Millimeter and Microwave Engineering for Communications and Radar

Critical Reviews Series: Recent Titles

Defining the Global Information Infrastructure: Infrastructure, Systems, and Services (CR56, S. F. Lundstrom)

Adaptive Computing: Mathematics, Electronics, and Optics (CR55, S.-S. Chen, H. J. Caulfield)

Millimeter and Microwave Engineering for Communications and Radar (CR54, J. C. Wiltse)

Glass Integrated Optics and Optical Fiber Devices (CR53, S. I. Najafi)

Handbook of Critical Dimension Metrology and Process Control (CR52, K. Monahan)

64- to 256-Megabit Reticle Generation: Technology Requirements and Approaches (CR51, G. Hearn, R. Naber)

Fiber Optics Reliability and Testing (CR50, D. K. Paul)

Diffractive and Miniaturized Optics (CR49, S. H. Lee)

Photorefractive Materials, Effects, and Applications (CR48, P. Yeh, C. Gu)

Optical Technologies for Aerospace Sensing (CR47, J. E. Pearson)

Optical Inspection and Testing (CR46, J. D. Trolinger)

Integrated Optics and Optoelectronics (CR45, K.-K. Wong, M. Razeghi)

Fiber Optic Sensors (CR44, E. Udd)

Optomechanical Design (CR43, P. R. Yoder, Jr.)

The Aviation Security Problem and Related Technologies (CR42, W. H. Makky)

Lens Design (CR41, W. J. Smith)

Optical Pattern Recognition (CR40, J. L. Horner, B. Javidi)

Infrared Thin Films (CR39, R. P. Shimshock)

Infrared Optical Design and Fabrication (CR38, R. Hartmann, W. J. Smith)

Standards for Electronic Imaging Systems (CR37, M. Nier, M. E. Courtot)

Machine Vision Systems Integration (CR36, B. G. Batchelor, F. M. Waltz)

Digital Optical Computing (CR35, R. Athale)

Critical Reviews of
Optical Science
and Technology

Volume CR54

Millimeter and Microwave Engineering for Communications and Radar

James C. Wiltse
Editor

Proceedings of a conference held
10–11 January 1994
San Diego, California

Sponsored and Published by
SPIE—The International Society for Optical Engineering

SPIE OPTICAL ENGINEERING PRESS

A Publication of SPIE—The International Society for Optical Engineering
Bellingham, Washington USA

Library of Congress Cataloging-in-Publication Data

Millimeter and microwave engineering for communications and radar : proceedings of a conference held 10-11 January 1994, San Diego, California, sponsored by SPIE—The International Society for Optical Engineering / James C. Wiltse, editor.
p. cm. — (Critical reviews of optical science and technology; v. CR54)
ISBN 0-8194-1505-7
1. Microwave integrated circuits—Congresses. 2. Telecommunication systems—Congresses. 3. Radar—Congresses. I. Wiltse, James C. II. Society of Photo-optical Instrumentation Engineers. III. Series.
TA1660.M555 1994 94-40112
621.381'32—dc20 CIP

Published by
SPIE—The International Society for Optical Engineering
P.O. Box 10
Bellingham, Washington 98227-0010 USA
Telephone 206/676-3290 (Pacific Time) • Fax 206/647-1445

Printed in the United States of America.

Contents

Preface

This Critical Reviews proceedings contains state-of-the-art summaries of numerous millimeter-wave topics, including technology developments and progress in several applications areas. Each manuscript was derived from an invited presentation given at the Millimeter and Microwave Engineering for Communications and Radar Critical Reviews conference held 10–11 January, 1994, in conjunction with the International Conference on Millimeter and Submillimeter Waves and Applications in San Diego, California. The idea for this Critical Reviews conference on millimeter-wave techniques was initiated by Mr. Hal Buscher.

The authors, all well-known specialists in their fields, have gathered the most current information on their subjects. The results described illustrate the rapidity of growth in the field of millimeter waves.

I would like to give special thanks to the speakers and authors who prepared these manuscripts. A significant effort was required, and the results are a major contribution to the field.

James C. Wiltse

Monolithic millimeter-wave integrated circuits, application and performance

Barry R. Allen

TRW Space & Electronics Group
One Space Park, Redondo Beach, CA 90278

ABSTRACT

Recent millimeter-wave three-terminal semiconductor device performance improvement enables monolithic millimeter-wave integrated circuits to be a viable option for many applications through 100 GHz. The low cost of the MMIC circuits, a result of circuit design and fabrication technique improvements, provides a path to low cost millimeter-wave systems for new commercial and military applications. Key technical advances that make the fabrication of high-performance monolithic millimeter-wave integrated circuits possible will be described. System applications and the present state of the art in monolithic millimeter-wave integrated circuits will also be discussed.

1. INTRODUCTION

Monolithic millimeter-wave integrated circuits combine passive circuit elements with semiconductor devices on a single semiconductor substrate. Since the techniques developed to support monolithic microwave integrated circuits are applicable to millimeter wave monolithic integrated circuits, MMIC will be used to refer to both types of circuits. Because the wavelength is shorter and the passive elements smaller, some circuit techniques are more practical at millimeter wave than at microwave, however, the difference is only fabrication or cost related. While frequency range that is addressed includes 30 to 300 GHz, some of the circuit and fabrication techniques are applicable above 300 GHz, the design trend in the submillimeter-wave region is toward designs based on quasi-optical techniques which will not be discussed in detail.

An encouraging aspect for the continued development of millimeter-wave MMICs is recent commercial interest in applications of millimeter-wave systems. Automobile radars, high data rate wireless networks, improved resolution all weather landing systems and other applications where the small antenna size, wide spectrum, and minimal interference provide millimeter wave systems a significant advantage over lower frequency systems. Millimeter-wave MMICs will be needed to provide the low system fabrication cost required to make these commercial applications feasible.

Monolithic millimeter-wave integrated circuits have been pursued since semiconductor devices were developed that operate at millimeter wave frequencies. The first monolithic millimeter-wave integrated circuits were simple matching structures connected to mixer diodes and were fabricated on both silicon and gallium arsenide substrates. In 1983, three-terminal devices were demonstrated that had performance capability suitable for millimeter-wave operation [1,2]. Since that time the development of monolithic millimeter-wave integrated circuits has been driven by improving device performance and circuit design techniques as evidenced by a chronology of papers published in the Microwave and Millimeter-Wave Monolithic Circuits Symposiums. In 1982, a single paper on a monolithic Ka-band diode mixer was published[3]; by 1987 there were two sessions on millimeter waves, including ten papers with the majority of the circuits based on three-terminal devices. In recent symposia, the trend has been to include millimeter-wave and microwave circuits together with division more by application than by frequency. Several companies have now demonstrated good performance from MMICs using three-terminal devices through 100 GHz [4,5,6,7], including an integration of a low noise amplifier and mixer on the same chip. The increase in MMIC operating frequencies will continue with improving device performance. Devices for 140 GHz amplifiers have been demonstrated[8] while 200 GHz amplifiers are likely to be developed within 2 years, and even 300 GHz operation is possible before the end of the decade. With advances in numerical EM simulation and the modeling of circuits, the development of MMICs should follow the device development by only a year or two.

2.0 APPLICATIONS

In the past, applications where system performance was paramount were the primary driving need in the development of millimeter-wave systems. However, as budgets continue to decline, cost performance of millimeter-wave systems will become of more importance. The inherent ability of MMICs to provide repeatable system performance at a lower recurring cost recommends their consideration for use in cost-sensitive millimeter-wave systems, including military, scientific and commercial systems.

Military systems to date have usually exploited one of three characteristics of millimeter-wave systems: a higher angular resolution from a given antenna aperature and a smaller aperature antenna for a given gain than in microwave systems, and attenuation due to atmospheric gases of the millimeter-wave radiation. The former property is extensively employed in smart munitions while the latter two properties are employed in military communication systems, including secure communication systems. As military budgets decrease and cost performance becomes a more prominent component of system design, MMICs will find application in military millimeter-wave systems to provide repeatable performance providing a lower recurring cost through higher levels of system integration .

Scientific systems to date have usually exploited the increased interactions of millimeter-waves with atoms and molecules compared to microwave interactions. Atmospheric remote sensing systems, whether looking up from the ground or down from high altitude, employ millimeter-wave systems to gain more accurate temperature or pressure vs. altitude measurements. In the future, these same properties will be used to monitor pollution in the atmosphere. In these systems, angular resolution is presently obtained by receiving millimeter-waves at multiple angles sequentially. MMICs provide a cost effective method to construct simultaneous receiving channels at multiple antenna angles for increased sensitivity.

Unlike military and scientific systems, commercial systems have been and will continue to be driven by cost performance. Recent advances in the cost performance as well as electrical performance and repeatability of millimeter wave MMICs has led to increased interest in millimeter-wave commercial systems. Compact chip designs with their small size and therefore, low cost are critical to low-cost millimeter-wave system development.

The commercial applications that have recently received the most attention are automotive sensors and communication systems[9,10,11]. The potential automotive applications include forward, side or rear looking radars at multiple millimeter-wave frequencies. In Europe, toll booths that use microwave links to automatically charge automobiles as they pass would employ millimeter-wave systems at 77 GHz to provide a higher data rate and possibly form the basis of a data transmission systems with automobiles. In Japan, automotive systems are being developed at 60 GHz because the high atmospheric attenuation reduces long range interference. In the U.S., systems are in development from 24 GHz to 94 GHz. The highest frequency requires the smallest antenna, but power output is more difficult to obtain as the frequency increases. Millimeter-wave MMICs usable for automotive radar applications have been presented in the literature[12,13] and provide the most viable approach to obtain the production cost needed for this application.

Potential communication applications include systems requiring wideband and small antenna size compared to competing microwave systems. Wireless data networks such as

Europe's proposed network at 38 GHz are an example of those types of systems which require low unit cost and will be enabled by millimeter-wave MMICs.

A few government long range communications systems or radar systems require more power, but most millimeter-wave applications require modest transmit power levels, usually well below 1 W. In general, terrestrial governments applications primarily use atmospheric transmission widows near 35 GHz, 94 GHz, 140 GHz, and 220 GHz with frequencies above 220 GHz not usually considered because the atmospheric attenuation rises rapidly. For atmospheric remote sensing systems or secure communications systems, high attenuation frequencies at 60 GHz (0_2), 118 GHz (0_2) and 183 GHz (H_20), are primarily used. To date, the development of millimeter-wave MMICs has been driven by the need for government systems utilizing either low or high atmospheric attenuation, with most of the work to date at 35 GHz, 60 GHz, or 94 GHz at moderate power levels.

3.0 DEVICE TECHNOLOGY -- PERFORMANCE AND FABRICATION

3.1 Introduction

Most of the recent monolithic millimeter-wave integrated circuit work has been with three terminal active devices. The device types include metal semiconductor field effect transistors (MESFET), high electron mobility transistors (HEMT), and heterojunction bipolar transistors (HBT). All of the early work was on GaAs substrates, but recently InP substrate HEMT and HBT MMICs have also been reported. The InP based devices have higher millimeter-wave gain and lower noise figure. Both GaAs and InP provide low dielectric loss substrates for integration of millimeter-wave active and passive devices.

3.2 Passive Component Fabrication

The fabrication of the passive components of a millimeter-wave MMIC is similar whether the active device has two or three terminals. The MMIC fabrication process includes the steps needed to produce the active devices with additional steps for passive circuit elements, such as thin-film capacitors and either thin-film or bulk resistors for coupling and bias networks. The interconnect metallization is used to form distributed transmission line elements or semi-lumped elements such as spiral inductors. Although most early millimeter-wave MMICs used fully distributed circuit techniques and did not necessarily require all the circuit components such as thin-film capacitors or spiral inductors, as the process, device and circuit modeling techniques mature, the trend is toward a more lumped element approach, resulting in smaller chip sizes.

A complete MMIC process includes the process steps needed to fabricate resistors and capacitors and two levels of interconnect metallization. Usually interconnect crossovers are fabricated using an air bridge to minimize capacitance and coupling between the transmission lines at the cross over. The typical cross section of the elements of the complete MMIC is shown below in Figure 1. The interconnect metallization is usually gold with a thin adhesion layer at the substrate surface. The most commonly used materials for the

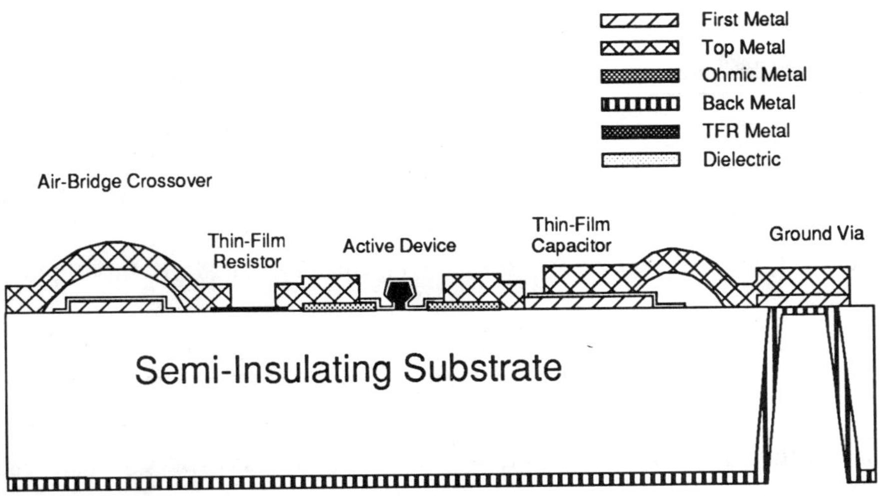

Figure 1. Cross section of typical millimeter-wave MMIC process

capacitor dielectric are Si_3N_4 or SiO_2. The details of the methods used to pattern the passive elements differ in detail among the various semiconductor manufacturers, but the results and process capabilities are similar. Line width accuracy is a few tenths micron with line widths and spaces usually five microns or greater. More linewidth precision and finer line pitch is possible, but not necessary for the majority of microwave and millimeter-wave applications.

The passive components are fabricated using photolithography and the common process techniques of etching, plating and lift-off. The etching process is the familiar photolithographic patterning used for most printed circuit boards. For the etch process of Figure 2, the material to be patterned is deposited uniformly over the substrate. A coating of photosensitive material (photoresist) is applied and exposed to light though a photomask. After development the photoresist forms a protective covering over the material to be patterned. The undesired material is removed by an etch and then the photoresist is removed to leave the desired circuit pattern. Etching is useful for both metal and dielectric patterns although the accuracy of the etch process is best for thin material since the amount of undercut that results from an isotropic etch is proportional the thickness of the patterned material. To increase metal thickness after a metal etch process, gold electroplating is sometimes used.

The lift-off process shown in Figure 3 is another major patterning process used in the formation of the passive elements. In the lift-off process, the photoresist is applied and devel-

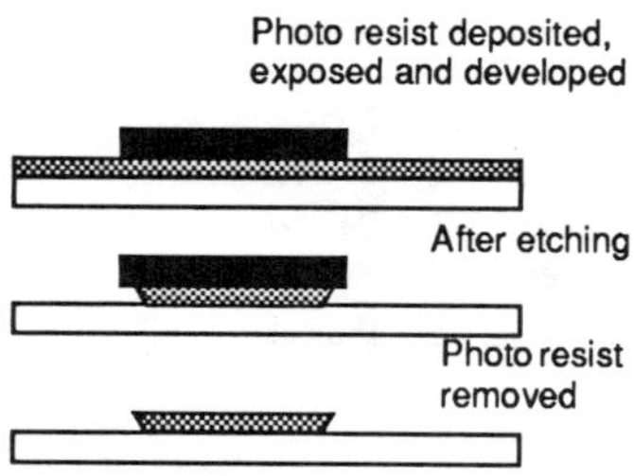

Figure 2. Typical MMIC etching process

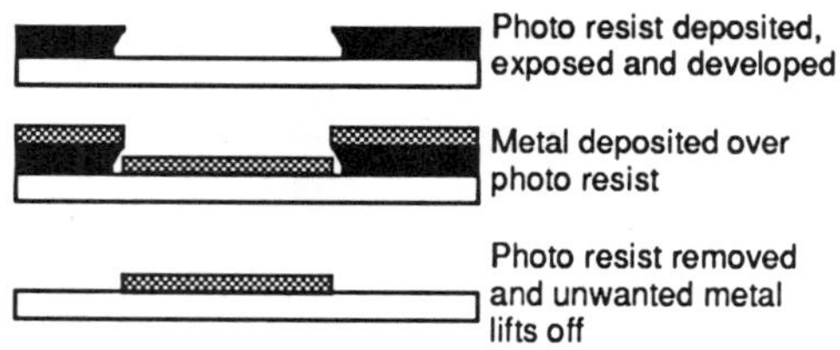

Figure 3. Typical MMIC lift-off process

oped first. The metal to be patterned is deposited uniformly on top of the photoresist. The overhang shown in the cross section of the resist is desirable so that there is a break in the metal at the pattern boundaries. After metal deposition, the resist is dissolved and the unwanted metal lifts off with the photoresist. The lift-off process provides accurate dimensional control, but the thickness of the metal can only be a fraction of the photoresist thickness to provide clean lift off.

The final process steps for passive elements are substrate thinning and ground via-hole formation. Wafer thinning and polishing is usually a combination of mechanical grinding and chemical polishing. A reactive ion etch (RIE) process or wet etch is used to form the ground via-holes. The RIE process is preferred since there is improved control of the via hole size and the side walls of the via are more vertical than with a wet etch process. These two process steps are needed for microstrip circuits, but are not used for coplanar waveguide circuits.

3.3 Two-Terminal Devices

The earliest device type used in monolithic millimeter-wave integrated circuits was the Schottky diode. For some applications such as mixers or efficient high power frequency multipliers[14], high performance diodes continue to be preferred. Also, above 120 GHz only monolithic diode circuits have been reported in the literature.

High quality Schottky diodes for millimeter-wave applications are usually fabricated on a GaAs substrate and require a highly doped n+ contact layer under a lightly doped n region. One of the main difficulties in fabrication of a high performance millimeter-wave diode for an integrated circuit is the contacting of the anode of the near planar diode without adding significant parasitic shunt capacitance. Several methods have been developed to accomplish this, with the most common using an air-bridge to the diode. A representative schematic of a millimeter-wave monolithic Schottky diode process is shown in Figure 4.

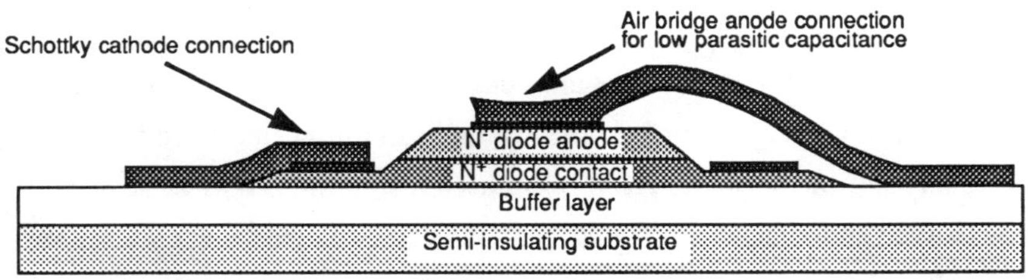

Figure 4. Typical MMIC millimeter-wave MMIC Schoktty diode cross section

With many diode circuits a single metal layer is sufficient, but an additional metal layer is usually fabricated to provide two levels of metal and thin film capacitors. The diode n+ contact layer can be used to fabricate a bulk resistor. With the improvement in device and circuit modeling, diode based monolithic millimeter-wave integrated circuits have application potential that has not been fully exploited, especially above 100 GHz.

3.4 Three-Terminal Devices

The earliest three-terminal devices suitable for millimeter-wave MMICs were GaAs MESFETs [1]. MESFETs have been successfully used for MMIC low noise amplifiers[15], power amplifiers[16], and VCOs[17] at 60 GHz or below. MESFETs for millimeter-wave MMICs are usually grown using molecular beam epitaxy (MBE) for the active channel. Gate lengths are 0.25 micron or less with a T cross section preferred for lower gate resistance, lower noise, and higher gain. The typical cross section of an MMIC MESFET is shown in Figure 5. A fabrication process similar to that described for the passive circuit

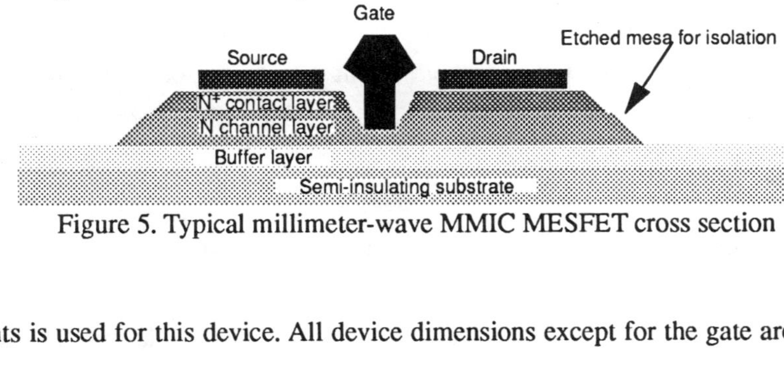

Figure 5. Typical millimeter-wave MMIC MESFET cross section

elements is used for this device. All device dimensions except for the gate are fabricated by optical lithography. The device is fabricated on an etched mesa to provide device isolation. The fabrication of the gate is usually done by direct write electron beam lithography (EBL). For this process step, the resist used to define the gate is sensitive to high energy electrons. An EBL machine is used expose the gate area, and the exposed resist is removed during development. The gate recess is etched and the gate metallization is deposited. The gate is usually formed in a lift-off process. To define a T-gate cross section, multiple resist layers and EBL exposures are usually required.

While MESFETs continue to be the most common active device for applications below 20 GHz, the HEMT has replaced the MESFET in most millimeter-wave MMIC applications. Both GaAs and InP substrate HEMTs are used for millimeter-wave devices. The earliest HEMTs were lattice matched $Al_xGa_{1-x}As/GaAs$ HEMTs with x about 25% to 30%. Most of the HEMTs on GaAs substrates are now $Al_{0.25}Ga_{0.75}As/In_xGa_{1-x}As/GaAs$ HEMTs with x about 12-25%. The devices are called pseudomorphic HEMTs because the $In_xGa_{1-x}As$ channel is not lattice matched to the GaAs substrate. There are many varieties of pseudomorphic HEMT on GaAs substrates which take advantage of the flexibility of MBE material growth to produce multiple quantum wells and other structures that increase the charge controlled by the gate. For a given gate length, the gain of the device increases with increasing InAs in the channel, but 30% InAs is about the practical upper limit at this time for a realizable device on a GaAs. Through proper device design, good low noise and power performance are obtainable from the same pseudomorphic HEMT device[18].

To take advantage of the improved gain with higher InAs content in the HEMT channel, recent device and millimeter-wave MMIC development has included InP substrate based HEMT devices. The InP substrate provides a lattice match to $In_xGa_{1-x}As$ with x=0.53. These devices provide higher f_T and f_{max} compared to the pseudomorphic HEMTs on GaAs. The lowest noise and highest gain millimeter-wave devices are InP substrate HEMTs. The operating voltage for millimeter-wave InP HEMTs have not been as high as the GaAs based pseudomorphic devices, but recent device developments have increased the operating voltage of lower frequency devices[19], and further work may improve the millimeter-wave power performance.

Except for the starting material, the basic fabrication of a HEMT device for an MMIC is similar to the fabrication of the MESFET. The HEMT gate etch is more difficult to control than the MESFET because of the thin channel and multiple material layers under the gate.

Mesa etching for device isolation is not always required if ion implant isolation is used. In any event, the HEMT process is more planar than a mesa etched MESFET process because of the thinner active layers.

Recently the performance of heterojunction bipolar transistors on both InP and GaAs has improved to the point where practical millimeter-wave circuits are now possible[20,21]. These devices are especially well suited to low noise oscillator and pulsed power applications. The basic device is smaller than a HEMT with the same current handling capability, leading to a compact device layout for high power, however, the thermal management of the HBT is more difficult than for the MESFET or HEMT. The development of millimeter-wave MMICs using HBTs will continue as the performance of the device continues to improve.

3.5 Multi-Function MMICs

Multi-Function MMICs provide more than a single circuit function on a chip. The development of the pseudomorphic HEMT, with good performance as both a low noise amplifier and a power amplifier, has made possible a variety of multifunction chip types with excellent performance. Although the gate diode of the HEMT provides reasonable mixer conversion loss through 100 GHz[6], an optimized mixer diode integrated with the HEMT will provide improved performance. Several techniques are in development to provide multiple millimeter-wave semiconductor types on a single wafer. One technique is to grow all the layers needed for two device types across a single wafer[7,17]. The material is etched to the proper set of device layers at each device site. The material structure for an implementation of this technique for Schottky diode and HEMT is shown schematically in Figure 6.

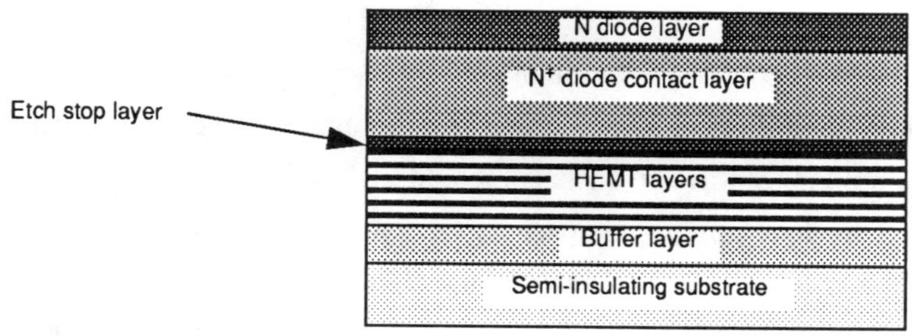

Figure 6. Material structure for integrated HEMT and Schottky diode

With a second technique called selective epitaxy, the required layers are grown only at the device sites need. The first device structure is grown uniformly across a wafer. The wafer is patterned with a mask and the unwanted layers are etched off at the sites for the alternate device. The alternate device structure is grown through the lift-off resist layer. After removal of the resist, the wafer has the proper device layers at each device site. With either approach, additional processing steps are needed to fabricate more than one type of active device, but many of the steps can be made common.

4.0 CIRCUIT TECHNOLOGY-- DESIGN AND TEST

The improvements in circuit simulation and testing have been as important to the development of MMICs as the improvements in semiconductor processing. With accurate device measurement and circuit simulation, it now possible to develop a new small signal circuit design from measurement of a new device with high confidence that the circuit will operate as simulated. The picture is not quite as good for high power circuits because of the present simulation limitations for both large passive circuits and nonlinear active circuits. Most present millimeter-wave MMIC designs use distributed networks with MIM capacitors for DC blocks. The compact design of millimeter-wave MMICs using semi-lumped elements is not as mature as at lower frequencies. More work is needed to develop the design techniques for compact designs and high yield, especially for low cost commercial applications. Design centering and statistical design techniques used to increase yield for MMICs are also applicable to millimeter-wave MMICs and will be applied more to millimeter-wave design as the technology matures.

4.1 Design

Microstrip is the most common transmission media for millimeter-wave MMIC circuit design. The microwave design techniques and modeling procedures established for microstrip design of MMICs provides a sound basis for millimeter-wave design with present commercial node based circuit simulators having reasonably accurate circuit models for isolated microstrip elements through about 100 GHz. However, the simulators largely neglect the effect of coupling between microstrip discontinuities which affect circuit performance more for millimeter-wave designs than for microwave designs. Accurate circuit modeling at millimeter-wave frequencies should include coupling effects among multiple discontinuities, especially for compact circuit layouts. The only feasible way to include coupling effects is numerical solution of the electromagnetic problem. Several commercial software products are now available to solve planar circuit EM problems. At the present time, the complexity of circuits amenable to numerical solution is limited. For complicated designs, the passive structures must be partitioned and solved separately assuming no coupling. This procedure works well, but care in circuit partioning is required.

For some applications, coplanar waveguide (CPW) offers advantages over microstrip design. For operation of microstrip based MMICs above 220 GHz, a substrate thickness of 50 μm or less would be needed to avoid strong coupling to the fundamental surface wave. Since the electric field is mostly tangential to the substrate surface, the coupling to the surface wave in CPW is weaker than in microstrip so the substrate thickness does not limit the operating frequency. CPW also provides reduced ground inductance compared to ground vias through a microstrip substrate, improving high frequency device gain. However, since the CPW substrate is thicker, thermal resistance through the substrate is higher than for microstrip, (although thermal transfer can be improved using flip-chip mounting or a thick ground plane). Circuit modeling for CPW elements is not as mature as microstrip element modeling, transmission losses are higher when compared to microstrip and large CPW chips need special attention when mounting to avoid high-Q substrate resonances. More work needs to be done on CPW for millimeter design, but the circuit technique will become more important, especially when combined with other transmission line techniques such as coplanar strip and slotline.

At this time most millimeter-wave MMIC designs have used techniques developed from microwave designs. The millimeter-wave signal is coupled on and off the chip through bond wires, or in a few cases by integrated waveguide transitions. In the future, especially at the higher frequencies, more chips will be developed using quasi-optical transmission and coupling techniques borrowed from optical and submillimeter systems. So far MMICs using quasi-optical techniques have been demonstrated as single channel submillimeter-wave circuits[22] and simple multiple channel millimeter-wave systems[23]. Quasi-optical techniques are well suited for efficient power combining and will be needed for efficient high power solid state millimeter-wave amplifiers. As the yield increases and the cost decreases for complex millimeter-wave MMICs, quasi-optical circuits will be increasingly used for high performance millimeter-wave systems.

4.2 Test

The development of on-wafer millimeter-wave testing has been one of the major factors in the rapid improvement in the performance and complexity of millimeter-wave MMICs. On-wafer testing is valuable for measurement of complete circuits and for accurate model development and verification. Using a ground-signal-ground coplanar interface between the probe tip and the chip under test, on-wafer RF probes through 110 GHz are now available[24,25]. Millimeter-wave on-wafer testing capability includes most tests that would have formerly required diced chips in test fixtures. On-wafer testing is especially useful for complex multifunction millimeter-wave chip designs to verify operation of individual circuit functions. With on-wafer testing, it is practical to measure many more chips and chip types than would have been feasible with fixture testing, providing important data on both the repeatability of the semiconductor process and the accuracy of the design models.

5.0 PERFORMANCE

The last few years the design procedures for MMICs have improved to the extent that the chip performance is approaching the limit set by the active device performance. The difference in circuit loss between hybrid integrated circuits and monolithic integrated circuits is less at millimeter-wave than at microwave frequencies because the nominal 100 μm substrate thickness is close to optimum for microstrip circuit designs near 100 GHz. The optimal substrate thickness occurs just below the thickness where coupling between the microstrip mode and other substrate modes becomes significant.

For systems that require accurate phase and amplitude matching such as phased arrays and power combined millimeter-wave amplifiers, the performance obtainable from MMICs is superior to hybrid designs due to the close matching performance of devices fabricated under identical conditions within the same circuit. The easier control of undesirable parasitics also makes the monolithic implementation of wideband millimeter-wave amplifiers and mixers less difficult than in hybrid designs. As advances in active device processing provides high yield active devices with good performance past 100 GHz, the performance advantage of monolithic circuit implementation will be even more evident due to the increased difficulty in hybrid circuit fabrication and assembly.

5.1 Low Noise Amplifiers

Low noise amplifiers were one of the first applications for active millimeter-wave MMICs. Initially noise figures were significantly higher than hybrid circuits using the same type devices. The modeling accuracy of both the device noise parameters and the on-chip passive matching networks has steadily improved, bringing the performance of monolithic LNAs closer to the performance of hand tuned hybrid LNAs which

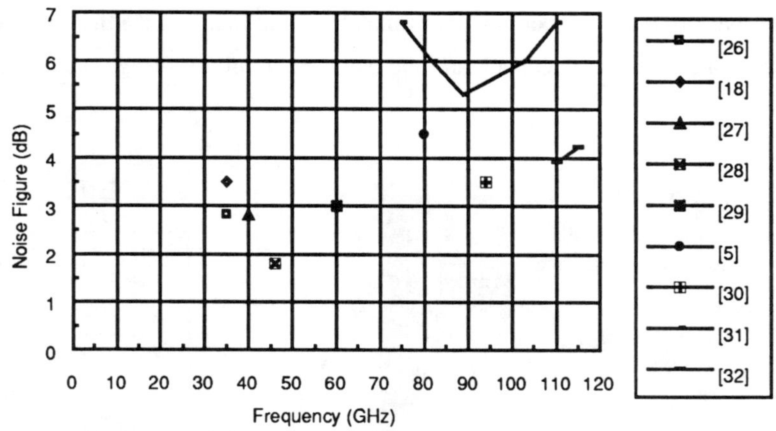

Figure 7. Millimeter-wave HEMT low noise amplifiers

have a slight performance advantage since the hybrid substrate parameters can be optimized to minimize circuit losses. Table 1 and Figure 7 show a representative list of some millimeter-wave MMIC LNAs and their measured performance. Two substrate type

Table 1: Millimeter-wave MMIC LNA performance

Frequency (GHz)	Noise Figure (dB)	Gain (dB)	Stages	Device (Channel/Substrate)	Reference
35	2.8	14	2	0.2 μm, AlGaAs/GaAs HEMT	[26]
35	3.5	17	2	0.25 μm InGaAs/GaAs HEMT*	[18]
40	2.8	21.5	3	0.15 μm InGaAs/GaAs HEMT	[27]
46	1.8	25	2	0.1 μm InGaAs/InP HEMT	[28]
60	3	24	3	0.1 μm InGaAs/InP HEMT	[29]
80	4.5	23	4	0.1 μm AlGaAs/GaAs HEMT	[5]
94	3.5	21	3	0.1 μm InGaAs/GaAs HEMT	[30]
75-110 (89)	6.8 (5.3)	20	4	0.1 μm InGaAs/InP HEMT	[31]
110/115	3.9/4.2	19.6/12	2	0.1 μm InGaAs/GaAs HEMT	[32]

amplifiers are represented in Table 1; GaAs substrate amplifiers and InP substrate amplifiers. A few example circuits are given below.

A typical millimeter-wave GaAs substrate pseudomorphic InGaAs channel HEMT MMIC LNA photograph and measured performance is shown in Figure 8[26]. The chip was one of the first designs to use spiral matching inductors to 40 GHz and is a balanced two-stage design using Lange 90° couplers to provide good VSWR. The second stage uses parallel RLC feedback to provide good interstage matching and to minimize the sensitivity to process variations. With the feedback and the balanced amplifier design, the gain is flat and

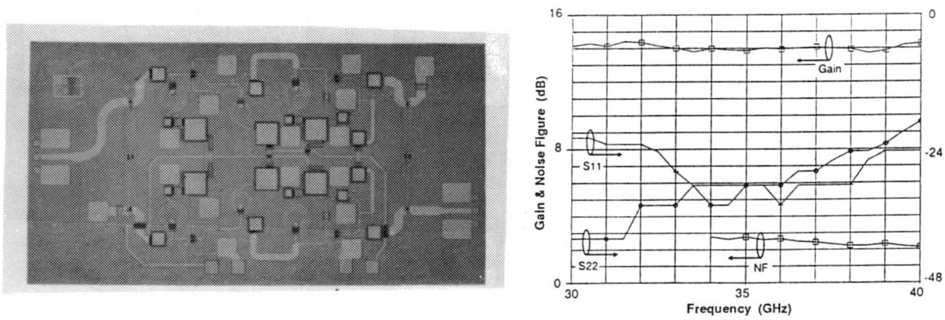

Figure 8. Ka-band pseudomorphic HEMT low noise amplifier photograph and performance. The circuit was built on a GaAs substrate.

the port VSWR is low across the full 26-40 GHz waveguide band. This approach sacrifices some gain and noise figure capability but provides repeatable performance and good process latitude.

Better noise figure, higher gain, and reduced DC power can be obtained from InP substrate HEMTs with a higher indium percentage in the channel than is possible on GaAs substrates. A two-stage Q-band InP substrate HEMT MMIC LNA photograph and measured performance is shown in Figure 9[28]. This amplifier uses series inductive feedback in the

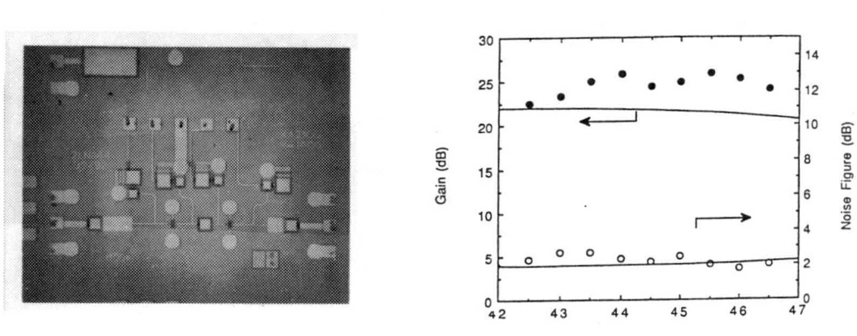

Figure 9. Q-band InP HEMT low noise amplifier photograph and performance

source of each HEMT to improve the input power match at the minimum noise figure match. The required drain bias for the active devices is about 0.8 V, which is lower than required for an equivalent pseudomorphic HEMT on GaAs. The low drain bias and high transconductance at low drain current results in a factor of three lower power dissipation than in an equivalent GaAs based MMIC. Compared to a pseudomorphic GaAs substrate HEMT MMIC, the power savings per dB of gain is closer to a factor of four, since fewer stages are needed for the InP substrate HEMT MMIC. This power savings is only for low signal level conditions. If high linearity is required, the only power savings that InP based devices will provide is due to the higher gain per stage.

At the higher millimeter wave frequencies, 0.1 μm gate length pseudomorphic HEMT devices on GaAs are capable of operation to slightly over 100 GHz. The photograph and measured performance of a pseudomorphic HEMT MMIC amplifier that is near the upper frequency limit for useful operation using this type of device is shown in Figure 10[32].

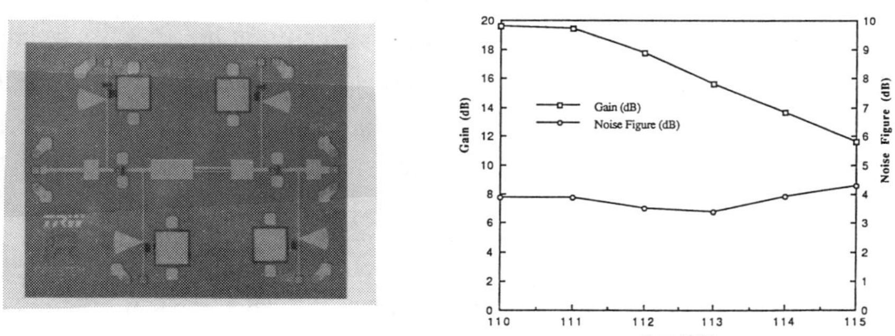

Figure 10. D-band pseudomorphic HEMT low noise amplifier photograph and performance

The amplifier is also operating near the upper limit of microstrip matching elements on a 100 μm thick GaAs substrate. The design incorporates simple matching networks with no on-chip thin-film capacitors or thin-film resistors required. Low frequency stability is established by bulk semiconductor resistors loading the drain and gate bias lines. Although resistive loading the bias lines results in additional DC dissipation, the wideband stability is excellent. Quarter-wave radial stubs are used on each bias line to short out the effects of the resistive bias networks in the band of interest. DC blocks are not needed at the input and output of the amplifier since the finline waveguide transition provides a DC open circuit; a quarter-wave coupled line is used for the interstage DC block. Although this amplifier is near the performance limit of 0.1 μm gate length GaAs devices, further device improvements are expected to move the frequency capability of MMICs even higher.

Present InP based HEMT devices should be capable of providing equivalent performance to 150 GHz or higher.

5.2 Power Amplifiers

The development of millimeter-wave MMIC power amplifiers has lagged somewhat behind the low noise development because of the increased difficulty in both device fabrication and circuit design for power amplifiers compared to low noise amplifiers. The devices for millimeter-wave power amplifiers require higher breakdown voltage and higher operating currents. The device physical parameters needed to meet these requirements usually conflict with the device requirements for high gain and high frequency operation. The higher power density results in higher operating temperature in a millimeter-wave power device compared to a low noise device. Circuit design usually involves matching to many device cells in parallel resulting in a higher Q, more complex matching network. Power amplifier design also involves thermal issues not associated with low noise amplifier design. Within the last three years significant progress has been made in the development of millimeter-wave MMIC power amplifiers. Although some of the early millimeter-wave MMIC power amplifiers were based on GaAs MESFETs, much of the recent progress in millimeter-wave power amplifier design has been the result of maturing HEMT processing and improvements in MBE material profiles. In Figure 11 and Table 2

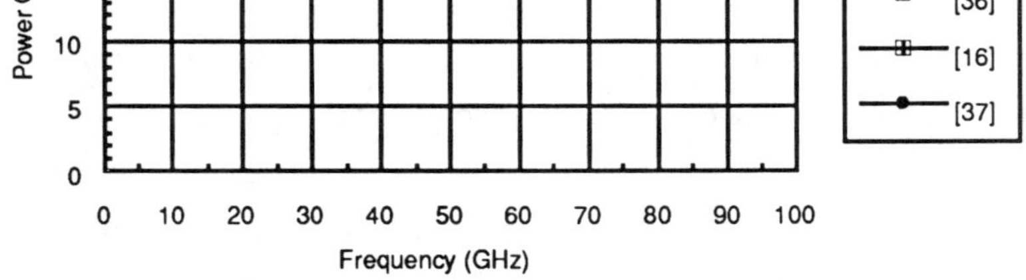

Figure 11. Millimeter-wave MMIC power amplifiers

are a representative list of some of the results for millimeter-wave power MMICs. The data in Table 2 includes both MESFET and HEMT results. The performance levels for

Table 2: Millimeter-wave MMIC power amplifier performance

Frequency (GHz)	Power Output (dBm)	Power Added Efficiency (%)	W/mm	Device (Channel/Substrate)	Reference
30	27	32.3	0.63	0.25 µm, InGaAs/GaAs HEMT	[33]
35	27	33	0.63	0.25 µm InGaAs/GaAs HEMT*	[18]
35	30.4	21	0.35	0.15 µm InGaAs/GaAs HEMT	[34]
44	24	33	0.56	0.15 µm InGaAs/InP HEMT	[35]
47	22.6	11	0.23	0.3 µm GaAs MESFET	[36]
58	19.8	11	0.32	0.3 µm GaAs MESFET	[16]
93.5	20	6	0.16	0.1 µm InGaAs/GaAs HEMT	[37

millimeter-wave power MMICs are expected to improve rapidly over the next several years as the process and design techniques mature.

A representative millimeter-wave power MMIC is shown in Figure 12[34]. This chip was

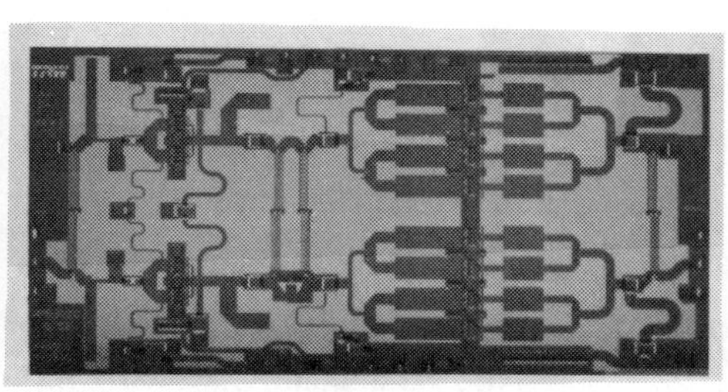

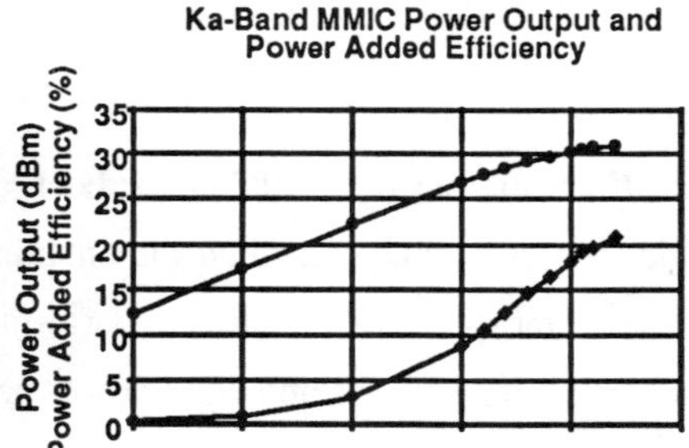

Figure 12. Ka-band 1 Watt MMIC

one of the first millimeter-wave power MMICs to surpass 1 W output power. The process is a 0.15 µm gate length pseudomorphic HEMT process. The output stage has a total device periphery of 3.2 mm and provides about 1 W output at 35 GHz. The design is a two-stage balanced amplifier with Lange couplers at the input, interstage, and output. Although the Lange couplers introduce some loss (about 0.3 dB), the improvement in VSWR and process sensitivity is significant compared to other design approaches. The design of this chip made extensive use of numerical electromagnetic simulation of the passive matching networks. The low input and output VSWRs provided by the couplers are useful in maintaining good phase matching when combining several chips for increased output power capability.

The chip is designed to be on-wafer tested at full power using low duty cycle pulsed measurements to avoid excessive heating. Care was taken in the bias network design to avoid low frequency instabilities under the expected terminations presented by the on-wafer DC probes. During pulsed testing, both the drain voltage and RF drive are pulsed. The output power from on-wafer pulsed measurement and CW fixtured measurements agreed to within 1 dB. The on-wafer pulsed power measurement system provides a low cost method to screen parts before installing in a higher level assembly.

The highest frequency useful power MMICs operate at is usually 10-20% lower in frequency than an equivalent device used in a low noise MMIC. Since a power MMIC operates partially saturated, additional gain is needed. Combining a large number of cells to obtain high power also introduces additional loss, requiring higher gain devices. Presently the upper frequency limit on power MMICs is about 94 GHz[37]. Power MMICs operating significantly above this frequency with ≥20 dBm output will require device improvements over those presently available although some increase in efficiency and power at 94 GHz should be possible through improved design and modeling techniques.

5.3 Multi-Function Circuits

Recently a number of MMICs have been developed that combine several functions on a single chip. A leading contributor to enabling increased integration is the improvements in the pseudomorphic HEMT process performance, yield and repeatability. It is now possible to obtain both low noise and good power performance from a single process[18]. The integration of low noise and power devices on the same chip will enable many low cost millimeter-wave communication and radar applications. The millimeter-wave chip integration is not yet to the level of some microwave MMICs, but the HEMT process capability has not yet been reached. At present, increased functional integration is more of a design problem than a processing problem.

One of the advantages of the MMIC process is the repeatability for circuits that require accurate phase and amplitude matching. An example of a multifunction chip that exploits the repeatability of the MMIC process is shown in Figure 13[38]. This chip includes both

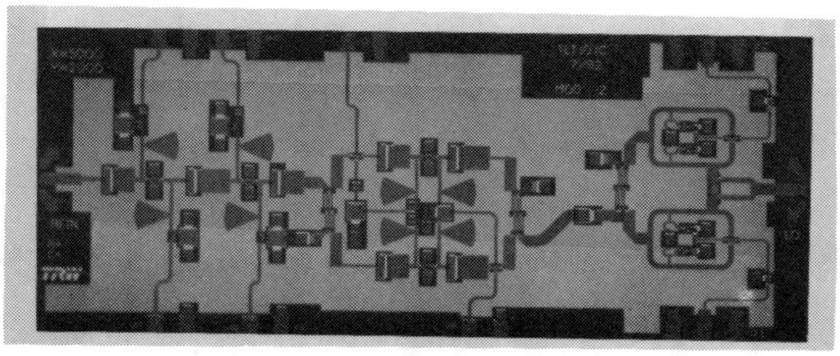

Figure 13. W-band image reject downconverter

a W-band low noise amplifier and an image reject diode mixer integrated on a single chip. The diode is formed by connecting the HEMT source and drain together for the diode cathode with the multi-finger gate metal used as the diode anode. This diode has good conversion loss performance, but suffers from excess $1/f$ noise at low IF. The downconverter chip provides about 6 dB SSB noise figure for a 50-500 MHz IF using an external IF 90 degree hybrid combiner. Due to the excess diode noise, the noise figure of the downconverter increases below about 50 MHz. The good phase and amplitude match between the in-phase and quadrature mixers provides 20 dB image rejection. The low IF noise performance of this type of downconverter could improved through processing techiques[7] that enable integration of a quality GaAs diode with the active HEMT device.

A final example of a multifunction millimeter-wave MMIC is shown in Figure 14[13].

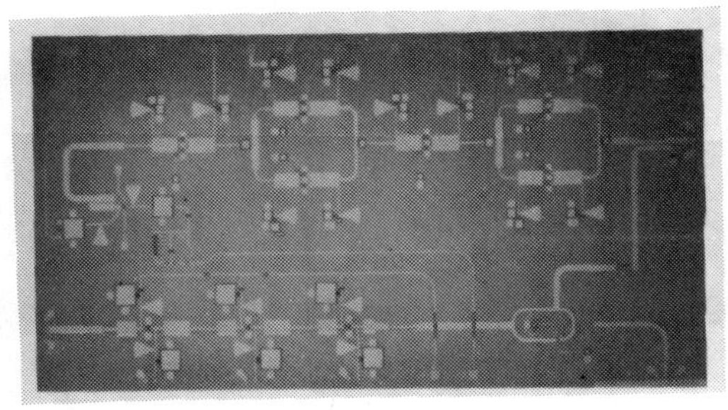

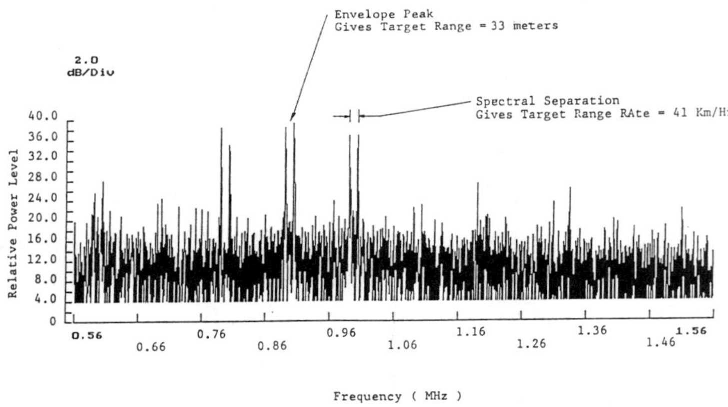

Figure 14. W-band single-chip FM-CW radar and detected target

This is the first chip to include all the active RF functions needed to implement a 94 GHz FM-CW radar on a single chip. The functions include a VCO, several amplifier stages for the transmit function, a low noise amplifier, a mixer, and the coupler to supply LO from the transmitter to the mixer. Several automotive radar test systems using this chip have been built and tested. The plot in Figure 14 shows the detection of a moving automobile at about 30 m range from one of the radar test systems. Although this chip has the functionality needed to implement an FM-CW radar, further work will be needed to decrease the chip size to provide a cost effective implementation for a W-band automotive radar.

6.0 CONCLUSIONS

The development of millimeter-wave MMICs has attained the state where high performance, low cost system applications are feasible. Progress in processing, design, and test has helped millimeter-wave MMIC design capability become comparable to microwave design capability. While diode based circuits are already capable of operation into the submillimeter-wave region, present active device circuits provide good performance for both low noise and power amplifier applications through about 100 GHz. Processing and design enhancement will continue to improve MMIC performance below 100 GHz, especially power output and power added efficiency. Further advancement in active device performance is likely to extend the performance of three terminal device MMICs to near 200 GHz within the next five years.

The design and fabrication cost of millimeter-wave MMICs will continue to decrease with the improvements in processing and design techniques. The design accuracy of millimeter-wave MMICs has improved greatly over the last few years, but much improvement is still possible, especially with the more compact designs. Distributed networks should become more compact with the size of a matching network decreasing at least linearly with wavelength lumped element millimeter-wave networks should further reduce the size of the circuits as the modeling of millimeter-wave components improves. At present the size of millimeter-wave and microwave MMICs are comparable, mainly because of the more extensive use of lumped element microwave networks compared to distributed elements used for most millimeter-wave networks. As the size of the matching networks is reduced the level of integration will be increased while chip cost is lowered.

The outlook for increased use of millimeter-wave MMICs is positive, with many of the near term needs coming from commercial applications. As the costs decline millimeter-wave systems that now appear too costly will be enabled. Further development of design and process techniques will also provide added millimeter-wave MMIC performance capability stretching to 200 GHz or more in the near future.

REFERENCES

1. E. T. Watkins, J. M. Schellenberg, L. H. Hackett, H. Yamasaki and M. Feng, "A 60 GHz GaAs FET amplifier", *IEEE MTT-S International Microwave Symposium Digest*, pp. 145-147, 1983.

2. K. Joshin, T. Mimura, M. Niori, Y. Yamashita, K. Kosemura, and J. Saito, "Noise performance of microwave HEMT", *IEEE MTT-S International Microwave Symposium Digest*, pp. 563-565, 1983.

3. A. Contolatis et. al., "Ka-band monlithic GaAs balanced mixers," *IEEE Microwave and Millimeter-Wave Monolithic Circuits Symposium*, 1982.

4. R. Majidi-Ahy, M. Riaziat, C. Nishimoto, M. Glenn, S. Silverman, S. Weng, Y. C. Pao, G. Zdasiuk, S. Bandy, and Z. Tan, "5-100 GHz InP CPW MMIC 7-section distributed amplifier", *IEEE Microwave and Millimeter-Wave Circuits Symposium Digest*, pp. 31-34, 1988.

5. M. W. Trippe, S. Weinreb, S.W. Duncan, A. Eskandarian, B. A. Golja, D. C. Martel, G. Mendenilla, B. Power, H. B. Sequeira, S. B. Southwick, S. P. Svensson, D-W. Tu, and N. E. Byer, "MM-wave MIMIC receiver components", *IEEE Microwave and Millimeter-Wave Circuits Symposium Digest*, pp. 51-54, 1991.

6. K. W. Chang, H. Wang, T. H. Chen, K. Tan, J. Berenz, G. S. Dow, A. C. Han, D. Garske, and L. C. T. Liu, "A W-band monolithic pseudomorphic InGaAs HEMT downconverter", *IEEE Microwave and Millimeter-Wave Circuits Symposium Digest*, pp. 55-58, 1991.

7. H. C. Huang, P. Laux, J. Bass, S. W. Chen, T. Lee, S. Tadayon, and O. Aina, "A W-band multi-function MMIC", to be published in *IEEE Microwave and Millimeter-Wave Circuits Symposium Digest*, 1994.

8. K. L. Tan, D. C. Streit, P. D. Chow, R. M. Dia, A. C. Han, P. H. Liu, D. Garske, and R. Lai, "140 GHz 0.1 μm gate-length pseudomorphic $In_{0.52}Al_{0.48}As/In_{0.60}Ga_{0.40}As/$ InP HEMT", *IEEE International Electronic Devices Meeting Digest*, pp. 239-242, December, 1991.

9. M. Kotaki, Y. Takimoto, E. Akutsu, Y. Fujita, H. Fukuhara, and T. Takahashi, "Development of millimeterwave automotive sensing technology in Japan", *IEEE MTT-S International Microwave Symposium Digest*, pp. 709-712, 1992.

10. H. H. Meinel, "Applications of microwaves and millimeterwaves for vehicle communications and control in Europe", *IEEE MTT-S International Microwave Symposium Digest*, pp. 609-612, 1992.

11. A. G. Stove, "Automotive radar at 80-90 GHz", *IEEE MTT-S International Microwave Symposium Digest*, pp. 613-616, 1992.

12. J. Berenz, M. LaCon, and M. Luong, "Single chip Ka-band transceiver", *IEEE Microwave and Millimeter-Wave Circuits Symposium Digest*, pp. 517-520, 1991

13. K. W. Chang, G. S. Dow, H. Wang, T. H. Chen, K. Tan, B. Allen, and J. Berenz, "A W-band single-chip transceiver for FMCW radar", *IEEE Microwave and Millimeter-Wave Circuits Symposium Digest*, pp. 41-44, 1993

14. S. Chen, T. Ho, F. R. Phelleps, J. L. Singer, and K. Pande, "Rigorous design of a 94 GHz MMIC doubler", *IEEE Microwave and Millimeter-Wave Circuits Symposium Digest*, pp. 89-92, 1993.

15. M.A.G. Upton, A. W. Jacomb-Hood, P. C. Chao, S. J. Nightingale, and J. D. Merrill, "EHF monolithic receive module technology", *IEE Military Microwaves Conference Digest*, pp. 173-178, July 1985.

16. H. A. Hung, G. M. Hegazi, T. T. Lee, F. R. Phelleps, J. L. Singer and H. C. Huang, "V-band GaAs MMICC low-noise and power amplifiers", *IEEE Transactions on Microwave Theory and Techniques*, vol. 36, no. 12, pp. 1966-1975, December 1988.

17. M.G. McDermott, C. Sweeney, M. Benedek, and G. Dawe, "Monolithic Ka-band VCO using quarter micron GaAs MESFETs and integrated high-Q varactors", *IEEE Microwave and Millimeter-Wave Circuits Symposium Digest*, pp. 103-106, 1990.

18. A. Kurdoghlian, C. S. Wu, W. Yau, J. Chen, M. Hu, C. Pao, and D. Bosch, "The demonstration of Ka-Band multi-functional MMIC circuits fabricated on the same PHEMT wafer with superior performance", *IEEE Microwave and Millimeter-Wave Circuits Symposium Digest*, pp. 97-98, 1993.

19. M. Matloubian, L. M. Jelloian, M. Lui, T. Liu, and M. A. Thompson, "Ultra-high breakdown high-performance AlInAs/GaInAs/InP Power HEMTs," *IEEE International Electron Device Meeting Digest*, pp. 915-917, 1993.

20. N. H. Sheng, W. J. Ho, N. L. Wang, R. L Pierson, P. M. Asbeck, and W. L. Edwards, "A 30 GHz bandwidth AlGaAs-GaAs HBT direct-coupled feedback amplifier", *IEEE Microwave and Guided Wave Letters*, vol. 1, no. 8, pp. 208-210, August 1991.

21. K. W. Kobayashi, L. T. Tran, S. Bui, J. Velebir, D. Nguyen, A. K. Oki, and D. C. Streit, "InP based HBT millimeter-wave technology and circuit performance to 40 GHz", *IEEE Microwave and Millimeter-Wave Circuits Symposium Digest*, pp. 85-88, 1993.

22. S. S. Gearhart, J. Hesler, W. L. Bishop, T. W. Crowe, and G. M. Rebeiz, "A W-band 760-GHz planar integrated Schottky receiver", *IEEE Microwave and Guided Wave Letters*, vol. 3, no. 7, pp. 205-207, July 1993.

23. K. D. Stephan, F. H. Spooner, and P. F. Goldsmith, "Quasioptical millimeter-wave hybrid and monolithic PIN diode switches", *IEEE Transactions on Microwave Theory and Techniques*, vol. 41, no. 10, pp. 1791-1798, October 1993.

24. E. M. Godshalk, "A W-band wafer probe", *IEEE MTT-S International Microwave Symposium Digest*, pp. 171-174, 1993.

25. S. M. J. Liu and G. G. Boll, "A new probe for W-band on-wafer measurements", *IEEE MTT-S International Microwave Symposium Digest*, pp. 1335-1338, 1993.

26. J. Yonaki, R. Carandang, B. R. Allen, M. Hoppe, W. L. Jones, D. C. Yang, and C. L. Brunnenmeyer, "35 GHz InGaAs HEMT MMIC downconverter", *IEEE Microwave and Millimeter-Wave Circuits Symposium Digest*, pp. 47-50, 1991.

27. K. H. G. Duh, S. M. J. Liu, S. C. Wang, P. Ho, and P. C. Chao, "High performance Q-band 0.15μm InGaAs HEMT MMIC LNA", *IEEE Microwave and Millimeter-Wave Circuits Symposium Digest*, pp. 99-102, 1993.

28. D. C. W. Lo, R. Lai, H. Wang, K. L. Tan, R. M. Dia, D. C. Streit, P. Liu, J. Velebir, and B. Allen, "A high-performance monolithic Q-band In-P-based HEMT low-noise amplifier", *IEEE Microwave and Guided Wave Letters*, vol. 3, no. 9, pp. 205-207, September 1993.

29. K. W. Chang, H. Wang, R. Lai, D. C. Lo, and J. Berenz, "A V-band monolithic InP HEMT downconverter", *IEEE GaAs IC Symposium Digest*, pp. 211-214, 1993

30. H. Wang, T. N. Ton, K. L. Tan, G. S. Dow, T. H. Chen, K. W. Chang, J. Berenz, B. Allen, P. Liu, D. Streit, G. Hayashibara, and L. C. T. Liu, "An ultra low noise W-band monolithic three-stage amplifier using 0.1μm pseudomorphic InGaAs/GaAs HEMT technology", *IEEE MTT-S International Microwave Symposium Digest*, pp. 803-806, 1992.

31. S. M. J. Liu, K. H. G. Duh, S. C. Wang, O. S. A. Tang, and P. M. Smith, "75-110 GHz InGaAs/GaAs HEMT high gain MMIC amplifier", *IEEE GaAs IC Symposium Digest*, pp. 273-276, 1993.

32. H. Wang, K. L. Tan, T. N. Ton, G. S. Dow, P. H. Liu, D. C. Streit, J. Berenz, M. W. Pospieszalski and S. K. Pan, "A high gain low noise 110 GHz monolithic two-stage amplifier", *IEEE MTT-S International Microwave Symposium Digest*, pp. 783-785, 1993.

33. H. Q. Tserng, P. Saunier, and Y. Kao, "High-efficiency broadband monolithic pseudomorphic HEMT amplifiers at Ka-band", *IEEE Microwave and Millimeter-Wave Circuits Symposium Digest*, pp. 51-54, 1992.

34. M V. Aust, B. Allen, G. S. Dow, R. Kasody, G. Luong, M. Biedenbender, and K. Tan, "A Ka-band HEMT MMIC 1 watt power amplifier", *IEEE Microwave and Milli-*

meter-Wave Circuits Symposium Digest, pp. 45-48, 1993.

35. W. Lam, M. Matloubian, A. Kurdoghlian, L. Larson, A. Igawa, C. Chou, L. Jelloian, A. Brown, M. Thompson, and C. Ngo, "High-efficiency InP-based HEMT MMIC power amplifier for Q-band applications", *IEEE Microwave and Guided Wave Letters*, vol. 3, no. 11, pp. 420-422, November 1993.
36. T. Ho, K. Pande, F. Phelleps, and J. Singer, "A 0.6-watt U-band monolithic MESFET power amplifier", *IEEE MTT-S International Microwave Symposium Digest*, pp. 531-534, 1993.
37. T. H. Chen, K. L. Tan, G. S. Dow, H. Wang, K. W. Chang, T. N. Ton, B. Allen, J. Berenz, P. H. Liu, D. Streit, and G. Hayashibara, "A 0.1-W W-band pseudomorphic HEMT MMIC power amplifier", *IEEE GaAs IC Symposium Digest*, pp. 71-74, 1992
38. H. Wang, K. W. Chang, T. N. Ton, M. Biedenbender, S. Chen, J. Lee, G. S. Dow, K. L. Tan, and B. R. Allen, "High-yield W-band monolithic HEMT low-noise amplifier and image rejection downconverter chips", *IEEE Microwave and Guided Wave Letters*, vol. 3, no. 8, pp. 281-283, August 1993.

InP Gunn diodes and millimeter-wave applications

J.D. Crowley

Litton Solid State, Santa Clara, CA 95054-3095

ABSTRACT

InP Gunn diodes provide low-cost, efficient, reliable solutions to power generation and amplification problems throughout the millimeter-wave range. InP has replaced GaAs as the material of choice for most millimeter wave Gunn diodes. It offers a significant efficiency advantage and has extended the frequency range of practical Gunn oscillators beyond 140 GHz. Device performance in applications ranging from 35 to 140 GHz will be described, including local oscillators, VCOs, power amplifiers, and power combiners. CW output powers exceed 500 mW at 35 GHz, 150 mW at 94 GHz and 50 mW at 140 GHz, while conversion efficiencies exceed 15%, 6%, and 2.5%, respectively.

1. INTRODUCTION

Gunn diodes are a commonly used device for generation of microwave and millimeter-wave power. Gunn oscillators are simple to operate, low-cost, reliable, and possess good amplitude and phase noise characteristics. Therefore, they are often a preferred source for local oscillators or medium power transmitters. The intrinsic negative resistance of a Gunn diode exists for approximately an octave bandwidth centered on the transit time frequency of the active layer. When embedded in the proper impedance matching circuit, non-oscillating, stable amplification can be achieved over a wide bandwidth. The bulk nature of Gunn diodes make them very robust and resistant to damage from ESD, EMI, large RF signals, or load mismatches.

Many types of solid state devices are finding use throughout the millimeter wave region of the spectrum. Impatt diodes, either GaAs or silicon, are used to meet the very high power requirements, pulsed or CW. However, the availability of Impatt diodes is growing increasingly scarce. Three terminal devices, e.g., HEMT and PHEMT MMICs, have shown rapid improvement in performance for low-noise and medium-power amplification. For the very high frequencies beyond 60 GHz, the technology to produce such devices is extremely costly. As a result, the availability of devices is quite restricted and the cost per device is extremely high. Only in high volume will the the cost per chip be affordable for millimeter wave applications. Gunn diodes are readily available for frequencies up to 140 GHz and are affordable even in small quantities. Their availability and ease of use has made possible the realization and demonstration of many millimeter-wave system concepts. Two very important commercial applications which have been lead by Gunn devices are millimeter-wave commercial radios and vehicle radars. Gunn diodes employ mature material , fabrication, assembly, and circuit technologies which can be applied to producing devices and components throughout the millimeter-waves range.

For frequencies below 35 GHz, GaAs Gunn diodes are the typical device employed in Gunn oscillators. For frequencies from 35 GHz to 140 GHz, InP Gunn diodes are now commonly used. GaAs Gunn devices became commercially available in the mid 1960's. However, around 1970 InP was recognized as having material properties which should yield improved performance of Gunn devices. In the early 1970's fundamental materials issues, e.g., high quality substrates and epitaxial layers, limited the realization of InP Gunn devices which had any significant performance advantages. As better quality material became available, fabrication technology advanced, InP Gunn devices began to show the performance which had been promised. Early work with InP was concentrated at lower frequencies and stressed pulsed operation. Exceptionally high efficiencies (22%) were demonstrated. The development and demonstration of millimeter-wave devices occurred in the late 1970's and early 1980's. Today it is as a millimeter-wave device that InP has found its important area of application.

The millimeter-wave advantage which InP has achieved is best explained by its electron velocity field characteristic of Figure 1 and its material properties of Table 1, where they are contrasted to GaAs. The larger peak-to-valley ratio of the electron velocity field characteristic in InP leads to a greater dc-to-rf conversion efficiency. The shorter time constants associated with electron energy transfer result in a cut-off frequency for the transferred electron effect which is at least twice as high in InP as it is in GaAs. Based on these properties InP should offer higher efficiency to higher frequencies. There are several other material properties which lead to important advantages for practical devices. The greater effective transit velocity results in thicker active layers, which is beneficial in the material growth process for high frequency devices. The temperature dependence of the electron velocity field curve is lower in InP. This translates into RF performance which is less dependent on temperature and oscillators with fewer cold-start problems. InP has a higher thermal conductivity which means lower thermal resistance for similar geometrical structures. The higher threshold field means higher operating voltages and higher RF impedances, which make impedance matching easier. One disadvantage which InP possesses is that its low-field mobility is lower than GaAs. This can lead to high internal positive parasitic series resistance in devices which are not fabricated in an optimum fashion. Many of the advantages of InP have been well documented and demonstrated[1,2,3,4].

InP Gunn diodes do exhibit high dc-to-rf conversion efficiency and low bias power consumption. Efficiencies of InP are typically a factor of two greater at 35 GHz than those of GaAs, and are often a factor of 5 times greater in W-band (75-110 GHz). CW output powers show a similar increase. In addition, the output power variation with case temperature can be extremely low, allowing excellent performance over wide temperature ranges. InP Gunn diodes operate at higher voltages and lower bias currents, which lead to higher device impedances. Circuit matching is easier; cold start problems are rare; and, bias circuit oscillations are not as large. Excellent reliability has been demonstrated, as well as good AM and FM noise performance.

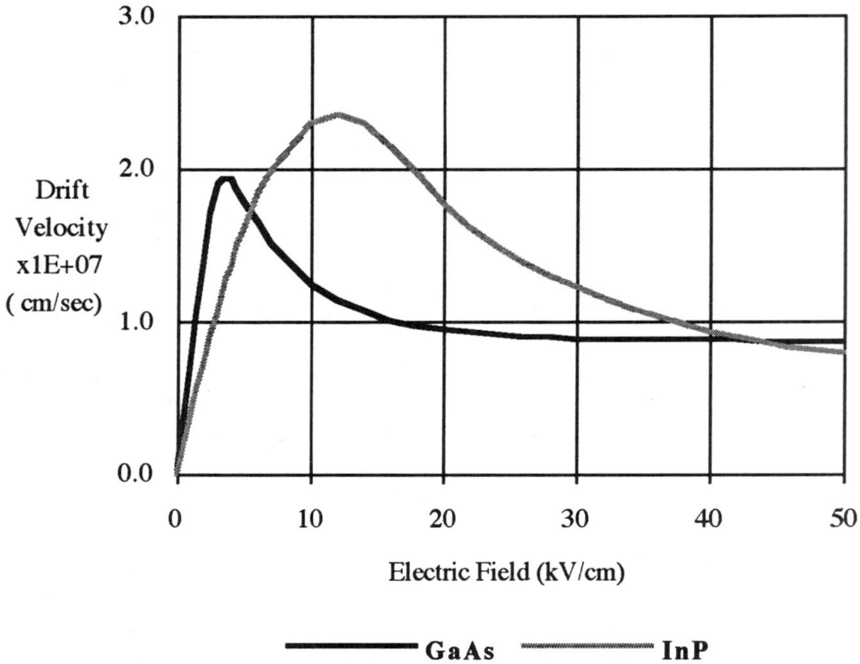

Figure 1. Electron velocity field characteristics for GaAs and InP

Table 1. Comparison of material properties.

Material Property	GaAs	InP
Peak to valley ratio (300 K)	2.2	3.5
Threshold field (kV/cm)	3.2	10.0
Effective transit velocity (cm/sec)	7.0e+6	1.2e+7
Temperature dependence of electron velocity(%/C)	-0.15	-0.1
Inertial energy time constant (psec)	1.5	0.75
Energy relaxation time (psec)	0.4	0.2
Energy gap at 300 K (eV)	1.43	1.34
Thermal conductivity (W/cm C)	0.54	0.68
Breakdown field at Nd=1 E+16 (kV/cm)	400	500

2. InP GUNN DIODE CONSTRUCTION

The physical construction of an InP Gunn diode is shown in Figure 2. A small diameter InP mesa is sandwiched between an integral gold heat sink and a top contact. The InP mesa consists of an active layer, which is adjacent to the integral heatsink, an N^+ buffer layer, and N^+ substrate material. The mesa diameter ranges from 50 to 100 microns depending on the frequency of operation and the required output power. The height of the mesa is 10 microns. Such a thin mesa is necessary to reduce the positive parasitic series resistance of the InP substrate material and to reduce skin effect losses at the very high frequencies. An integral heatsink device configuration is used for its thermal characteristics, as well as its advantages for handling in the diode assembly process.

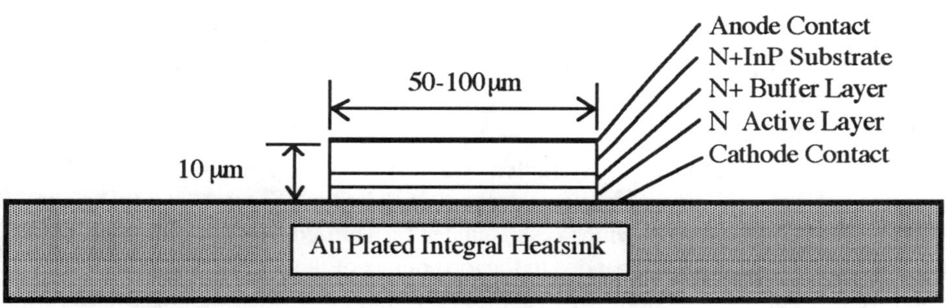

Figure 2. Physical construction of InP Gunn diode.

The epitaxial layers which are the heart of the device are produced using either vapor phase epitaxial (VPE) or metal-organic chemical vapor deposition (MOCVD) growth methods. Both approaches have produced state-of-the-art material with low background doping levels (mid 10^{13} cm^3) and high liquid nitrogen mobilities (100,000 cm^2/V-sec). There are a variety of InP Gunn epitaxial doping profiles which have been successful in producing good RF performance. Two of the most common profiles are shown in Figure 3. The two-layer profile consists of an N^+ buffer layer followed by an N-type active layer. The three layer profile has an additional layer, an N^+ contact layer. Each has advantages and disadvantages. The two-layer profile is simpler to produce with a higher yield. It has yielded the highest efficiencies and lowest operating temperatures. Since it uses a direct-metal cathode contact to the active layer, its performance is not as reproducible from wafer to wafer. The three-layer profile yields lower efficiencies, lower output powers, and operates at higher active layer temperatures, but is very reproducible from device to device and wafer to wafer.

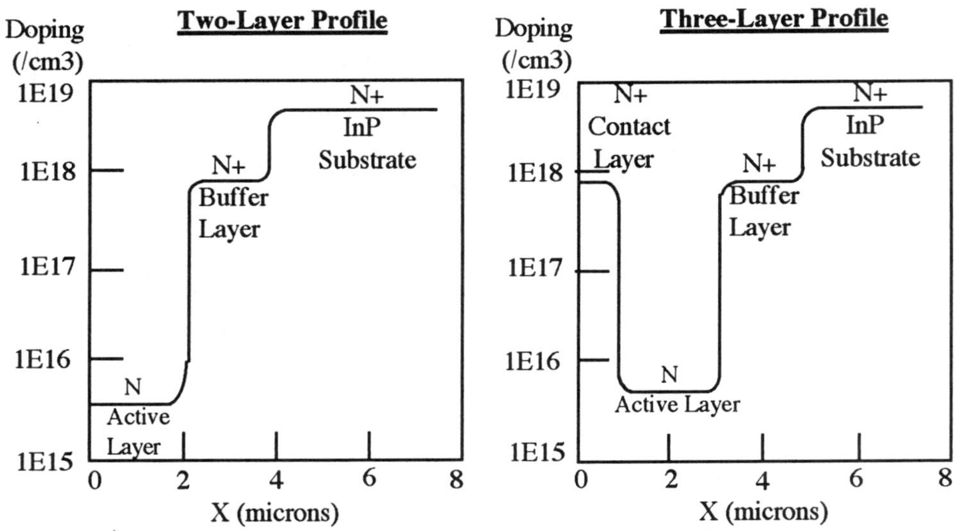

Figure 3. Two very common epitaxial layer doping profiles used to construct InP Gunn diodes.

The InP die of Figure 2 is assembled into an hermetically sealed package. The cross section of a typical millimeter-wave package is shown in Figure 4. The InP die is bonded to a gold plated pedestal which is surrounded by a brazed ring of alumina or quartz. Preformed gold leads are bonded to the top contact of the die and the top of the ring housing. A gold plated copper lid is solder-brazed to top of the ring housing to form a hermetic seal. Small diameter alumina ring housings have been successfully used up to 110 GHz. However, several quartz ring housing designs are commonly used for frequencies above 75 GHz to reduce package parasitics. Quartz ring housings are used for W-band microstrip assemblies and for all diodes above 110 GHz.

The millimeter-wave package and the methods used to assemble the device into the package have a significant effect on the performance of a Gunn diode. The equivalent circuit model of Figure 5 represents a typical Ka-band (26-40 GHz) InP Gunn diode assembled into an alumina ring housing package. The InP mesa is represented by a negative resistance in parallel with a capacitance. The other six remaining elements in the model are parasitic elements associated with the package and the assembly methods. These include inductances associated with the lid, the leads, and the pedestal to which the device is bonded, and shunt capacitances associated with the ring housing and lid. As operation is attempted at higher frequencies the impact of these parasitics becomes more significant. The net result is to lower the effective net negative resistance at the packaged device terminals. Due to these parasitic elements, special low parasitic quartz ring packages have been developed for operation at the very high frequencies, e.g. 140 GHz.

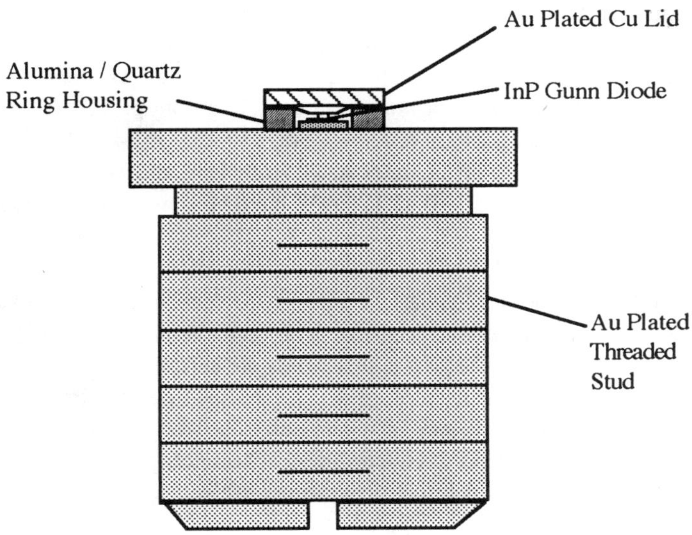

Figure 4. Cross section of packaged InP Gunn diode.

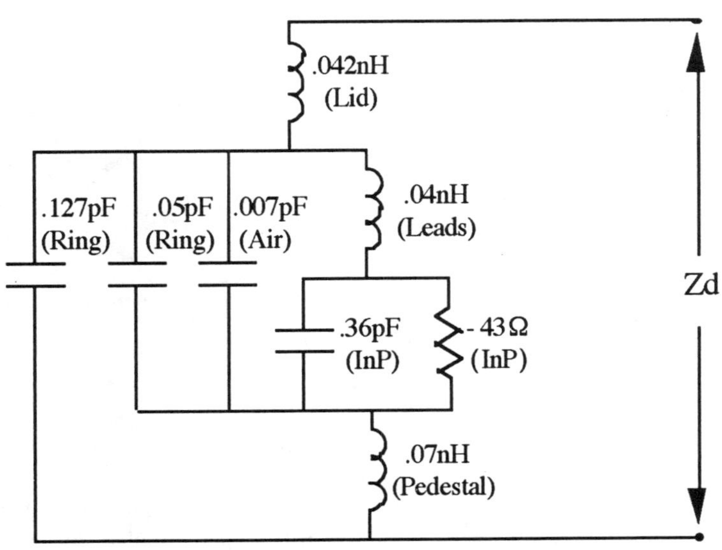

Figure 5. Model for typical packaged Ka-band InP Gunn diode.

3. POWER and EFFICIENCY

InP Gunn diodes have been available commercially since 1985. Standard catalog products can be obtained for frequencies ranging from 35 GHz to 140 GHz. Figure 6 summarizes the current performance limits for power and efficiency of CW InP Gunn diodes throughout the millimeter-wave range. CW output powers exceed 500 mW at 35 GHz, 350 mW at 44 GHz, 250 mW at 56 GHz, 150 mW at 94 GHz and 50 mW at 140 GHz. DC-to-RF conversion efficiencies range from 15% at 35 GHz, to 7% at 94 GHz to 2.5% at 140 GHz. Power levels of 175 mW have been measured at 94 GHz and 65 mW at 138 GHz.

The demand for simple solid state sources of power at frequencies beyond the currently available 140 GHz continues to grow. The fundamental limits of InP Gunn devices have not yet been reached. Performance limits are currently constrained by fabrication, packaging, and circuit technologies. The dashed lines of Figure 6 show the projections of performance up to 190 GHz, where output powers of 5 mW and 1% conversion efficiency are expected.. It is very likely as technology progresses that InP Gunn diodes will eventually be capable of generating usable power up to 220 GHz.

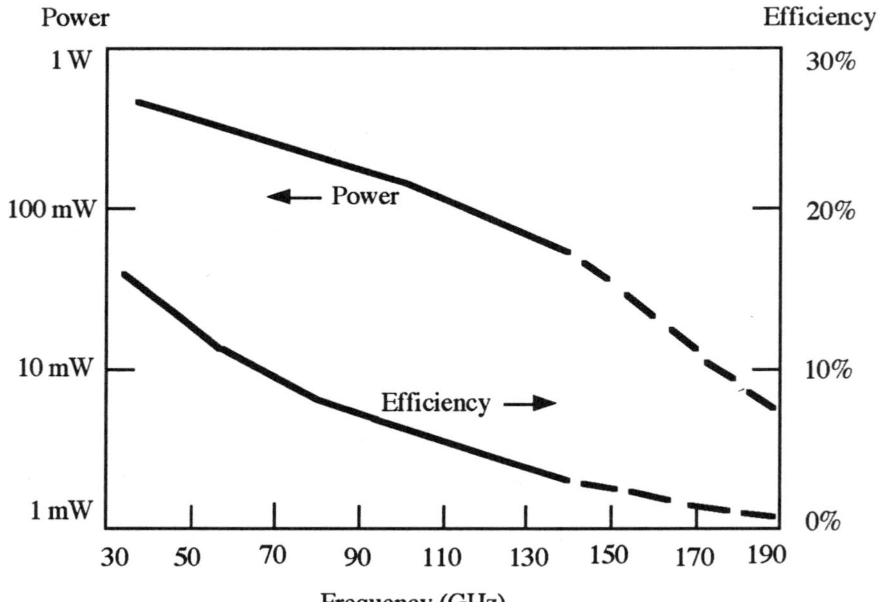

Figure 6. Power and efficiency vs. frequency for CW InP Gunn diodes.

Pulsed operation of InP Gunn diodes can yield even higher peak powers and efficiencies. Pulsed operation at duty cycles less than 10%, and at pulse lengths less than 1 υsec can alleviate thermal limitations and allow operation at higher operating voltages, higher current densities and larger operating currents. At 35 GHz a peak power of 1.25 W with an efficiency of 20% has been obtained. At 94 GHz a peak power of 325 mW with an efficiency of 7% has been obtained. Optimized pulsed devices could yield even higher peak powers.

4. MILLIMETER-WAVE APPLICATIONS

High efficiency and high output power have made InP Gunn diodes very attractive to millimeter-wave designers. Table 2 lists some of the major applications for frequencies from 35 GHz to 180 GHz and include both military and non-military uses. InP diodes are used in low-noise local oscillators for high frequency radiometers, transmitters in point-to-point radio links, VCOs in FMCW radars for both smart munitions and vehicle radars, drivers for varactor multipliers, drivers for pulsed Impatt transmitters, and exciters for electron tube transmitters.

Table 2. Millimeter-wave applications of InP Gunn Diodes

Fo (GHz)	APPLICATION
35	Smart munitions / fire control radars
38	PCN Base stations
44	Satellite communications up link
60	Secure / inter satellite communications Vehicle communications / automotive radars
75	Weather satellite radiometers
77	Vehicle radars
80	Smart munitions
92	Weather satellite radiometerS
94	Smart munitions
	Vehicle radars
140	High resolution radars
180	Weather satellite radiometers

5. OSCILLATORS

5.1 Cavity Oscillators

A wide variety of cavity-type oscillators have been constructed for many applications over the years. The complexity of cavity oscillators range from simple bias tuned sources to the more complex mechanically tuned and varactor tuned oscillators. The advantage of cavity-type oscillators is their low-loss, high Q nature which results in good phase noise. Their disadvantages are cost, size and reproducibility. Repeatable performance from unit to unit requires very tight control of mechanical parameters, which becomes more and more costly as the frequency of operation increases. Despite these disadvantages, cavity-type oscillators are still the most practical means of constructing millimeter-wave sources up to 140 GHz. A number of examples of cavity-based oscillators will be described.

PCN base station transceivers require a frequency tunable oscillator for the 37-40 GHz band. Individual oscillators are required to cover approximately a 300 MHz portion of this band and provide an output power of 100 mW or +20 dBm over a wide temperature range from -20 to +65 °C. A number of options exist for meeting this requirement including MMIC VCOs/power amplifiers or low frequency VCO/multiplier/power amplifier. For reasons of cost, complexity and performance, Gunn VCOs are currently being used to meet this requirement. Gunn oscillators have a significant advantage over MMIC VCOs in phase noise performance. The performance of an InP Gunn VCO which is used in a 38 GHz transceiver is shown in Figure 7. Frequency and power tuning curves are shown for three case temperatures -20, +25, and + 65 °C. Tuning bandwidths of approximately 500 MHz are achieved. Output power is approximately +23 dBm and varies less than 1 dB over temperature and tuning. The measured phase noise of this InP VCO is shown in Figure 8. At 100 kHz offset phase noise is -87 dBc/Hz. InP provides a significant performance advantage because of its higher power, good efficiency, and low variation with temperature.

A wide variety of W-band InP Gunn oscillators have been constructed for local oscillators and transmitter applications. Typical performance of a 94 GHz InP Gunn oscillator over a very wide temperature range, from -35 °C to +70 °C, is shown in Figure 9 . The output power at the temperature extremes is virtually the same, 90 mW. Low frequency drift with temperature was measured, 2 MHz /°C. Typically, values of 3 to 4 MHz/°C are easily obtained.

Vehicle radars are currently being demonstrated and tested using millimeter wave radars. Millimeter wave radars offer a practical solution to the problems which have prevented vehicle radars from becoming a reality. Small antenna size, narrow beamwidths, and high resolution requirements have forced system designers to turn to millimeter waves. One of the most common frequencies of operation is at 77 GHz. Many of the system approaches utilize some form of an FMCW radar. Typical system requirements call for a frequency tunable source with an output power of 50 mW and tuning bandwidths of 200 MHz. Operation is needed over a wide temperature range, -30 to +70 °C. W-band InP diodes can be used to easily construct a low cost bias tuned VCO which readily meets the system requirements. Typical performance of a 77 GHz VCO is shown in Figure 10. Frequency tuning and output power characteristics are shown for temperatures of -30, +20, and +70 °C. Output powers exceed +17.5 dBm under all conditions.

The plots of Figure 5 illustrate the characteristics of a 140 GHz InP Gunn oscillator which is now commercially available. This device possesses well-behaved characteristics as a function of bias, and low sensitivity to temperature. Power levels of 40 mW CW with conversion efficiencies greater than 2% are readily achieved.

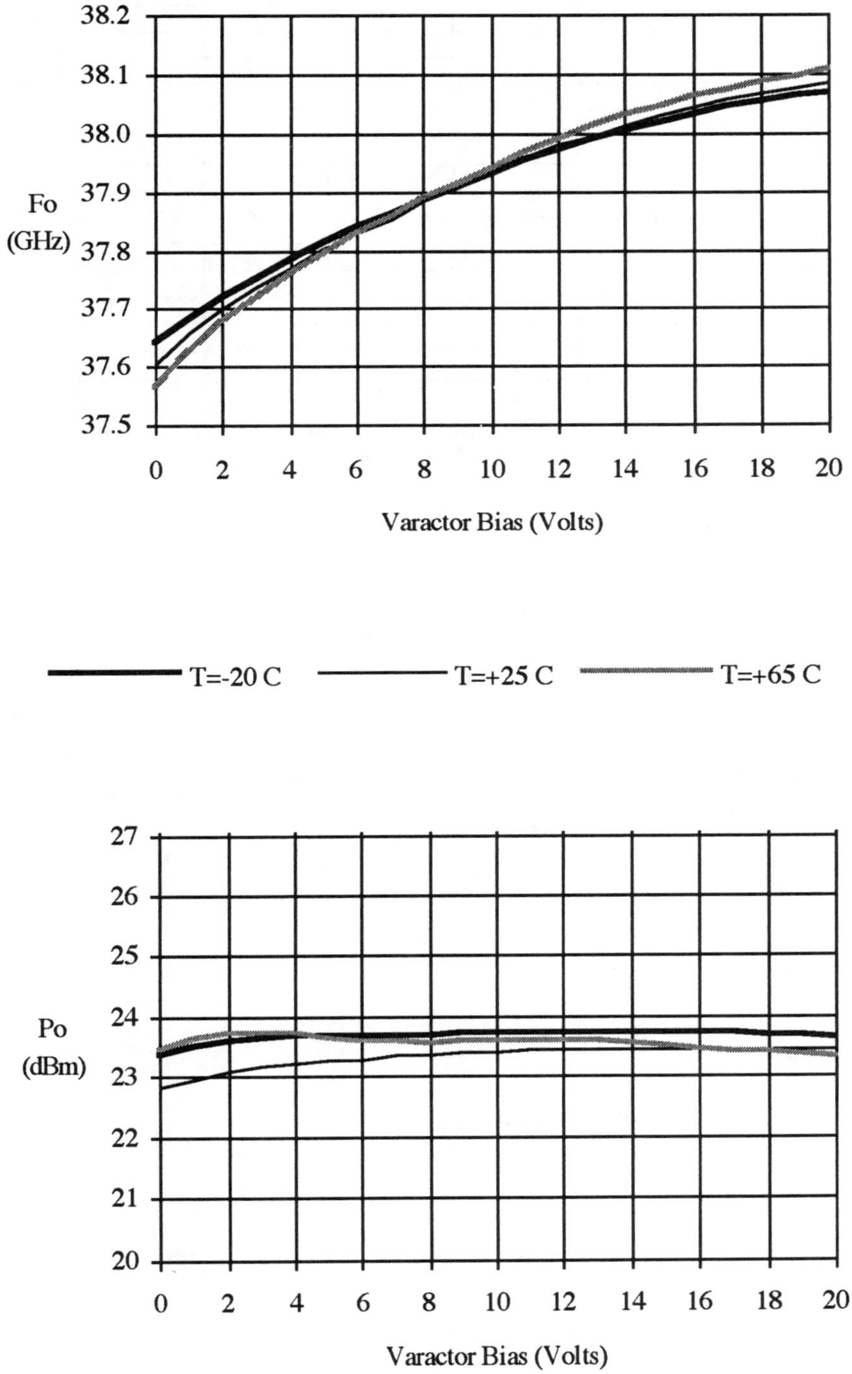

Figure 7. Frequency and power tuning characteristics of a 38 GHz VCO

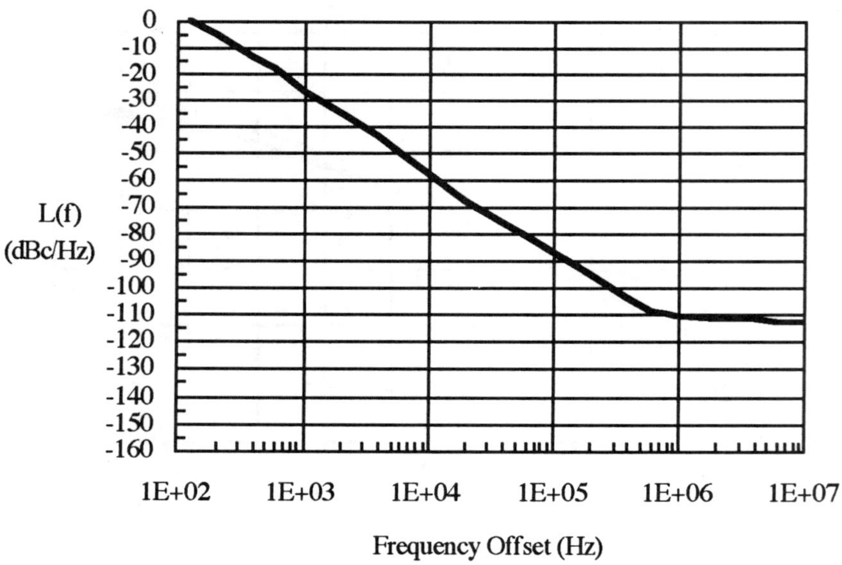

Figure 8. Measured phase noise of 38 GHz InP VCO.

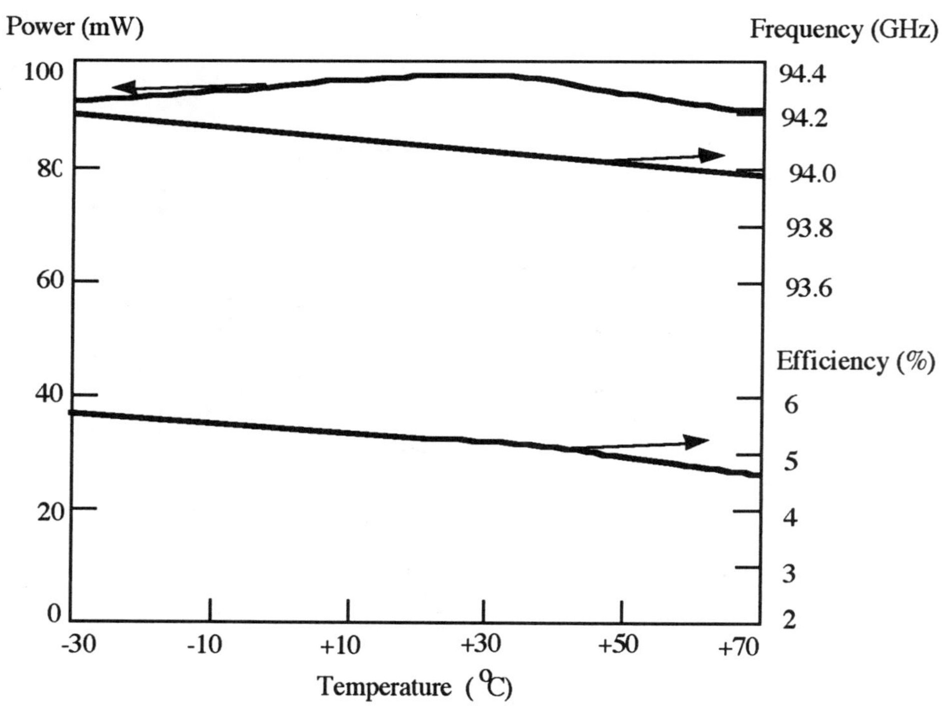

Figure 9. Performance of 94 GHz InP oscillator over temperature.

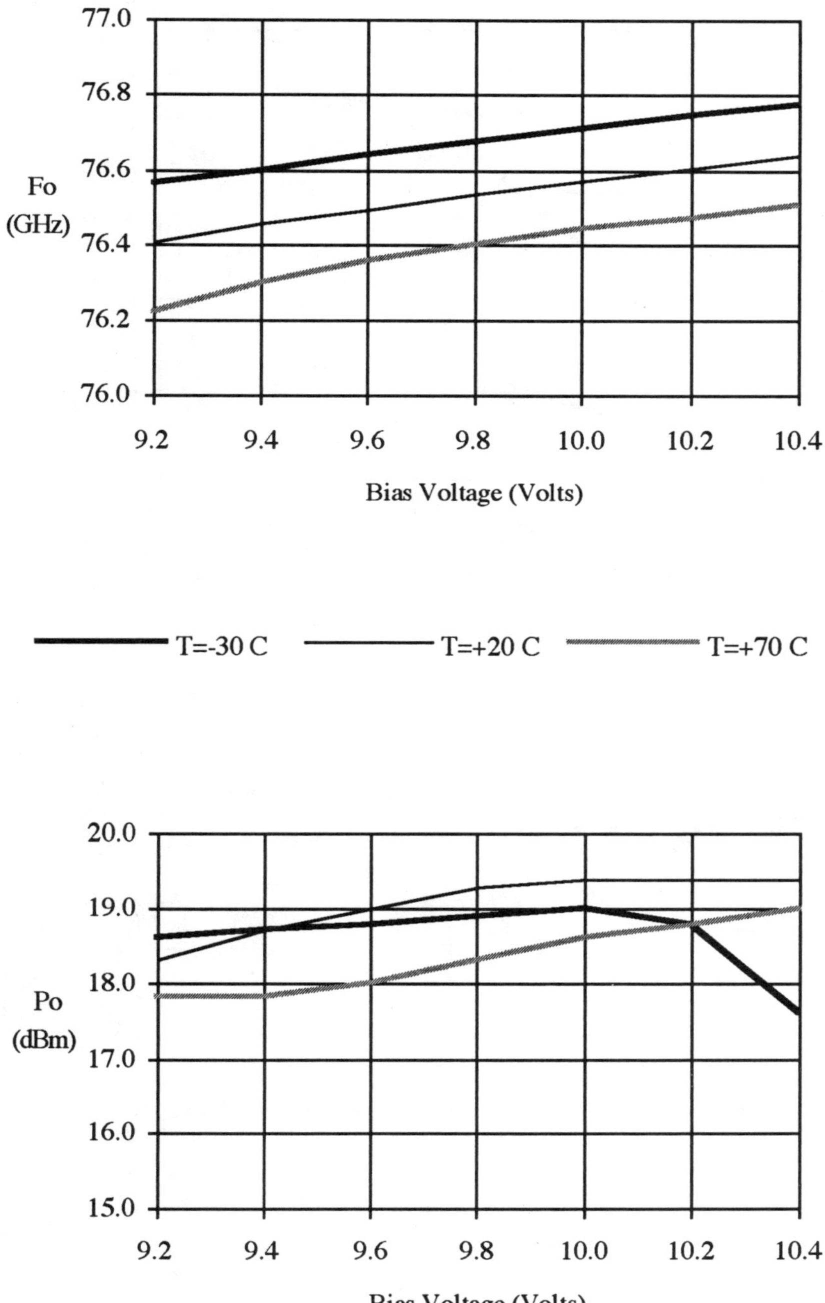

Figure 10. Bias tuned frequency and power characteristics of a 77 GHz InP oscillator for vehicle radar transceivers.

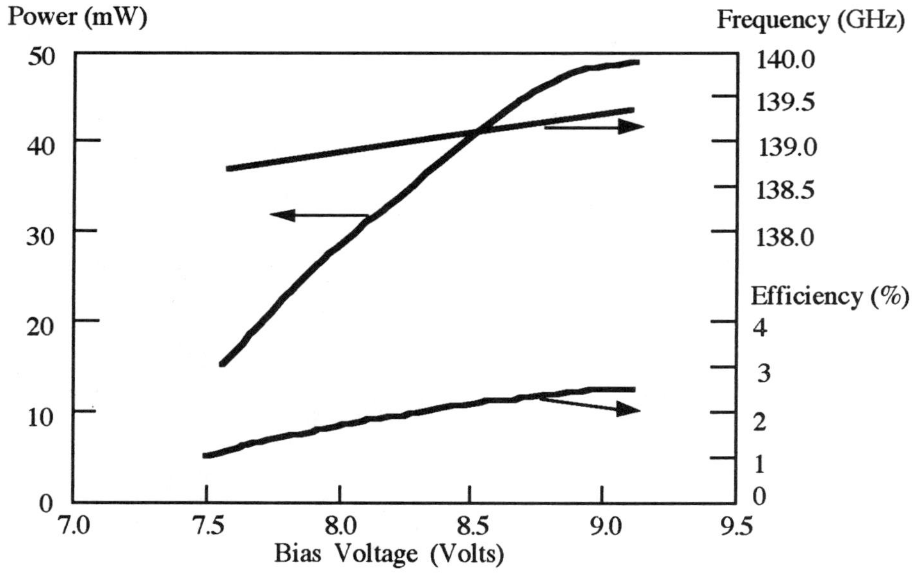

Figure 11. Bias characteristics of a 140 GHz InP Gunn oscillator.

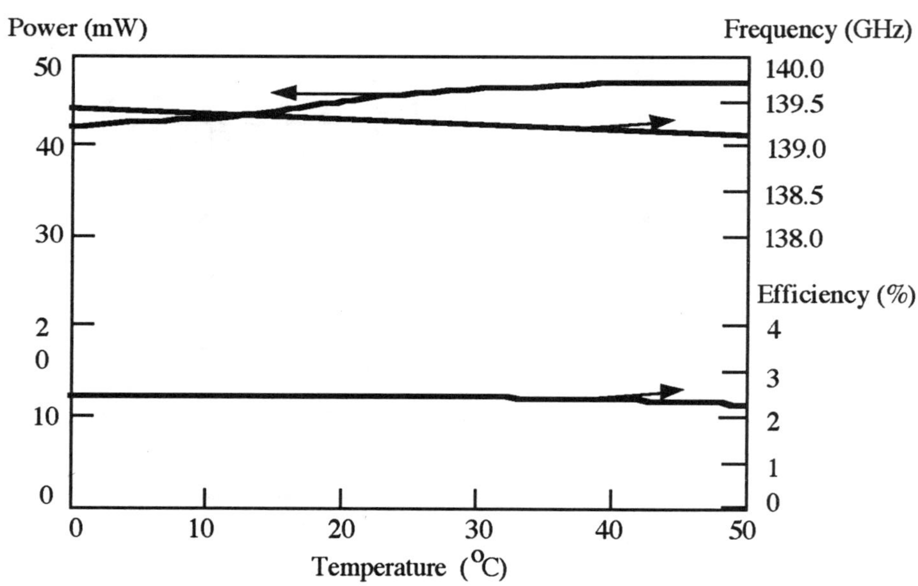

Figure 12. Performance of an InP 140 GHz oscillator over temperature.

5.2 Power Combining

InP Gunn diodes are readily combined in a variety of power-combining circuits to increase output powers. High combining efficiencies, approaching 100%, are routinely achieved. Two diodes have been combined to yield 500 mW at 60 GHz. Two- and four-diode combiners have been built at 94 GHz using radial resonators coupled in an in-line waveguide configuration. Output powers of 175 mW were achieved with two diodes and 250 mW with four diodes.

5.3. Microstrip Oscillators

The implementation of millimeter-wave systems demands increased integration to lower cost and to reduce size. Microstrip varactor-tuned oscillators have been developed and demonstrated in both Ka-band (26-40 GHz) and W-band (75-110 GHz) and show the potential for integrating InP oscillators into a microstrip assembly. Microstrip oscillator configurations make it significantly easier to implement series varactor tuning and achieve larger tuning bands. Tuning bandwidths of 1 GHz were obtained in Ka-band while tuning bandwidths of 1.5 GHz were obtained at W-band.

A typical microstrip circuit layout is shown in Figure 13. This circuit layout was used with 10-mil Duroid substrate. The InP Gunn diode is packaged in a standard diode package of the type shown in Figure 3. A beamlead varactor is used in a DC and RF series configuration in the center of an open-circuited stub. The matching circuit interfaces with a 50 ohm output line which contains a DC block and a finline transition to a waveguide output. The finline transition would not be present if the VCO were a part of an integrated assembly. Measurements were made at the waveguide port following the finline transition. The tuning characteristics of a typical 35 GHz microstrip VCO are shown in Figure 6. Output powers greater than 200 mW were achieved over the full tuning range. These types of oscillators have very good performance over wide temperature ranges and do not suffer from cold start problems. The phase noise has been measured and it was found to be within several dB of the results shown in Figure 8. Microstrip 38 GHz oscillators will provide a significant cost and operational advantage for use in transceivers for PCN base station applications.

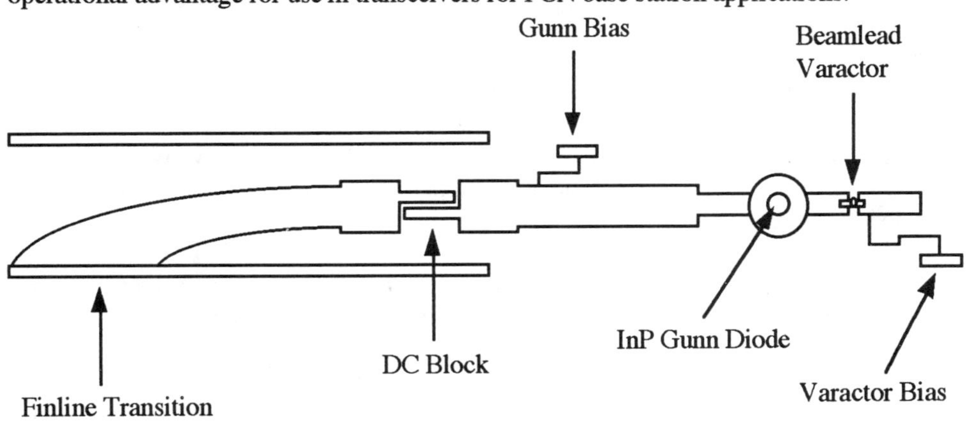

Figure 13. Microstrip VCO circuit layout.

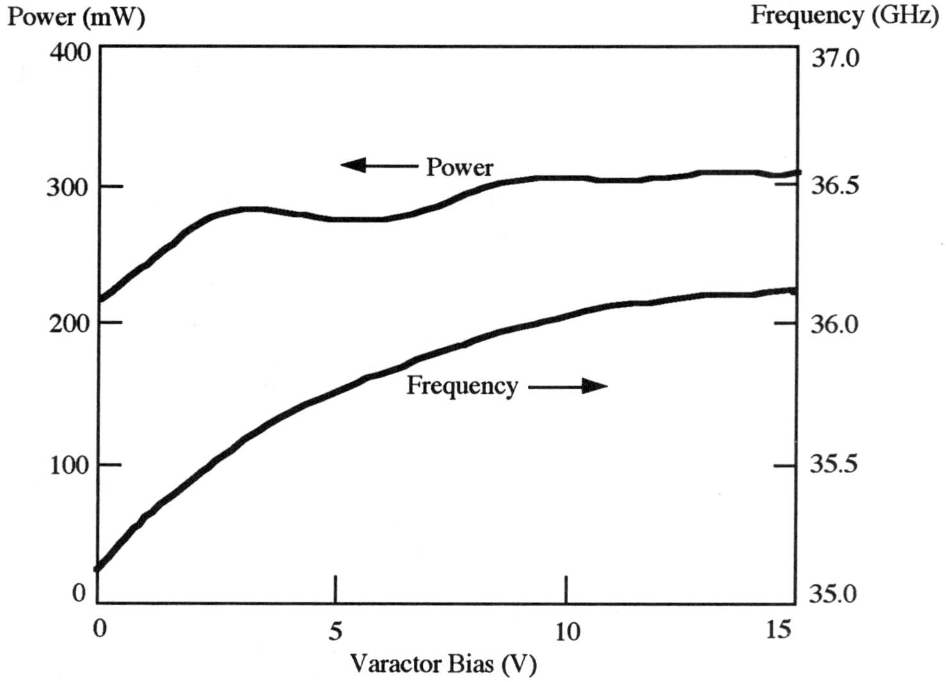

Figure 14. Tuning characteristics of an InP microstrip VCO.

6. MM-WAVE AMPLIFIERS

The intrinsic negative resistance of an InP Gunn diode can be used to provide stable, non-oscillating power amplification to very high frequencies. Millimeter-wave power amplifiers have been developed for frequencies from 35 GHz through 110 GHz using InP Gunn diodes as the amplifying device. Instantaneous bandwidths of 20 GHz have been observed in W-band. Multistage amplifiers typically produce small signal gains of 30 dB with noise figures in the 16 dB range. Saturated output powers range from 0.5 watt at 35 GHz to 100 mW at 94 GHz. Power combining techniques have been developed which use two diodes in the final stage of amplification, allowing output powers to be doubled. Amplifiers have been demonstrated which produce one watt at 44 GHz and 200 mW at 94 GHz.

Throughout the millimeter-wave portion of the spectrum there exists a need to provide power amplification. Electron tube amplifiers and their associated power supplies offer a complex and costly solution to the problem. Solid state options are simpler, less expensive, and more reliable, but are also limited in output power capability. However, for many system applications, the power that can be achieved with solid state amplifiers (100 mW to one watt) is sufficient to to be useful and may lead to system advantages in the areas of size, cost, and reliability. Three-terminal devices offer a cost effective solution for low power requirements. MMIC amplifiers using HBT and PHEMT technology offer promise for future millimeter-wave applications. Their limited availability leaves many millimeter-wave system developers with few options for current system development and implementation. A solid state amplifier option for medium power requirements which is available today for frequencies up through 100 GHz uses indium phosphide Gunn diodes.

InP Gunn amplifiers have been demonstrated and are commercially available at all of the common millimeter-wave bands centered at 35, 44, 56, and 94 GHz. Designs have been optimized for high output power, for performance over wide temperature ranges, and for performance independent of input and output load conditions. Amplifiers with small signal gains of 30 to 40 dB are quite typical. As these amplifiers are driven into their large signal regime, their gain slowly compresses until their saturated output power is reached. For all drive conditions, from small to large signal, these amplifiers remain stable, non-oscillating negative resistance amplifiers.

The most common use of these InP amplifiers is as a post amplifier following an upconverter as shown in Figure 15.. The InP amplifier is used to increase the signal level from the 0 dBm level to the +20 dBm level, where it provides the final transmit power, or is used to drive an even higher power transmitting device. In some cases the InP amplifier has been used to drive pulsed silicon Impatt diodes, and in many others it was used to drive high power electron tubes. InP amplifiers have been cost effective in systems where they preceded either CW or pulsed extended interaction amplifiers (EIA's) or traveling wave tubes (TWT's). In these situations, the overall system cost and complexity were reduced significantly by lowering the gain requirement for the high power tube. Numerous millimeter-wave instrumentation systems have utilized InP amplifiers to increase system dynamic range.

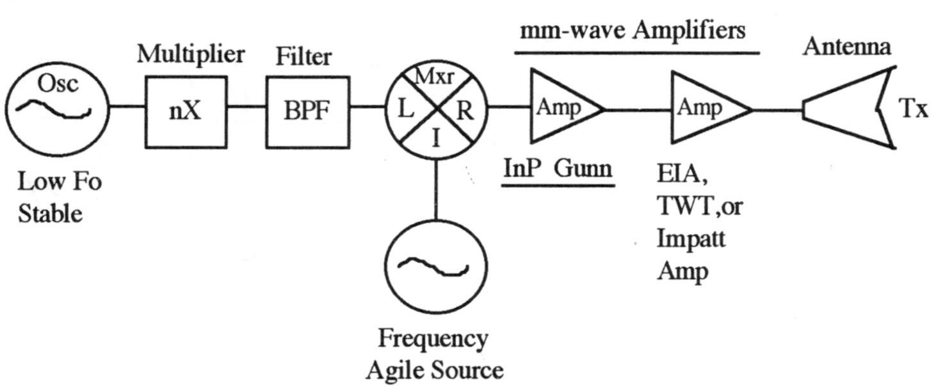

Figure 15. Millimeter-wave upconverter using an InP Gunn amplifier.

6.1 Reflection Gain Modules

When biased to approximately twice the threshold field level for onset of the Gunn effect, an InP Gunn diode possesses a significant negative resistance for frequencies in the vicinity of the transit-time frequency. It is this negative resistance which is used to provide reflection gain. To achieve stable, non-oscillating amplification using a Gunn diode, proper selection must be made of the active layer doping, length, and cross sectional area, as well as the cathode electrical characteristics, and matching circuit impedances.

A usable gain stage requires the proper negative resistance diode, impedance matching circuit, and ferrite circulator as shown schematically in Figure 16. The reflection gain is given by the following expression [5]

$$G = |\Gamma|^2 = |(Z_c - Z_d^*)/(Z_c + Z_d)|^2$$

Using impedance de-embedding techniques, the small signal negative impedances of packaged InP Gunn diodes have been measured. A typical measured diode impedance locus is shown in Figure 17 along with a measured matching circuit impedance locus for a Ka-band application. For stable amplification, the real part of the circuit impedance must be maintained greater than the diode negative resistance and the circuit reactances must not be allowed to cancel. If both these conditions are maintained, the gain at any particular frequency is approximately inversely proportional to the separation of the two impedance points. In other words, in order to achieve stable gain, the circuit impedance locus is brought close to the diode impedance locus, but is not allowed to cross it. This will insure stable amplification and prevent oscillations. The packaged Gunn device

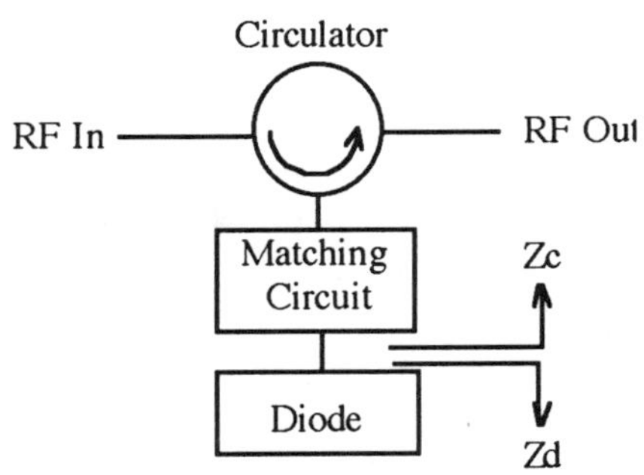

Figure 16. Reflection gain module

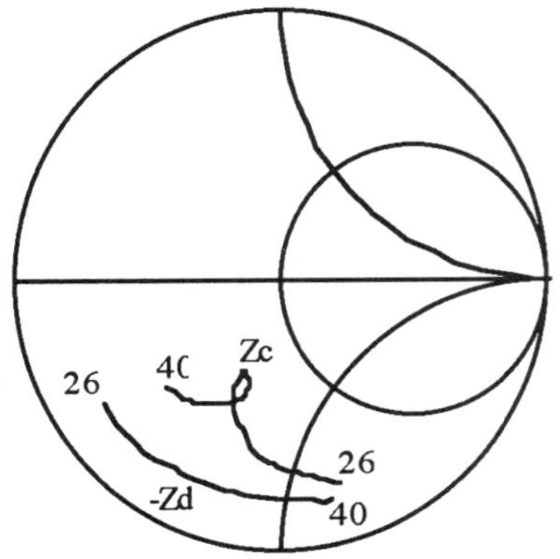

Figure 17. Impedance vs. frequency (GHz) for InP diode (Zd) and matching circuit (Zc).

impedance locus is a function of frequency, bias voltage, active layer temperature and rf voltage levels. The circuit impedance locus must be far enough removed from the diode impedance locus that, over temperature and a wide range of rf drive levels, the oscillation condition is never met. Typically, the negative resistance component of the device impedance is in the 5 to 10 ohm range. For all of the millimeter-wave bands, coax-to-waveguide impedance matching circuits have been developed. The characteristic impedance of the coaxial section, its length, and coupling to the waveguide fields are used to adjust the location of the circuit impedance locus and to control the gain level and bandwidth. At millimeter-wave frequencies, performing this transformation becomes a limiting factor, particularly as device impedance levels become lower, as is often the case with high power devices.

InP Gunn diodes have been produced which possess negative resistances over frequency ranges wider than most waveguide bands. Realization of impedance matching circuits limits the usable bandwidth for stable reflection gain. Figure 18 illustrates a measured reflection gain response in W-band. A gain of 8 dB was achieved over more than 20 GHz of bandwidth.

The impedance locus of properly designed InP Gunn diodes is relatively stable with temperature. Figure 19 shows the measured reflection gain response of an InP Gunn diode at -35 ^{o}C and +60 ^{o}C. The magnitude of the diode negative resistance changes very little. There is a slight shift in the center frequency of the gain response, indicating that the device reactance is changing slightly. Stable amplification has been easily achieved over wide temperature ranges without oscillations.

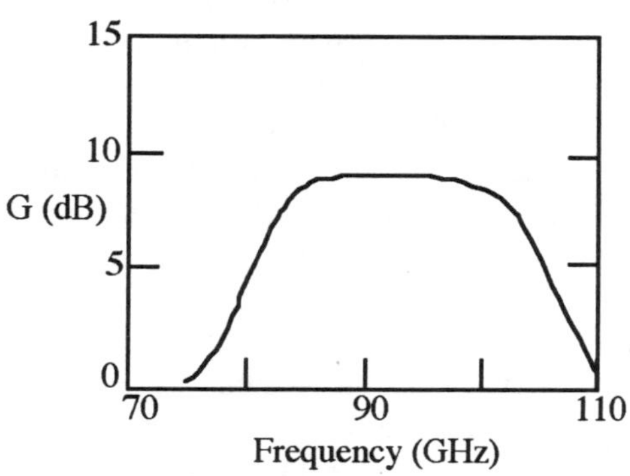

Figure 18. Broadband gain response of an InP Gunn diode.

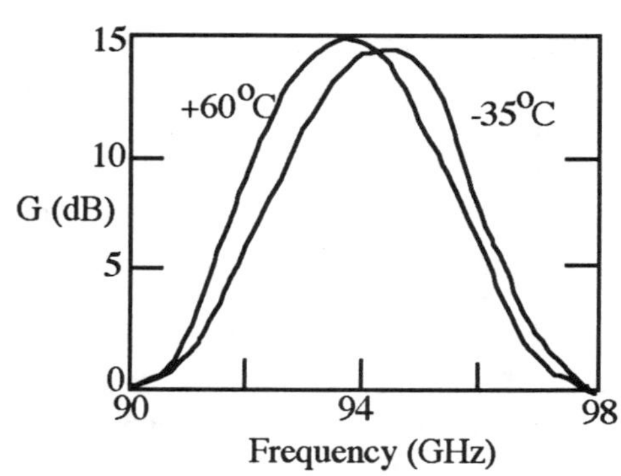

Figure 19. Effect of temperature on reflection gain response.

6.2 Multistage Amplifiers

To implement an amplifier, a ferrite circulator is used to separate incident and reflected signals from the reflection gain module. Successive gain stages are cascaded to provide multistage gain. A typical four-stage amplifier configuration is shown in Figure 5. For each reflection gain module there is one circulator junction. In addition, interstage junction isolators are used to prevent oscillations resulting from potential interaction of adjacent gain stages. Input and output junction isolators are used to provide good input match and to allow operation independent of load conditions. The number of ferrite junctions used in an amplifier with N gain stages is 2N+1.

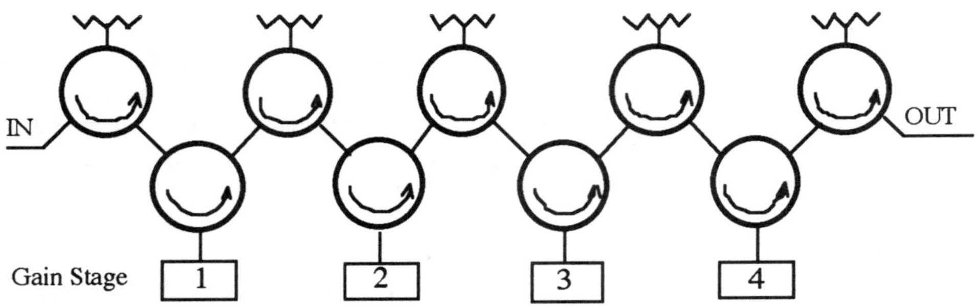

S.S.G. (dB)	8	8	7	7	=	30
L.S.G. (dB)	8	5	4	3	=	20

Figure 20. Four-stage amplifier.

For an amplifier stage having 8 to 10 dB of gain, the usable circulator bandwidth is defined by the bandwidth over which its individual port return loss is greater than 20 dB. The difference between the circulator port return loss and the reflection gain level determine the magnitude of the gain ripple across the amplifier bandpass. Excessive gain ripple is introduced when values of return loss minus reflection gain are below 10 dB.. This makes it extremely difficult to cascade multiple stages. Excellent circulators have been developed over the years which are well suited for use in Gunn amplifiers throughout the millimeter wave range. The insertion loss per pass through a single circulator junction is typically 0.3 dB. As frequencies increase, the operational bandwidth of ferrite junction circulators decreases, and the ferrite circulator becomes the bandwidth-limiting element in the amplifier. For most amplifiers, the cascaded circulators are manufactured using a single multijunction housing. Multijunction housings have been built at W-band with five, nine, and eleven junctions in one housing. This provides a cost effective and compact approach to building multistage amplifiers. Typical small signal and large signal gains of each stage of a four stage W-band amplifier are also shown in Figure 20. The output stages are tuned for lower small signal gain than the input stages. Total small signal gains as high as 40 dB have been provided. Typically the large signal gain of a four stage amplifier at its rated output power is 20 dB.

The output power capability of a multistage amplifier approaches that of an oscillating InP Gunn diode at the same frequency. Figure 21 shows the output power limit which has been achieved with InP amplifiers throughout the millimeter-wave range. Typical units can produce output powers of +28 dBm at 35 GHz, +27 dBm at 44 GHz, +25 dBm at 56 GHz, and +20 dBm at 94 GHz. Power combining is used to increase the output power capability beyond these limits. Techniques have been developed using two diodes in the output stage to double the output power, while maintaining stable amplifier characteristics. With two matched amplifier stages, excellent combining efficiency is obtained and output powers are increased by 3 dB. Output power of 1 watt has been achieved at 44 GHz, and 200 mW at 94 GHz. This dual diode approach has increased the current output power limit of InP amplifiers as shown in Figure 21.

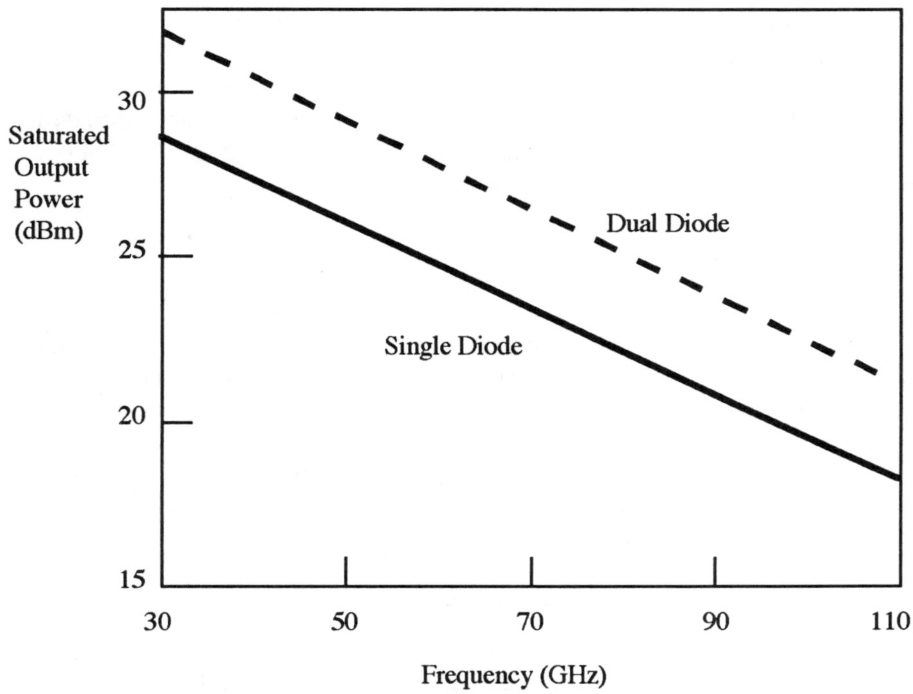

Figure 21. Saturated output power of InP amplifiers with single-diode and dual-diode output stages.

The performance of InP amplifiers will be enhanced by developments in several areas. The output power capability of the amplifying device can be increased by approximately 3 dB through improvements in device packaging and heatsinking techniques. Distributed area devices (e.g., annular ring structures) and embedded diamond heatsinks would allow thermal resistances of the packaged device to be significantly reduced, permitting higher bias powers to be dissipated. Operation at higher voltages and larger areas would then be possible, yielding higher output power.

The highest frequency at which this type of amplifier has been demonstrated is currently 110 GHz. However, fundamental mode operation of InP Gunn diodes has been demonstrated to extend to at least 140 GHz. It is possible then to construct negative resistance InP Gunn amplifiers up through 140 GHz. Manufacturing of the InP Gunn diode for these frequencies has already been demonstrated. The limiting factors will be construction of matching circuits and the availability of low loss broadband circulators. It is predicted that InP Gunn amplifiers can be constructed at 140 GHz with output powers in the 25-50 mW range with bandwidths of 2-4 GHz.

For most laboratory or ground-based applications, the current size and weight of amplifiers constructed in waveguide media are acceptable. However, implementation of microstrip amplifiers would make the use of these amplifiers much more feasible for air-borne or guided munition applications. The low cost and the potential of integration with other elements is greatly increased with a planar circuit media. Preliminary development work has been completed at W-band to demonstrate the feasibility of constructing gain modules in a microstrip format. Reflection gain using microstrip matching circuits has been demonstrated. The limiting factor at this time is the availability of ferrite microstrip circulator junctions with sufficient bandwidth and low insertion loss. It is anticipated that W-band amplifiers with 50 -100 mW output power capability could be constructed in a microstrip format.

6.3 InP Power Amplifier at 60 GHz

The 60 GHz amplifier shown schematically in Figure 22 is an example of a typical InP millimeter-wave power amplifiers. It consists of four gain stages and has includes input and output isolators. The power transfer curve measured at 61.5 GHz shows that the amplifier has a small signal gain of approximately 30 dB and it slowly compresses as the input drive is increased. The "soft" saturation is characteristic of InP Gunn amplifiers. This amplifier ultimately saturates at an output power >+27 dBm. At an input drive of +5 dBm the amplifier reaches an output power of +25 dB. Usable bandwidth of this unit is 4 GHz.

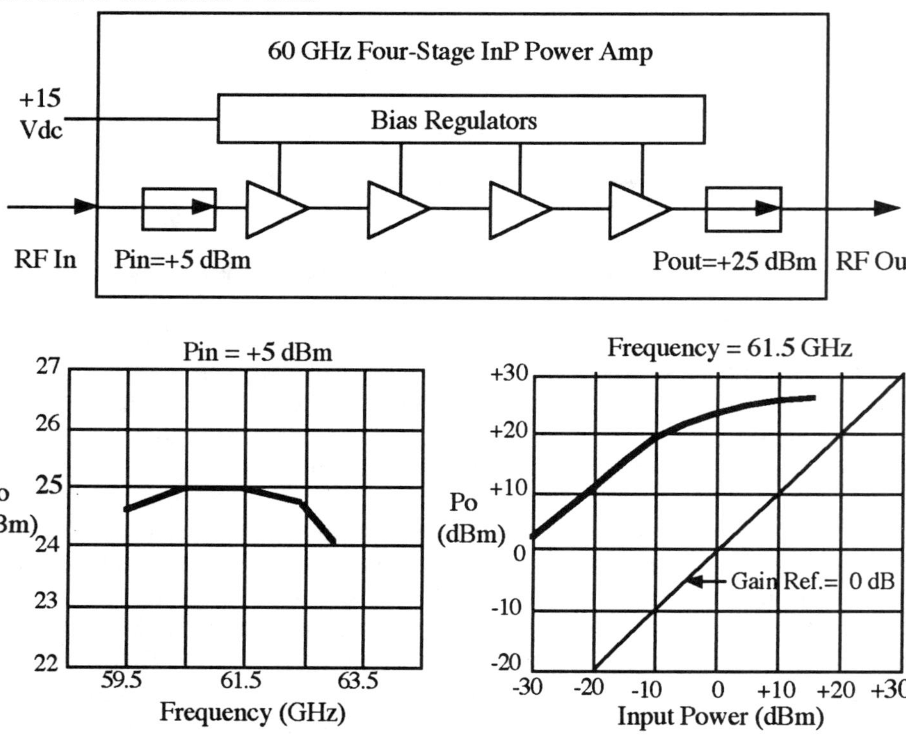

Figure 22. Measured output power and power transfer characteristic of 60 GHz four-stage InP power amplifier.

6.4 InP Power Amplifier at 95 GHz

A multi-stage 95 GHz amplifier with a dual-diode output stage is described in Figure 23. It was designed to operate with an input drive of - 5 dBm and provide a minimum output power of +22 dBm over a 2 GHz bandwidth from 94.5 to 96.5 GHz. This amplifier was used to amplify the output of an upconverter and drive a high power TWT transmitter in a 95 GHz missile tracking radar system. It consists of a five-stage driver chain followed by a dual-diode, hybrid coupled output stage. A 3 dB, 90° hybrid coupler was produced using the narrow-wall directional coupler design described by Tanaka [6]. Two closely matched high power reflection gain modules were installed on the coupled ports, and their reflection power was combined at the output port. To achieve good combining, the gain modules must be matched in amplitude and phase over frequency, temperature, and drive level. The measured data in Figure 23 shows a power transfer curve measured at 95 GHz, as well as, the output power versus frequency at an input drive of -5 dBm. Output powers of +23 dBm were achieved. Amplifier bandwidth exceeded 2 GHz. This approach has also been used at 44 GHz where one watt of output power was obtained..

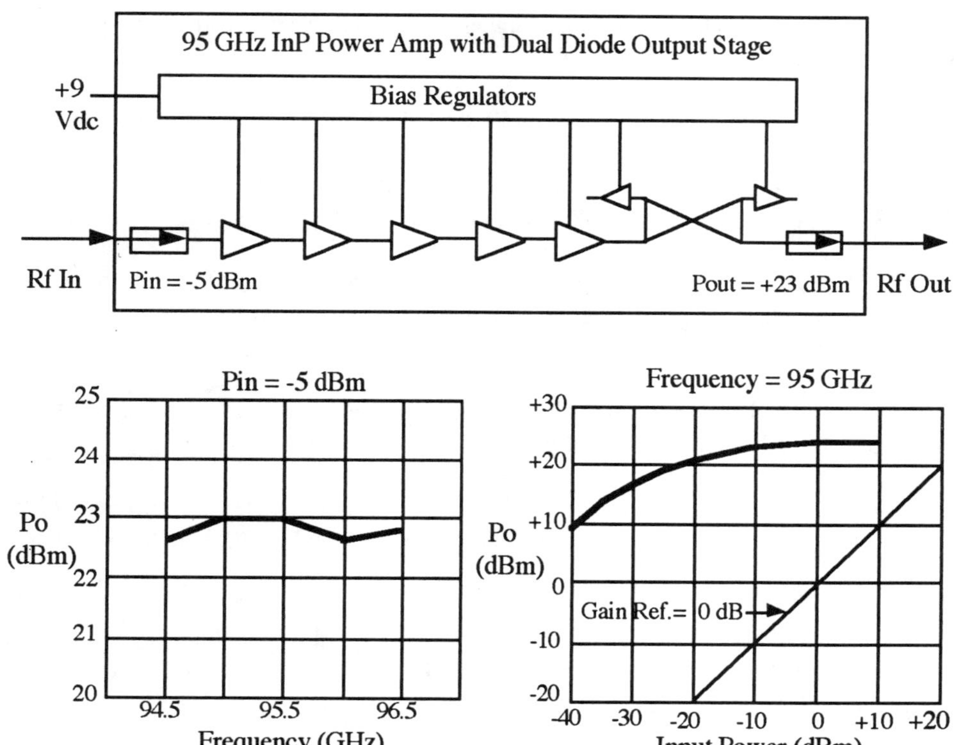

Figure 23. Measured output power and power transfer characteristic of a 95 GHz amplifier with dual diode output stage.

7. SUMMARY

InP Gunn diodes have unique properties which make them well suited to the task of efficient power generation and amplification throughout the millimeter-wave range. Diodes are commercially available for frequencies ranging from 35 GHz to 140 GHz. Output powers range from 500 mW at 35 GHz, to 150 mW at 94 GHz, to 50 mW at 140 GHz. Efficient, simple, reliable oscillators are now possible to very high frequencies and are being used in a wide range of application as power sources for radar systems, communication links, and instrumentation application

Many of the medium power requirements for millimeter-wave amplifiers can be met using InP Gunn diode reflection amplifiers. InP Gunn diodes have unique characteristics which make them suitable for millimeter-wave reflection amplifiers. Multistage amplifiers have been developed for all of the common millimeter-wave bands centered at 35, 44, 56, and 94 GHz. Four stage amplifiers with small signal gains of 30 dB and large signal gains of 20 dB are typical. Power levels of 500 mW at 35 GHz and 100 mW at 94 GHz are readily produced. Output powers can doubled by using two diode power combining in the output stage. Using power combining techniques, power levels of 1 watt at 44 GHz and 200 mW at 94 GHz have been demonstrated. Extension of this technology to even higher frequencies., higher powers and to a microstrip format are now feasible.

8. REFERENCES

1. B.K. Ridley: "Anatomy of the Transferred-Electron Effect in III-V Semiconductors", Journal of Applied Physics, Vol.48, No.2, February 1977, pp. 754-764.

2. M.R. Friscourt, P.A. Rolland, A. Cappy, E. Constant, G. Salmer: "Theoretical Contribution to the Design of Millimeter-Wave TEO's", IEEE Trans. ED, Vol.ED-30, No.3, March 1983, pp.223-229.

3. J.D. Crowley, J.J. Sowers, B.A. Janis, F.B. Fank: "High Efficiency 90 GHz InP Gunn Oscillators", Elect. Lett., 1980, Vol.16, No.18, pp.705-706.

4. M.A. di Forte-Poisson, C. Brylinski, G. Colmer: "High Power, High Efficiency LP-MOCVD InP Gunn Diodes for 94 GHz", Elect. Lett., 1984, Vol.20, No. 25, pp.1061-1062.

5. B.S. Perlman, C.L. Upadhyayula, R.E. Marx: Wide-Band Reflection-Type Transferred Electron Amplifiers", IEEE Trans. Microwave Theory Tech., Vol. MTT-18, No. 11, November 1970, pp. 911-921.

6. T. Tanaka: "Ridge-Shaped Narrow Wall Directional Coupler Using TE_{10}, TE_{20}, and TE_{30} Modes", IEEE Trans. Microwave Theory Techs., Vol. MTT-28, No. 3, March 1980, pp. 239-245.

Millimeter-wave power HEMT technology

Huan-Chun Yen
K. Tan, R. Lai, C. H. Chen
M. Biedenbender, D. Streit, M. Wojtowicz

Electronics System and Technology Division
Space and Electronics Group
TRW Inc.
One Space Park, Redondo Beach, CA 90278

ABSTRACT

Recent improvements in material structure, device layout, fabrication process, and device matching have resulted in simultaneous improvements in power output, power gain, and power-added efficiency from large size GaAs-based pseudomorphic AlGaAs/InGaAs HEMTs at millimeter-wave frequencies. Consequently efficient watt-level building blocks are now available to build solid state power amplifiers capable of 1 to 20 watt power output for next generation military and civilian systems. This paper describes a state-of-the-art 0.15 μm power HEMT technology suitable for applications in this frequency range. Examples of MIC and MMIC power amplifier circuits are given to illustrate the achievable power performance. Preliminary study of the power HEMT reliability is also discussed.

1. INTRODUCTION

Two terminal devices, especially the impact avalanche and transit time (IMPATT) diodes have been the primary high power solid state sources at the millimeter-wave (MMW) frequencies for some time. Depending on the operating frequencies, both silicon and GaAs IMPATTs have been demonstrated to produce one to several watts in the CW modes and greater than ten watts in the pulsed modes from a single device[1]. Despite these impressive power levels and their ability to meet some basic system performance requirements (e.g. seeker applications), they are gradually replaced by the emerging three-terminal power transistor technologies for a number of reasons: 1) being a two-terminal device, the IMPATT amplifier requires a non-reciprocal device such as circulator to separate the input and output; 2) being a negative resistance device, the IMPATT amplifier requires a high Q circuit to operate which for the best power performance usually means a metallic waveguide circuitry with very tight mechanical tolerance and labor intensive circuit tuning; 3) the IMPATT amplifier has limited waveform versatility. In most practical cases, it is dominated by the frequency-modulated signals; 4) the IMPATT amplifiers have low DC to RF conversion efficiency, typically only a few percent. As a result, extensive DC supply and heat dissipation management are required; 5) the IMPATT amplifier construction does not easily lend itself to the microelectronics fabrication techniques for the advantages of monolithic integration as well as higher yield and reduced cost.

The three-terminal devices that have both high power and low noise capabilities, on the other hand, offer an attractive alternative. Not only these transistor amplifiers can be fabricated with the well-established microelectronics fabrication techniques, but multiple circuit functions can be integrated on a single chip, achieving significant reductions in size and weight at the same time. Furthermore, sophisticated waveforms can be amplified and transmitted while highly desirable functions such as adaptive power management or automatic gain control can be easily implemented. However, until recent, diminishing power output from a single transistor at millimeter-wave frequencies has prevented us from building practical, affordable high efficiency solid state power amplifiers (SSPA) with 1 to 20 watts power output. Such SSPAs are planned for several next generation civilian and military systems for communications, seekers, and radar applications. Low power output from a single device requires extensive power combining to achieve a useful power level of one watt or higher, an approach invariably accomplished at the expense of the overall SSPA efficiency and reliability as well as at an increased cost.

Among the various three-terminal power transistor technologies, the high electron mobility transistor (HEMT) has been demonstrated to be the most viable candidate for efficient power generation at the millimeter-wave frequencies[2,3]. Recent results from large size (>300 μm) GaAs-based pseudomorphic AlGaAs/InGaAs power HEMTs have shown that high power output can be achieved at high efficiency[4]. Coupled with the demonstrated reliability, it is considered as the choice for the near term insertions at this frequency range. Some recent efforts were reported on the emerging InP-based power HEMT technology[5,6]. The initial results, although promising, still lag behind what can be achieved with the GaAs-based power HEMTs. Since the InP power HEMT technology is still in active development, we will not discuss it further in this paper. The heterojunction biploar transistor (HBT) is another power transistor technology actively pursued by the industry, especially for frequencies below 30 GHz using GaAs/AlGaAs HBTs[7,8]. However, using the advanced materials such as InP, InGaAs, GaInP, InGaAlAs in the HBT profile designs, we expect that the advanced power HBT design could become a competing alternative at this frequency range in the mid- to far-term.

In the following, we will concentrate on a 0.15 μm gate length GaAs-based pseudomorphic power HEMT design that is capable of efficient high power generation from 20 to 65 GHz. Excellent power performance was realized through advancements made in the areas of III-V compound semiconductor material growth, device layout, fabrication process, device characterization, and device-circuit matching designs. Tradeoff was made to achieve both power performance and reliability simultaneously. Representative examples of both MIC and MMIC power HEMT circuits based on this process are presented to show the attainable power performance up to 65 GHz. Watt-level power modules can be constructed with a small number of these types of chips. If a higher power output is desired, the modules can be further combined in a low loss combiner circuit such as a radial line combiner.

2. POWER HEMT DESIGN AND CHARACTERISTICS

A common GaAs-based 0.15 μm power HEMT microcell design was developed and demonstrated to be capable of excellent power performance from 20 to 65 GHz. Since a large number of planned applications operate in this frequency range, the common design approach was chosen to reduce the development and recurring cost without compromising the performance. Furthermore, a single common design makes the fabrication process more stable and robust due to the frequent exercising of the same process on a fabrication line. The power HEMT process was optimized to achieve the required performance and reliability simultaneously. Two power HEMT structures that are commonly used for power generation at these frequencies were investigated: doped channel and double heterostructure. Their respective device cross sectional views are shown in Figure 1.

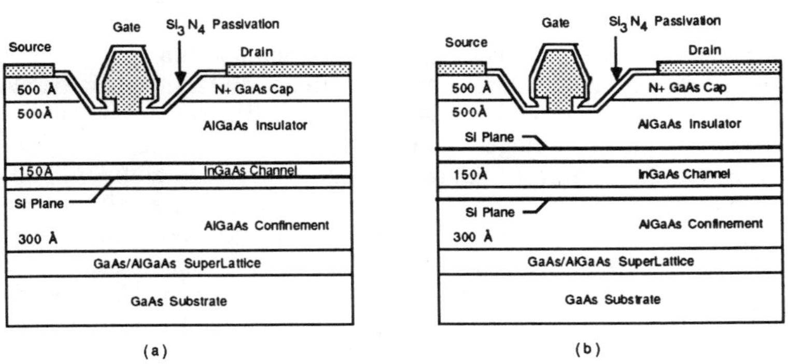

Figure 1. Cross Sectional View of AlGaAs/InGaAs Pseudomorphic Power HEMT Structures: (a) Doped Channel, (b) Double Heterostructure

Both structures are very similar except the way carriers in the channel are generated. The doped channel scheme has an electron-supplying doping layer directly embedded inside the channel, while the double heterostructure depends on the electron transferred from the two adjacent doping layers outside the channel. The doped channel HEMT structure although capable of higher current, has a lower cutoff frequency, typically in the range of 65 to 80 GHz compared to 75 to 90 GHz for the double heterostructure due to less favorable channel transport property caused by the additional carrier scattering from the channel doping. Operationally, we have found comparable power performance from both material profiles up to around 40 GHz beyond which the performance from the doped channel approach starts to lag behind that of the double heterostructure. The device design and the performance data discussed later are all based on the double heterostructure since it is needed to cover the high end of the frequency spectrum.

Silicon planar dopings of 5E12 /cm^2 were respectively inserted into the AlGaAs insulator and the AlGaAs confinement layers resulting in a sheet charge of 3.5 E12/cm^2 in the channel. Typical mobility of about 5,000 cm^2/V-sec at 300 °K was obtained. The undoped InGaAs channel thickness was kept below 150 Å to prevent the strained InGaAs layer from being relaxed which is essential to retain the superior electronic transport properties inherent to the InGaAs material. This structure is realized with molecular beam epitaxy technique with its precise doping and thickness control as well as sharp interfaces between neighboring layers that are prerequisites for good RF performance. The GaAs/AlGaAs superlattice buffer layer, although not directly participate in the power generation, plays an important role in achieving good device efficiency. By optimizing the band gap discontinuity at the channel/confinement layer and confinement layer/buffer layer interfaces, we can reduce the undesirable substrate current component and improve the channel carrier confinement. Experimentally, minimize this substrate current component to less than few percent of the total drain current can improve the device power-added efficiency by 3 to 5 percent at 60 GHz.

The basic power HEMT cell design has a total 400 μm gate periphery. High power output is obtained by combining several of these 400 μm cells in parallel at the output stage as illustrated later with the sample circuits. The cell as laid out in Figure 2 comprises of 8 T-shaped gate fingers grouped into two subsets separated by a source pad. Each finger is 0.15 μm in length and 50 μm in width. In addition, 320 μm cell was also used in some MMIC power amplifier designs for gain tailoring as well as power output stage. These cells have the same layout as that shown in Figure 2 except the gate width was reduced to 40 μm. The drain to source spacing is 2 μm. Ground vias were etched onto the three source pads with air bridges interconnecting the sources at the front side.

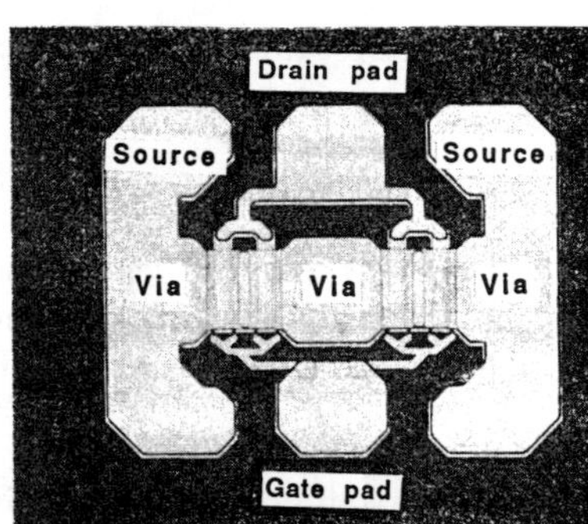

Figure 2. Layout of a 400 μm Power HEMT Device

The power HEMT was fabricated with a fairly standard HEMT process used in the industry. The processing sequence is shown in Figure 3. The discrete device was fabricated with the same flow except without the thin film resistor step. The device active areas were passivated with up to 2000Å thick silicon nitride dielectric for device stability and protection. Device passivation is required for this class of power HEMT because the exposed AlGaAs surface in the gate recess region is potentially reactive in

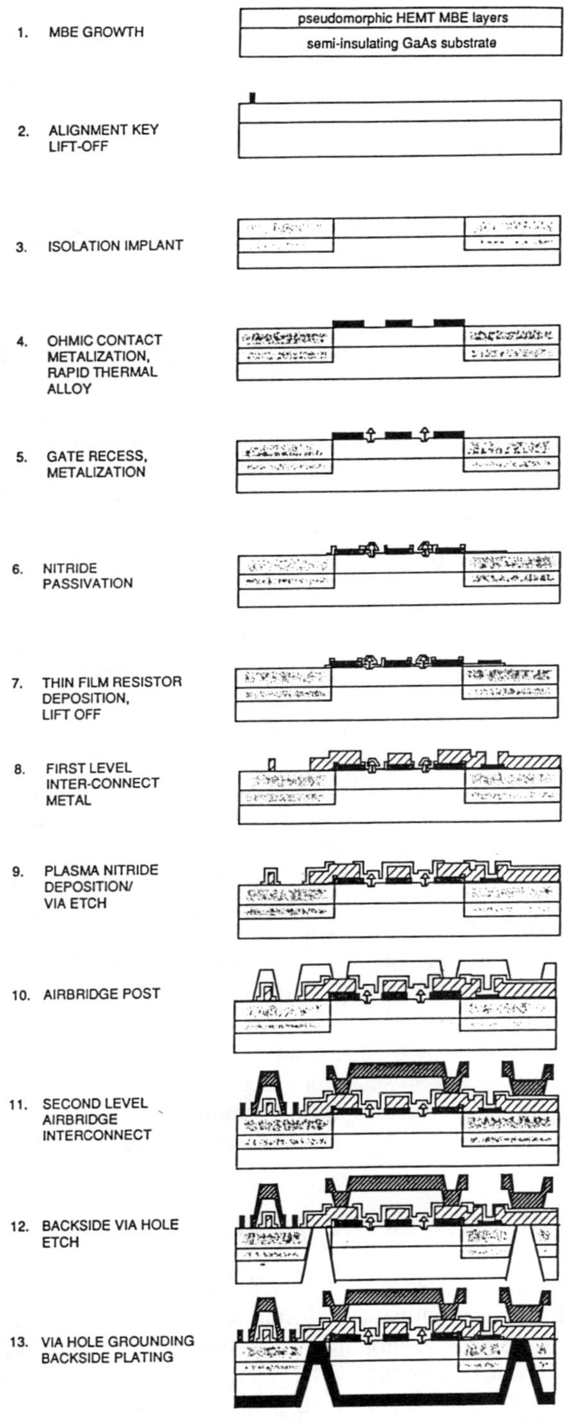

Figure 3. Power HEMT Process Sequence

the presence of oxygen which can cause the device characteristics to change in time. The 0.15 μm T-shaped gate was defined using electron beam lithography and Ti/Pt/Au metal system was evaporated onto the wafer for the gate formation. After completing the front side processing, the starting GaAs substrates were lapped down to 1.5 to 2.0 mils thick for the discrete devices to achieve a lower thermal impedance and lower source inductance. For the MMIC, the GaAs substrate was thinned down to 4 mils thick as determined by the considerations among the requirements of low device thermal resistance, low transmission line loss and the desired impedance level.

The maximum transconductance and channel current density of the resultant power HEMTs are generally greater than 500 mS/mm and 600 mA/mm respectively. The power HEMTs with the highest power output tend to be in more depletion mode with a pinchoff voltage less than -0.5V. The extent that power matching to the device was achieved is also a critical factor in determining the final observed power performance. The cutoff frequency for this class of devices generally exceeds 80 GHz. A minimum gate to drain diode breakdown of 9 to 10 volts (@ 0.1 mA/mm) for a single gate recess power HEMT is maintained to ensure both high power output and long term reliability. With a knee voltage of around 0.5 V, the power HEMTs could be biased at 4.5 to 5.0 volts and operated in deep Class AB for high power and high efficiency.

These device parameter values were experimentally found to be necessary to achieve good power performance, especially at V-band frequencies. They were obtained only after the trade-off among several competing process requirements such as high breakdown voltage, high cutoff frequency, optimal pinchoff voltage, and small gate lag was made. The gate lag measured the percentage of the total drain current that can be turned on instantaneously when the power HEMT was suddenly pulsed on from a pinchoff condition. It provided us a qualitative assessment of the power HEMT performance potential. Severe degradation was observed when the gate lag was much greater than 20%, most likely due to the charges trapped in the device surface or impurity states. Besides the intrinsic device characteristics, the actual power performance is also critically dependent on external factors such as the bias condition, device/circuit matching, and device operating temperature. Since the power devices are operated at a higher temperature and under a much larger RF field stress compared to the small signal devices, the reliability considerations will impose additional requirements such as the minimum breakdown voltage and maximum leakage gate currents. To evaluate power performance consistently, we have imposed the conditions that the averaged gate leakage current shall not exceed 0.1 mA/mm in the reverse direction and 1 mA/mm in the forward direction during nominal operation to ensure long life. These restrictions may be modified as more reliability data are accumulated in the future.

3. POWER HEMT PERFORMANCE

The 400 μm discrete power HEMT devices were embedded in a Ka, or Q, or V-band hybrid amplifier circuit for performance evaluation. These circuits have similar constructions: the device is mounted on a center carrier that has built-in prematched elements. The carrier is sandwiched between two end pieces which contain the finline transitions for transforming the microstrip media into the standard waveguide for measurements. The transition loss was determined by measuring the back-to-back insertion loss of a pair of end pieces. The one-sided loss ranges from 0.4 dB at 35 GHz to 0.55 dB at 60 GHz. The power performance reported here has been corrected for the transition loss at the output side. Table 1 summarizes the power performance from a 400 μm discrete device at 35, 44 and 60 GHz. The device was biased at 5.0 V drain voltage and tuned for maximum power output. With the base plate temperature of 30°C, the channel temperature was estimated to be around 90 to 100 °C. Both the drain and gate voltage were then adjusted to maximize the power-added efficiency (with a minimum power output of 150 mW) and the results are also given in Table 1. The best efficiency was obtained with a reduced drain bias of 4.5 V and the device operated at a deeper class AB mode as expected.

Frequency (GHz)	Optimized for Power Vds=5.0 V	Optimized for Efficiency Vds=4.5 V
35	250 mW (625 mW/mm) 40 % PAE 7.0 dB Gain	178 mW (445 mW/mm) 46 % PAE 6.8 dB Gain
44	252 mW (631 mW/mm) 36 % PAE 6.0 dB Gain	216 mW (540 mW/mm) 38 % PAE 6.3 dB Gain
60	205 mW (510 mW/mm) 25 % PAE 5.1 dB Gain	170 mW (425 mW/mm) 29 % PAE 4.4 dB Gain

Table 1. Power Performance of a Discrete 400 μm Power HEMT

The power output of 320 μm devices generally scale with the device size except that the power gain is slight higher (~0.3 dB) and the efficiency is better by about 4% due to easier circuit matching, less input signal attenuation along the gate fingers and better overall phase coherence across the cell.

Both the MIC and MMIC power amplifiers were developed at these frequencies based on the 320 and 400 μm cell designs described above. At Ka-band, one watt monolithic power amplifier was developed to operate over 34 to 36 GHz with greater than 9 dB power gain and >20% power-added efficiency[9]. The chip schematic and photograph are shown in Figure 4 and the associated power transfer characteristics is shown in Figures 5. The driver stage consists of four 320 μm cells and the output stage combines eight 400 μm cells. This design uses the Lange couplers extensively to achieve excellent input, output and inter-stage isolation and matching. Because of this conservative design approach, this amplifier is highly reproducible with an accumulated RF yield of 25.8% from a 60-wafer run demonstrated on a manufacturing line[10].

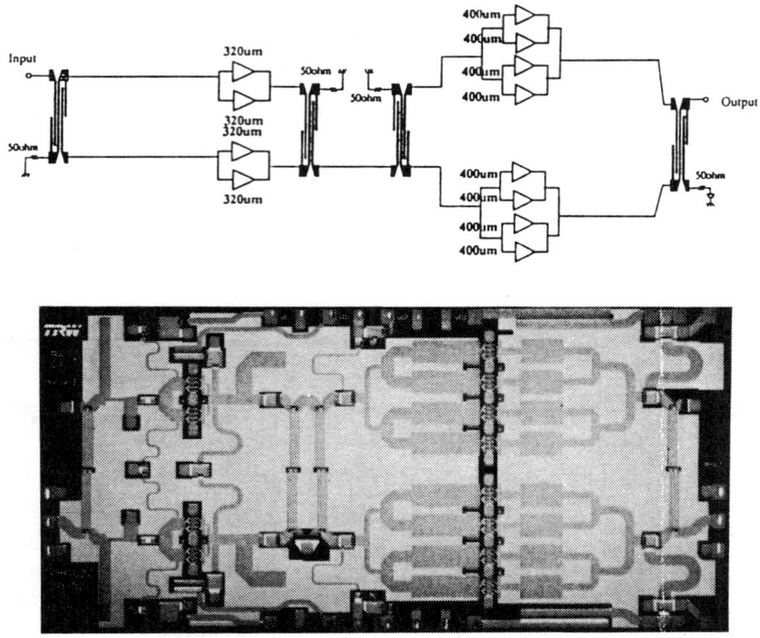

Figure 4. Block Diagram and Chip Photo of a Ka-Band MMIC Power Amplifier

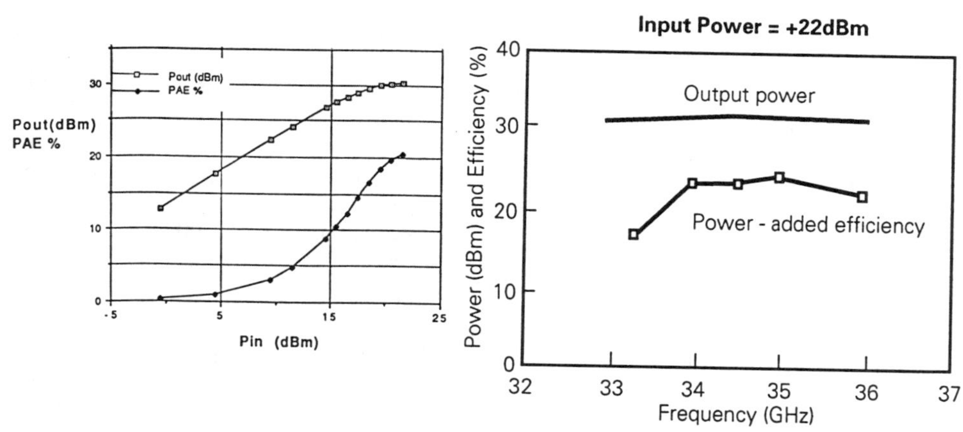

Figure 5. Performance of a Ka-Band Power Amplifier at 35 GHz

At 44 GHz, a more aggressive design was executed and shown in Figure 6. This design uses high impedance line corporate feed approach to combine 16 cells, each with four fingers of 45 μm wide at the output to obtain more than 820 mW power output over 4 GHz bandwidth as shown in Figure 6. The overall power-added efficiency is greater than 18%.

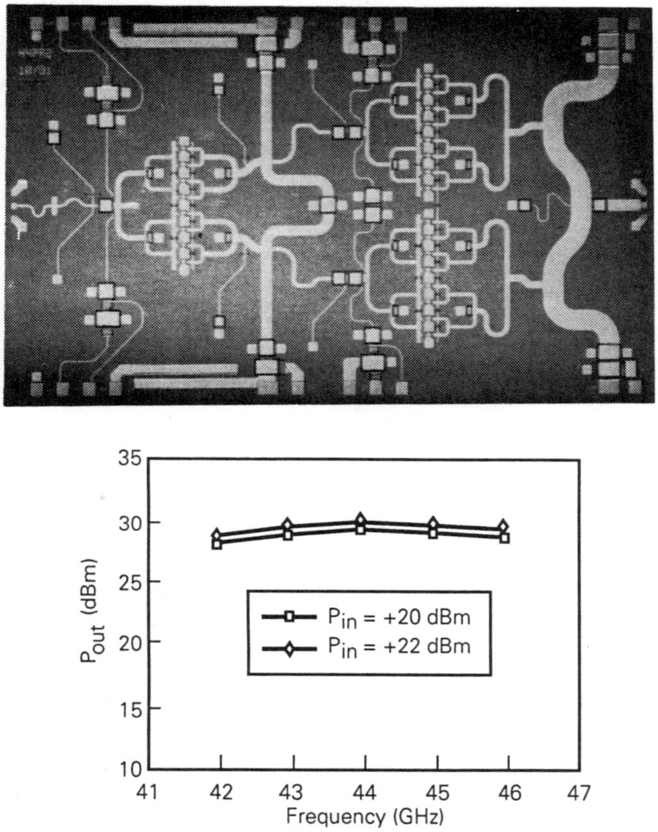

Figure 6. 44 GHz MMIC Power Amplifier and its Performance

At 60 GHz, broad band (> 4 GHz BW) MIC and MMIC power amplifiers based on the 0.15 μm power HEMT design were developed. A two-stage MIC amplifier (Figure 7) consisting of a 400 μm driver stage and a power stage with two 400 μm devices combined through a Wilkinson power splitter/combiner network produced 316 mW power output is shown in Figure 8. The saturated power gain of 7.8 dB and an overall power-added efficiency of 17% were obtained from this design. The matching network printed on separate alumina substrates was designed based on an extrapolated large signal device model and had greatly reduced the amount of labor required to align the amplifier over the needed bandwidth. However, despite demonstrated feasibility at 60 GHz, the MIC power amplifier approach is not cost effective at these frequencies.

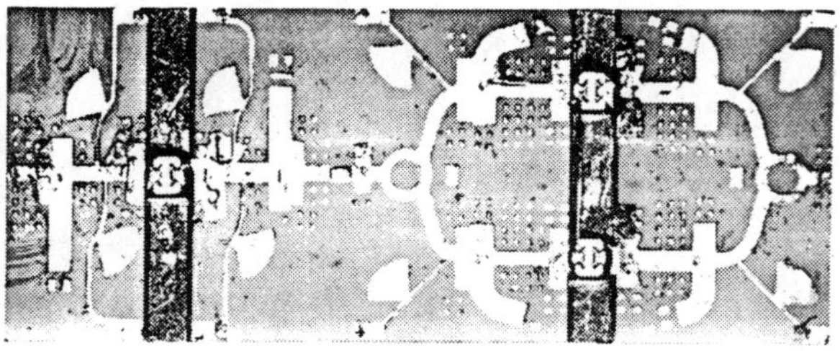

Figure 7. Photo of a Broadband V-Band Two-Stage MIC Power Amplifier

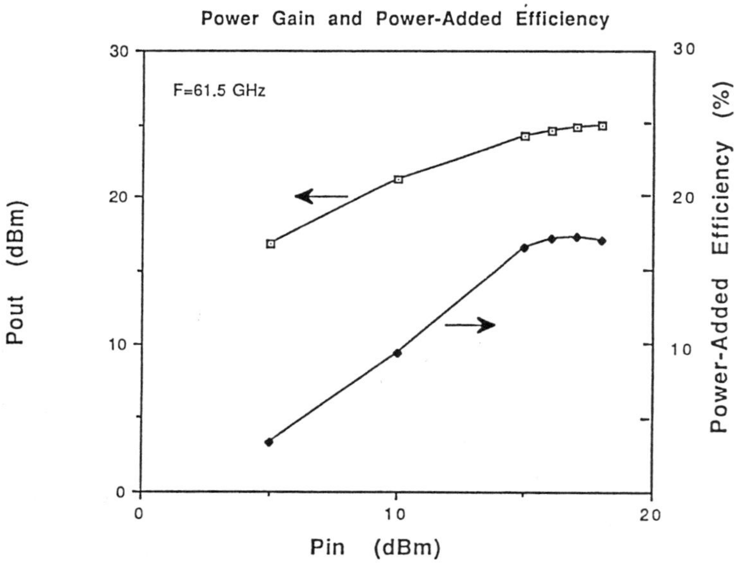

Figure 8. Power Transfer Characteristics of the V-Band MIC power Amplifier

A monolithic V-band power amplifier with similar bandwidth (Figure 9) was developed. It is a single-ended 2-stage design with two 320 µm feeding four 320 µm devices at the output stage. The amplifier circuit was designed with unpassivated power HEMT model at 60 GHz. However, both the unpassivated and passivated versions were fabricated and measured for comparison.

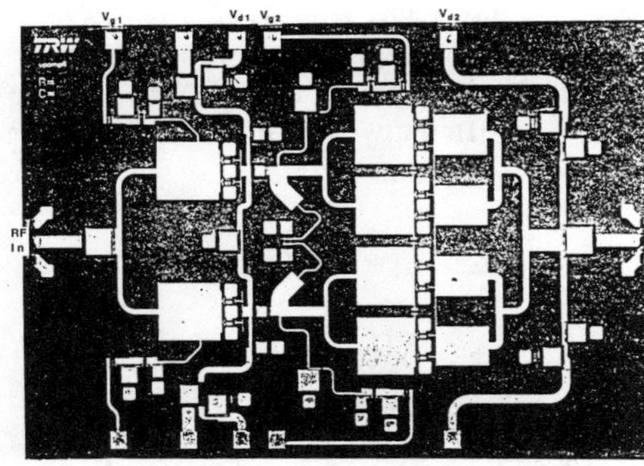

Figure 9. A Two-Stage Monolithic V-Band Power Amplifier

The RF performance results are summarized in Figure 10. Output power of 450 mW, 8.5 dB saturated gain and 14.5% PAE were achieved with the unpassivated, while 350 mW, 6.5 dB saturated gain, and >10% PAE were achieved with the passivated design. The lower performance from the passivated version indicates that the passivation had sufficiently affect the device/circuit matching that a redesign should easily correct this problem. Nevertheless, both the passsivated and unpassivated MMIC results given above represent the state-of-the-art power performance at 60 GHz.

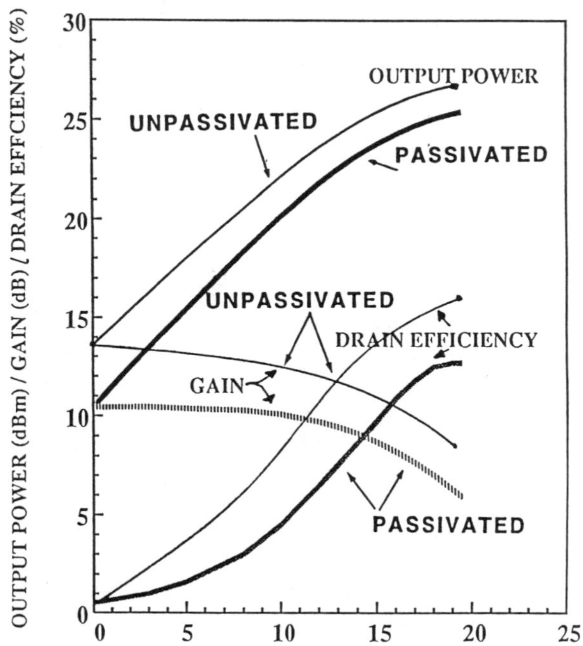

Figure 10. Measured Performance of Unpassivated and Passivated V-Band HEMT MMIC Power Amplifier at 60 GHz.

V-band power module using four of these passivated MMIC chips have been fabricated. The driver stage is a single MMIC chip followed by a power stage with three such chips combined through a three-way slitter/combiner network as shown in Figure 11. Averaged power output from 12 modules is 725 mW with >11 dB power gain over 4 GHz bandwidth at 60 GHz. Highest power output of 812 mW was obtained with 13 dB gain and 6.3% power-added efficiency. The module uniformity was excellent that a multi-watt power amplifier has been demonstrated[11].

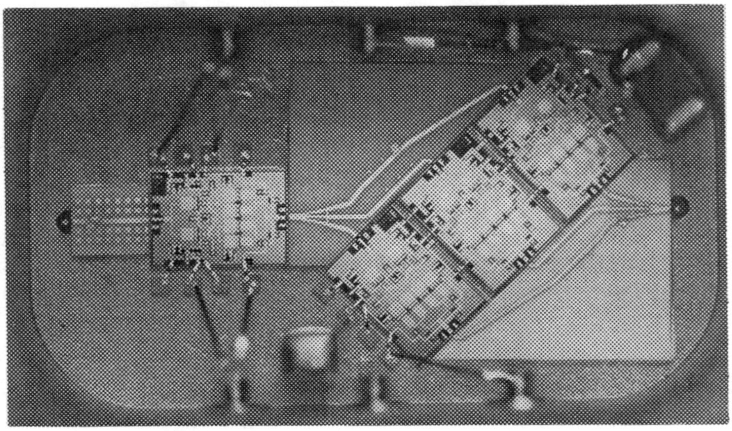

Figure 11. V-Band Power HEMT Module Employing Four MMIC Power HEMT Chips

4. POWER HEMT RELIABILITY

Reliability is an inseparable design consideration for power devices because the presence of large amount of DC currents, bias voltages, RF fields, and dissipated heat. They, individually and collectively impose restrictions on the device design and the operation range to meet the required reliability. For the 0.15 µm power HEMT design, we have studied its reliability based on the high temperature stabilization bake data and RF-stressed life test results. The stabilization bake is carried out at 240 °C ambient as a baseline fabrication step. Up to 5% changes were generally observed in the device DC characteristics between the pre- and post-temperature bakes. Devices with thicker passivation showed smaller changes in DC parameters. The gate-drain breakdown voltage generally improves by about 0.5 V after the bake. Even though these changes were well within the screening criteria of 10% changes in key DC parameters, we have found that RF burn-in is generally required to ensure the long term device reliability.

To provide a consistent basis for comparison and to ensure high reliability, the maximum reverse and forward gate leakage current of 0.1 mA/mm and 1 mA/mm respectively were imposed. Consequently, the 0.15 µm power HEMT design typically operates at a power density of 500 to 600 mW/mm at room temperature. RF-stressed accelerated life tests at 210 and 225 °C channel temperatures were performed on discrete devices to assess the power HEMT reliability characteristics. Ten sample amplifiers at each temperature were assembled and mounted on a test station. The amplifiers were tuned for the highest power output consistent with the gate leakage current constraints and were typically operated at 2 dB compression point. The failure criterion is a degradation of 1 dB in the output power at the given input power when measured at room temperature. Using this criterion, we have obtained an activation energy of 1.6 eV and a projected mean-time-to-failure of 1E7 hours at 125°C channel temperature[12].

Based on the rate of power gain degradation over the duration of the life tests along with other failure analyses, the failure mechanism appears to be a combination of the gate metal sinking and the surface oxidation of the AlGaAs insulator layer with the latter being the dominant factor during the initial phase only. We hypothesized that at the beginning of the life test, the AlGaAs surface of the insulator layer may still retain some of the chemical activity despite the stabake and RF burn-in cycles. The AlGaAs reacted with the residual oxygen present in the interface of the passivation layer and the AlGaAs insulator layer under high temperature and high RF field. Surface states were created which can trap the charges and suppress the channel current resulting in lower power output. This surface oxidation process was more pronounced at the beginning and stabilized within a couple of hundred hours corresponding to a faster degradation rate observed during the initial phase of the life test. Once the AlGaAs surface is stabilized, the ever present gate metal diffusion mechanism becomes the dominant failure mechanism which has a known characteristic activation energy of 1.6 to 1.8 eV.

5. CONCLUSIONS

A common 0.15 µm pseudomorphic AlGaAs/InGaAs power HEMT design capable of state-of-the-art power performance up to 65 GHz was presented. The demonstrated power performance from discrete devices as well as selected MIC and MMIC power amplifier circuits indicate that the GaAs-based pseudomorphic AlGaAs/InGaAs power HEMT technology today is capable of producing 1 watt power MMIC at 35 GHz, 0.8 watt at 44 GHz and 0.5 watt at 60 GHz. Coupled with the demonstrated reliability, this technology is the best choice for the near term insertion at these frequencies. However, there are areas that can be further improved before this process can be considered fully matured. For example: 1). methods for accurate large signal device characterization and modeling are still lacking, 2) circuit design techniques and advanced process improvements are still needed to increase the power MMIC efficiency by at least 10% and 3) further understanding of the power HEMT device and MMIC reliability characteristics and positive failure mechanism identification are highly desirable to ensure the long operating life.

6. ACKNOWLEDGMENT

The authors thank R. Dia and N. H. Nguyen, and L. Callejo for technical assistance, and D. Wang and E. Rezek for stimulating discussions. TRW's generous support for this work is also greatly appreciated.

7. REFERENCES

1. For Example, 1990 HUGHES MILLIMETER-WAVE PRODUCTS Catalog.
2. M-Y Kao *et al.*, "Very High Power-Added Efficiency and Low-Noise 0.15 µm Gate-Length Pseudomorphic HEMT's", IEEE Electron Device Lett., vol. 10, no. 12, pp 580-582, Dec. 1989.
3. K. L. Tan *et al*, "High-power V-band Pseudomorphgic InGaAs HEMT", IEEE Electron Devices Lett., vol. 12, no. 5, pp 213-214, May 1991.
4. H. C. Yen, *et. al.*, "Pseudomorphic InGaAs Power HEMT at 30-65 GHz", Workshop on Millimeter-Wave Power Generation and Beam Control, Huntsville, AL, 1993.
5. M. Matloubian, *et. al.*, "High Power V-Band AlInAs/GaInAs on InP HEMTs", IEEE Electron Devices Lett., vol. 14, no. 4, pp 188-189, April, 1993.
6. A. Kurdoghlian, *et. al.*, "High-Efficiency InP-Based HEMT MMIC Power Amplifiers", 1993 GaAs IC Symposium Digest, pp 375-377.
7. N. L. Wang, *et. al.*, "4 W, 7-12 GHz Common Base HBT MMIC Power Amplifier", 1992 GaAs IC Symposium Digest, pp 301-303.
8. F. Ali, *et. al.*, " An 8-15 GHz, 1 W HBT Power MMIC with 16 dB Gain and 48% Peak Power Added Efficiency", 1993 GaAs IC Symposium Digest, pp. 363-366
9. M. V. Aust, *et al.*, " A Ka-Band HEMT MMIC 1 Watt Power Amplifier", IEEE 1993 Microwave and Millimeter-Wave Monolithic Circuits Symp. Dig. pp 45-48.
10. M. Biedenbender, *et. al.*, "A Power Production Process for High-Efficiency Ka-Band MMIC Power Amplifiers", 1993 GaAs IC Symposium Digest, pp 341-344.
11. A. K. Sharma, *et. al.*, "Millimeter-Wave High power Amplifiers Using Pseudomorphic HEMTs", Submitted to 1994 IEEE MTT-S, International Microwave Symposium, San Diege, CA.
12. C. H. Chen, *et. al.*, "Reliability Study on Pseudomorphic InGaAs Power HEMT Devices at 60 GHz", Submitted to 1994 IEEE MTT-S, International Microwave Symposium, San Diego, CA

Quasi-Optical Power Combining Techniques

Robert A. York

Department of Electrical and Computer Engineering
University of California, Santa Barbara, CA 93106

Abstract

Quasi-optical power-combining arrays overcome some limitations of traditional power combining schemes, and are expected to facilitate development of compact and reliable millimeter-wave systems. This paper presents an overview of 1993 quasi-optical power combining technology, focusing on approaches for direct DC-to-RF conversion and beam amplification. Arrays for frequency up- and down-conversion, and beam control are briefly reviewed, and possible applications and future research efforts are identified.

1. Introduction and Motivation

High-speed device technology has advanced rapidly during the past decade, to the point where 300 GHz transistors and 1 THz diodes (Schottky, RTD) can be routinely fabricated. Much of this technology in the U.S. is a result of the ARPA MMIC program. However, advanced millimeter-wave devices have not yet found widespread use, due to extremely limited power-handling capacity (figure 1) and the expense associated with fabrication of conventional hybrid circuits at these frequencies. As a result, there is a lack of compact, reliable, high power sources, which has slowed or prevented the development of many millimeter-wave systems, despite increasing demand for such systems.

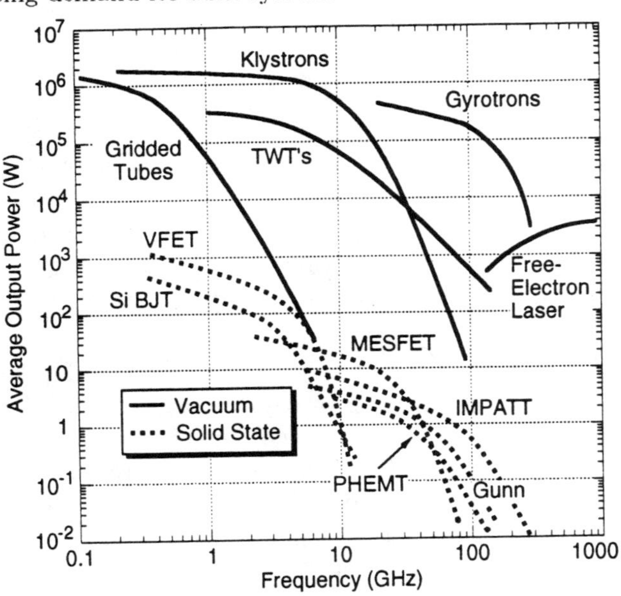

Figure 1 — Power handling capacity (continuous) from a variety of millimeter-wave sources (data adapted from Sleger *et al.*[1]).

Vacuum electronics are still the dominant technology for power in the millimeter-wave region, but solid-state electronics are generally more desirable in terms of size, weight, reliability, and manufacturability. The cost of production is a major concern today, and this also favors solid-state systems which can be mass produced using modern IC technology. To achieve the kind of power that is needed for many applications, the power from many individual devices must be added coherently. Over the years many power-combining techniques have been devised[3], but these perform poorly at millimeter-wave frequencies where large numbers of devices must be combined.

In 1986, Mink[5] published a classic paper which suggested the use of quasi-optical techniques for solid-state power-combining. Although this was not the first paper to suggest such methods (see Wandinger and Nalbandian,[4] for example), it is generally credited with providing the impetus for the increased research activity that followed. Quasi-optical power combining (QOPC) methods solve many of the problems which plague other combiners, and is a logical approach when viewed from the perspective of optical systems and the similarity to a well understood optical power combiner—the laser. At the same time, QOPC techniques introduce some new problems of their own. To better appreciate the issues involved, it is useful to first understand why conventional power-combining techniques fail at millimeter-wave frequencies.

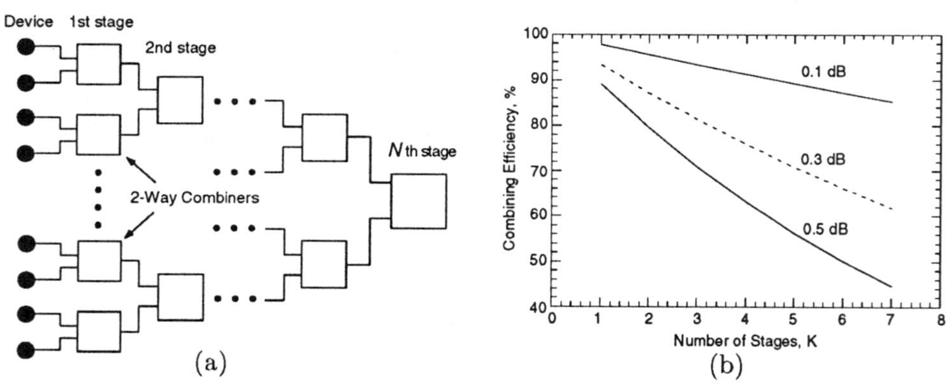

Figure 2 — (a) A "corporate" combiner, employing cascaded stages of two-way adders such as Wilkinson combiners. (b) Theoretical combining efficiency of the corporate structure for different levels of insertion loss for each two-way adder.

A very common combiner network that is used at both the chip level and component level is the "corporate" combiner shown in figure 2a. A number of devices are successively combined using two-way adders. The number of individual devices is $N = 2^K$, where K is the number of stages. The net output power is found to be $P_{\text{out}} = P_o 2^K L^K$ where P_o is the output power per device and L is the insertion loss of each stage. The combining efficiency is therefore $\eta = L^K$. This is plotted in figure 2b for three examples of insertion loss. For a large number of devices the power-combining efficiency can become unacceptably low. Furthermore, resistive, radiative, and dielectric losses make it increasingly difficult to maintain a low insertion loss at millimeter-wave frequencies, leading to an additional decrease in efficiency.

A more successful technique at microwave frequencies (for narrowband systems) has been the Kurokawa waveguide cavity combiner[2], shown in figure 3. This combiner uses a resonant waveguide cavity that is periodically loaded with devices, each with it's own impedance matching network and bias supply. The devices are typically spaced by $\lambda_g/2$, where λ_g is the guide wavelength. This method does not suffer as badly from waveguide losses as the corporate combiner. However, if the dimensions of the cavity are increased to accommodate more devices, the natural resonances of the structure become closer together, and the system is then in danger of "mode-hopping" or "multi-moding". This problem is thus a result of widely distributing the devices along a resonant dimension. Furthermore, this technique is not economically feasible for frequencies at or above 94 GHz because of the difficulties in machining the tiny waveguide dimensions and the labor-intensive method of construction.

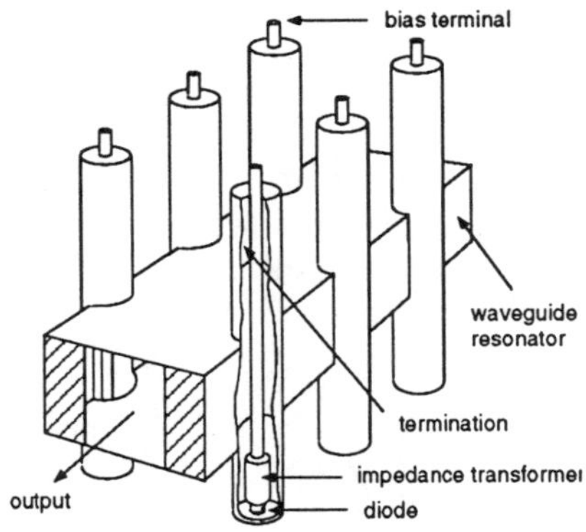

Figure 3 — Kurokawa single-cavity multiple device combiner.[2] For large numbers of devices, this technique suffers because the devices are widely distributed along the resonant dimension (see text).

The problems of waveguide losses and multi-moding are a result of extrapolating conventional transmission-line and circuit techniques to a regime where optical beam-guiding techniques are more suitable. Consider a simple gas laser system, as shown in figure 4a. Each gas molecule (when properly excited) delivers energy to an incident wave in much the same way as a negative resistance injection-locked amplifier. The individual contribution of each "device" is microscopic, but a large amount of power is obtained from the ensemble and carried in a propagating beam whose transverse dimensions are set by the external focusing elements (the mirrors). This fantastic power combiner differs from the previous examples in two important respects: the active devices are densely distributed over the cross sectional area of the beam, and the combined power is not guided by lossy metallic boundaries. As pointed out by Mink, this suggests the "quasi-optical" scheme shown in figure 4b. A planar array of solid-state devices is integrated into a radiating structure, which is designed to efficiently couple energy from the devices to a propagating beam in free-space (or vice-versa). The gain medium in this case is a two-dimensional device array, as opposed to the three-dimensional array in the laser. Since the devices are spread out transverse to the direction of beam propagation, the resonator (if present) can be operated in a low-order mode to avoid multi-moding problems. High combining efficiencies are anticipated, since there is little parasitic loss in the system. In addition, the arrays should degrade gracefully in the event of a device failure.

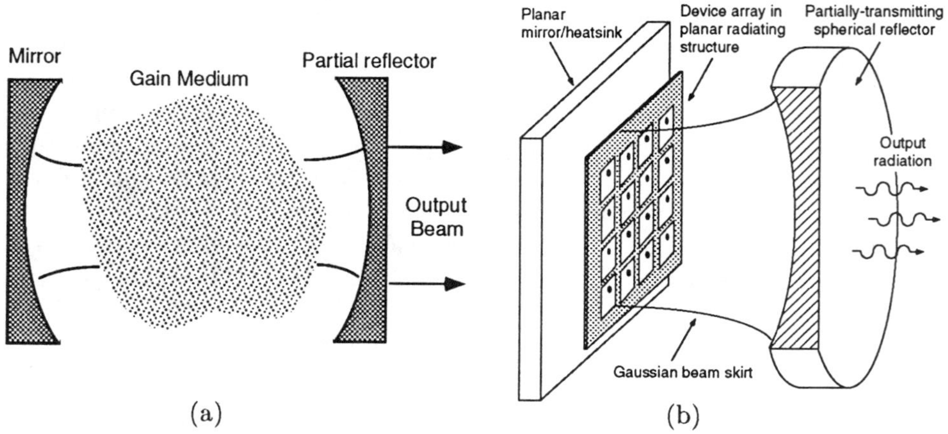

Figure 4 — (a) A simple gas laser. The gain medium is a collection of tiny amplifying devices with integral antennas (excited gas molecules), with feedback provided by external mirrors. (b) Quasi-optical laser analogy using a planar, solid-state millimeter wave array in a semi-confocal resonator.

In the laser system of figure 4a, the molecules interact with the propagating beam very efficiently. This occurs because (a) the spacing between the "devices", d, is a small fraction of a wavelength, and (b) the cross-sectional diameter of the "array" is large compared with a wavelength. That is, $d \ll \lambda \ll D$. Ideally a QOPC array would come as close to the optical limit as possible, not only to reduce diffraction losses, but also to pack more power into a smaller volume. However there are certain engineering issues which must be considered. The upper limit in achievable density of devices will almost certainly be governed by heat removal limitations rather than manufacturing limits, and this has at least two important consequences: (a) Heat removal considerations may strongly favor *reflecting* or uni-directional arrays which have an integral groundplane or heatsink on the backside of the substrate, rather than *transmission*-mode or bi-directional arrays for which the output beam can travel through the substrate; (b) The maximum power achievable per unit area scales with the *efficiency* of the active device rather than with the power per device. If the power per device is low, more power can be produced by increasing the device density, or the array size if thermal limits are reached.

For amplifying arrays, efficiency considerations will also govern the size of the array unit cell, such that for certain device technologies an electrically large cell size may be required. Another issue is that the active devices must be supplied with DC bias, ideally in such a way that a failure of one device does not lead to a catastrophic failure of all of the devices (as might be the case if the devices were biased in parallel).

These and other issues discussed below have been the focus of increasing research activity during the past few years. Some of the earliest work on QOPC arrays focused on the use of two-terminal device arrays for frequency multiplication and beam-steering.[79–83] By far the most effort has gone into the development of quasi-optical oscillator arrays, and more recently amplifier arrays. Two distinct approaches have emerged. The first conforms most closely to the ideal optical combiner discussed above, and has been labeled the "grid" approach. The grid arrays were first developed by Prof. Rutledge and co-workers at the California Institute of Technology, and are currently under investigation by other groups at University of Colorado at Boulder, University of Virginia, and Georgia Institute of Technology.

The second approach is an extension of classical antenna array systems. Each cell of the array contains a conventional planar antenna structure integrated with the appropriate circuitry, and the cell size is typically a half-wavelength. This approach is referred to as "spatial power combining" or "active array techniques", and have been developed by several groups at Cornell University, University of Massachusetts at Amherst, University of California at Santa Barbara, UCLA, and the University of Central Florida. These arrays are typically larger than the CalTech grids and accommodate fewer devices, but are generally somewhat simpler to design and can make use of extensive work on planar antenna structures[88–96].

This paper will emphasize current work in oscillator and amplifier arrays, since the bulk of the research effort has concentrated in these two areas, and since the near-term application of the technology will likely make use of these components. A brief overview of other types of arrays for mixing, frequency multiplication, parametric up-conversion, and beam-switching will be given. The work in active array oscillators has led to new beam-scanning techniques and a mode-locking technique for pulse generation analogous to a mode-locked laser, and these will also be discussed.

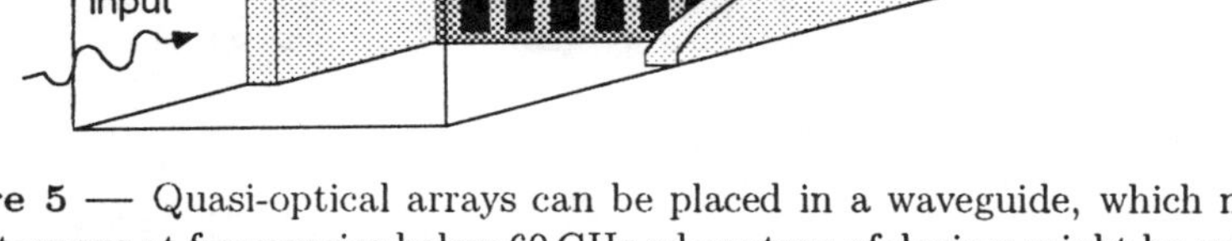

Figure 5 — Quasi-optical arrays can be placed in a waveguide, which may be advantageous at frequencies below 60 GHz where tens of devices might be sufficient and where off-the-shelf waveguide components are economically attractive.

It should be pointed out that this "new" power combining technique does not entirely preclude the use of metallic waveguide, as shown in figure 5. This may be attractive at frequencies below 100 GHz for several reasons: diffraction losses and focusing errors are minimized, since all of the energy is confined by the waveguide walls; the metal walls provide a convenient heatsink for the arrays in transmission; waveguide components are readily available in this frequency range and may be more economical than quasi-optical components; and using QOPC arrays in this way allows them to be retrofitted into existing waveguide systems. This technique may have somewhat lower combining efficiency than the open beam system due to waveguide losses, and cannot accommodate as many devices unless the waveguide cross-section is enlarged (in which case higher order modes may be excited). Furthermore, the nonuniform field profile in the waveguide means that edge elements on the array do not couple effectively to the waveguide mode. A possible solution is shown in figure 5, where a uniform field (approximating a plane wave) is artificially created in the guide by dielectric loading of the side walls.

Freq. (GHz)	Array size	Device type	Power (mW)	ERP (W)	Efficiency (%)	Reference
3.1	6×6	FET	220	3.0	22.0	[11]
5.0	10×10	FET	550	21.0	20.0	[12]
6.0	4×4	FET	—	28.2	—	[31]
7.3	3×3	FET	282*	6.1	31*	[39]
8.2	4×4	FET	184	10	26.0	[27]
8.3	3×6	Gunn	123	—	—	[13]
8.4	1×10	FET	1050*	10.5	10*	[49]
9.6	4×4	Gunn	415	22	1.0	[27]
9.7	5×5	FET	464	20.7	14.5	[10]
9.8	10×10	FET	10300	657	23.1	[21]
9.9	10×10	HBT	72*	6.3	3.9*	[72]
10.03	4×4	FET	163	17.4	—	[34]
11.6	4×4	FET	335	15.0	20.0	[16]
17.0	6×6	FET	235	3.3	7.0	[16]
34.7	6×6	HBT	—	0.12	—	[19]
37.0	2×3	PHEMT	5	—	—	[9]
60.0	2×4	IMPATT	2200	23.0	3.6	[32]

* Estimated from measured ERP and theoretical directivity.

Table 1 — Comparison of several reported quasi-optical sources. ERP is the effective radiated power (total power×directivity). Note that nearly all are proof-of-concept arrays; larger, more powerful arrays are expected in the near future.

2. Quasi-Optical Power Sources

Generating millimeter-wave power with quasi-optical arrays can be accomplished in several ways: by combining a large number of low-power sources to form an oscillator array; amplifying a low-power (single-device) source with an amplifier array; or following a lower frequency power source with a frequency multiplier array. Direct DC-to-RF conversion with an oscillator array has the advantage of fewer components and possibly better conversion efficiency. Many groups have concentrated on developing this technique, as shown in Table I. Some of the early work listed in this table has been previously summarized[6], while the most recent results have been published recently in a special issue of the *IEEE Transactions on Microwave Theory and Techniques*[7].

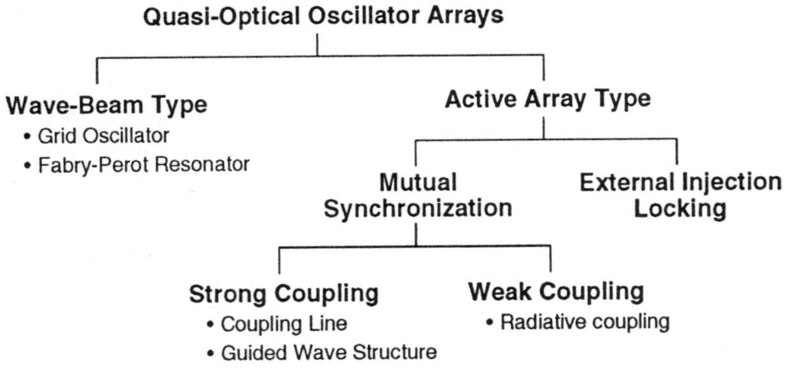

Figure 6 — Classification of quasi-optical sources (after Itoh[22]).

Oscillator arrays share two common requirements: all the devices must operate at the same frequency, and with a prescribed phase relationship (usually all-in-phase for broadside beamforming). Several approaches to oscillator arrays have been developed, and these have been classified by Itoh[22] as shown in figure 6. The most mature of these technologies (at least chronologically) is the grid approach, which has been demonstrated with arrays of up to 100 elements, and in both monolithic and hybrid (discrete device) assembly. All of the active array combiners have been fabricated using hybrid assembly and off-the-shelf packaged devices. It is important to point out that the work to date has been primarily proof-of-concept work, concentrating on developing an understanding of device-device and device-circuit interactions, and methods for reproducibly ensuring that the two key requirements for coherent combining are satisfied.

One parameter that is conspicuously absent from Table 1 is the combining efficiency, which is touted as a key advantage of QOPC arrays. This parameter is difficult to measure accurately, since it requires a knowledge of the total radiated power (often estimated from principal pattern measurements) and the total RF power produced by each device (including radiated power, power lost to substrate modes, and ohmic losses). Many of the published results for the active antenna arrays have estimated the combining efficiency by measuring the conversion efficiency of a single isolated element and assuming it does not change in an array environment. In all cases, the results have confirmed high combining efficiency (close to 100%), but this is not a good test of the concept because of the small size and low operating frequencies of the prototype arrays (where circuit losses are not ordinarily a problem). Furthermore, combining efficiencies greater than 100% have occasionally been estimated, which is a clear indication that the array elements behave differently in an array than when isolated.

2.1 Grid oscillators

The grid concept has been pioneered by the Rutledge group at CalTech. In a sense, every QOPC array is a "grid", meaning an ordered set of devices and radiators, however the term is currently understood to mean a certain type of array where the periodicity is small compared to a wavelength (in the surrounding medium). As such the grid represents a close electrical analogy to the distributed gain medium of a laser system. Ideally, the grid dimensions would be at least a wavelength or more on a side to interact efficiently with an incident plane wave. If the devices are terminated in such a way as to create a negative resistance at the frequency of interest, the grid can then be viewed as a quasi-continuous conductive sheet with a negative sheet resistance, and is hence capable of amplifying an incident electromagnetic beam. If the grid is placed before an external mirror to provide positive feedback, oscillation can occur.

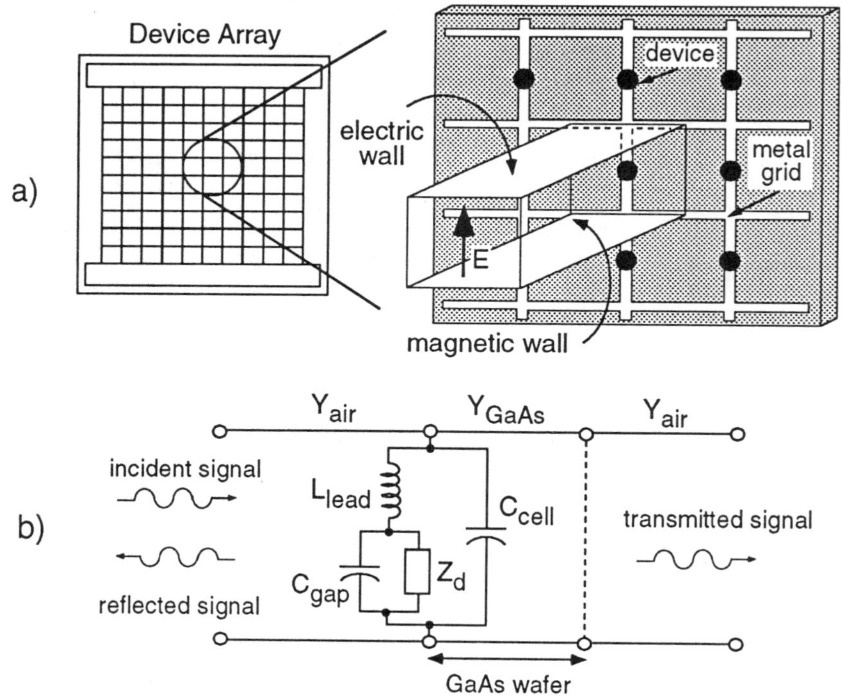

Figure 7 — The grid concept. (a) The symmetry of a dense device-loaded metallic mesh permits analysis by an equivalent parallel-plate waveguide circuit, as shown in (b). This grid employs a two-terminal device, but the concept extends easily to three-terminal devices.

If the grid is assumed infinite in extent and is driven by a uniform plane wave, the symmetry created by the periodic metal structure and the use of identical devices can be exploited as shown in figure 7. It is only necessary in this case to consider a unit cell of the array with the appropriate boundary conditions imposed at the walls of the cell. For excitation by a linearly polarized (vertical) plane wave, short circuit boundaries (perfect electric conductors) are required on the top and bottom walls, while open circuit conditions (perfect magnetic conductors) are required on the side walls. The response of the array is therefore equivalent to the response of a unit cell placed in a parallel plate waveguide, whose characteristic impedance depends on the aspect ratio of the unit cell. Since the cell is electrically small, only the TEM mode propagates in this equivalent waveguide circuit, and the current can be assumed uniform on the metallic grid.

When considering the unit cell it should be noted that the disadvantages normally attributed to electrically small antennas radiating in free space (poor radiation efficiency, difficult matching problems, limited bandwidth) are not relevant, since the dipole is radiating into a TEM waveguide. The waveguide effectively reproduces the mutual coupling effects between the elements in the array, and mutual coupling between such closely spaced antennas can have a dramatic effect on the driving-point impedance. Since the grid cell must be small compared with a wavelength, the range of impedances that can be synthesized is limited. There is little room for additional matching elements on the substrate, so external tuning elements (dielectric plates) are often used. In principle the device size (and hence impedance) could then be tailored to improve the match, but this also affects the power performance, bandwidth, and thermal design.

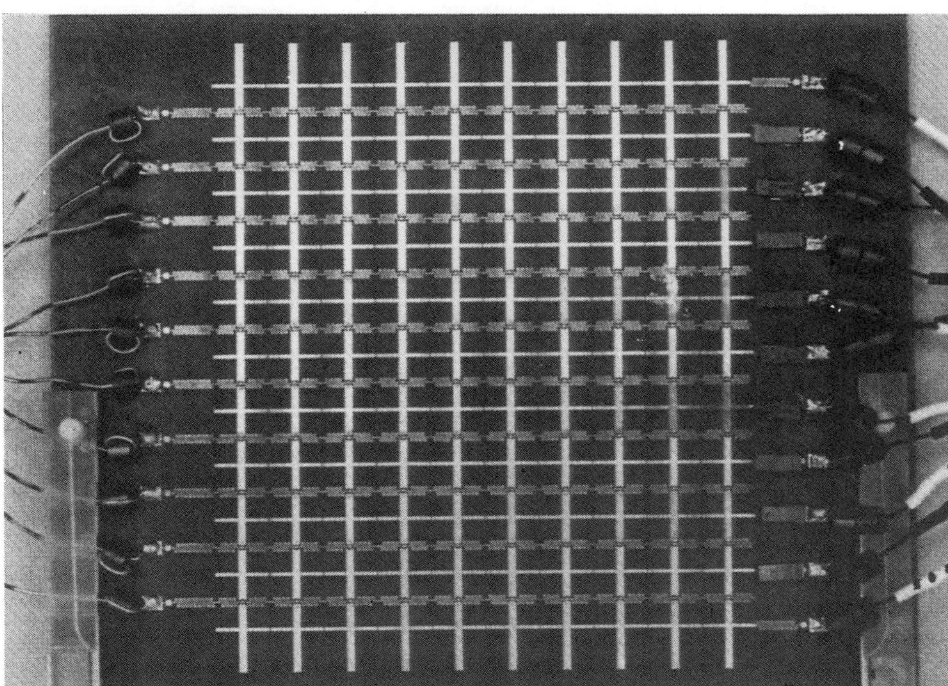

Figure 8 — Photograph of a 10 W, 100-element grid, operating at 9.8 GHz[21].

An elegant theory based on the induced EMF method and modal analysis has been developed[17] to calculate the equivalent circuit parameters shown in figure 7b, and has been verified on measurements of passive grid structures. This method works well for large grids comprised of simple geometric shapes for which the current distribution can be guessed at or approximated accurately. Given a certain device impedance, the cell and external reflector elements can then be designed so that the oscillation criteria are satisfied. In this way, the symmetry conditions are "built in" to the design, and it is argued that the entire array should then oscillate coherently and in-phase to produce a broadside beam. Active grids designed using this theory have conformed well to the predictions for low dielectric constant substrates, but there have been some discrepancies with higher dielectric constants for reasons which are not yet clear.[18] Recently a measurement scheme has been developed based on the equivalent waveguide concept, which allows the equivalent circuit elements to be measured directly.[15] This may facilitate the development of grids with more complicated metallic unit-cell structures that are difficult to model, as well as grids on substrates of high dielectric constant.

Several grid oscillators have been reported,[10–12,16,19] evolving rapidly toward the most recent version, a 100-element power MESFET design.[21] This array, shown in figure 8, produced a total radiated power of over 10 Watts, and an Effective Radiated Power (ERP) of 657 Watts. The array was fabricated on a Duroid substrate ($\epsilon_r = 2.2$). It was designed for operation at 10 GHz, and oscillated at 9.8 GHz. The grid used a Fujitsu FSX052 medium power FET device capable of providing 300 mW of power. The measured power per device was approximately 100 mW at 23% DC-to-RF efficiency. This was the fourth iteration of a design intended to extract maximum power from the device. Some problems with device failures prevented the array from being biased at the correct voltage and current required for maximum power. This was attributed to unexpectedly large RF currents flowing in devices on the edges of the array. The reasons for these large currents is not yet understood, but could be a result of either parasitic bias-circuit oscillations, or a manifestation of the electromagnetic edge condition. The device failures were not attributed to overheating, as the array was cooled by forced air and the device case temperatures were monitored with a thermocouple.

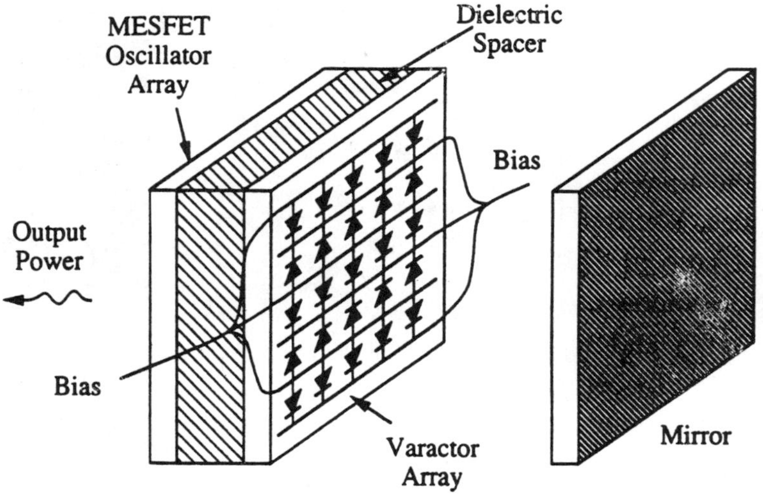

Figure 9 — A quasi-optical voltage-controlled grid oscillator, using a varactor phase-shifting grid (after Mader *et al.*[20]).

A 6 × 6 monolithic grid oscillator—fabricated at the Rockwell Science Center in collaboration with CalTech—has also been demonstrated at 35 GHz using HBT devices.[19] This array is noteworthy as the first successful monolithic demonstration of a quasi-optical oscillator array. Again, this was a proof-of-concept design. The induced EMF theory predicted an oscillation frequency of 39 GHz, and the first iteration oscillated at 35 GHz. The difference was attributed to the assumption of an infinite array in the modelling, as significant edge effects would normally be expected in such a small array. The measured E-plane pattern did not display a well defined main beam, which could be a result of the packaging and/or surface-wave resonances in the substrate; similar effects have been observed in other planar arrays on high dielectric constant substrates.[95,96]

A quasi-optical voltage-controlled oscillator (VCO), which is a combination of a grid oscillator and a varactor diode grid has been reported recently.[20] This is illustrated in figure 9. The varactor grid is designed to operate as a phase shifter, electronically controlling the feedback due to the mirror. A 10% tuning bandwidth was measured using a MESFET oscillator grid and a varactor tuning grid with a bow-tie metallization pattern.

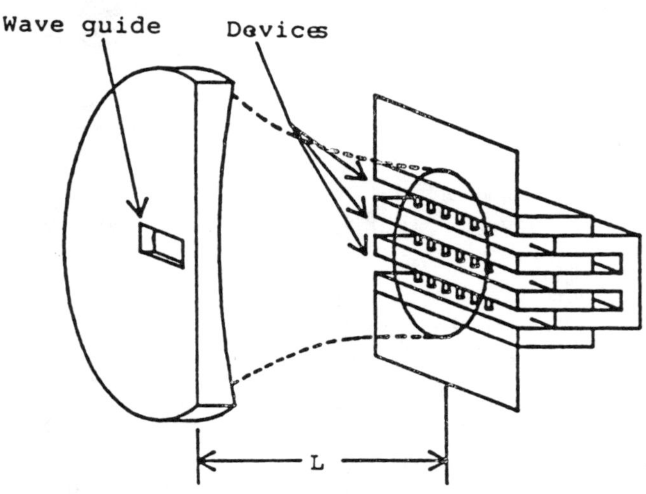

Figure 10 — A quasi-optical combiner in a Fabry-Perot cavity, using two-terminal devices in a "grooved" mirror (After Nakayama *et al.*[13,14])

Although not as well suited to monolithic implementation as the CalTech grid approach, similar work has been reported by Mizuno and co-workers using a grooved mirror loaded with active devices.[13,14] This active mirror forms part of a semi-confocal resonator, with power coupled through an aperture in the spherical mirror. Within the grooves and behind the devices are sliding backshorts for mechanical tuning. Using this technique, a 3 × 6 array of Gunn diodes has been developed at X-band,[13] and a 3-diode array at 50 GHz.[14]

2.2 Coupled-oscillator arrays

In 1986, Stephan[23] published a paper describing the use of coupled oscillator circuits for power combining, and developed a theory using Kurokawa's oscillator model[26] which could be used to observe the time evolution of amplitudes and phases of the oscillators in an array. In this technique, the oscillators are coupled together through a coupling network described by common N-port circuit parameters. This coupling allows the oscillators to mutually synchronize through the phenomenon of injection-locking,[25,26] which Stephan termed "inter-injection-locking". Each oscillator feeds an integrated planar antenna so that the power is combined quasi-optically. Note that external mirrors are *not* required. One possible advantage of this approach is that the oscillators and coupling network can be designed and tested independently. The design challenge is ensuring the proper steady-state phase relationship between the oscillators.

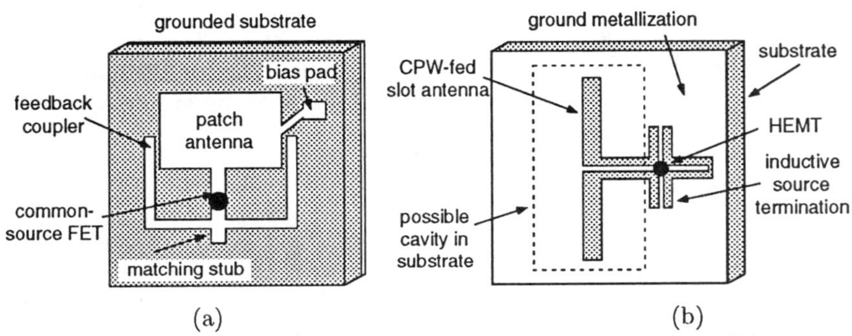

Figure 11 — Two examples of active antenna oscillators for arrays. (a) Feedback oscillator using patch antenna.[59] (b) FET oscillator using a slot antenna.[61,62]

An enormous variety of active antenna configurations have been developed which are suitable for coupled-oscillator arrays. Designs using both two and three terminal devices have been reported based on the patch antenna,[51–57,59] coplanar-waveguide-fed slot antenna,[58,61,62] and tapered-slot or notch antenna.[50,60] Two of the more common examples using patch antennas and cpw-fed slots are shown in figure 11. The design of such elements is straightforward. The antenna itself is often used as the resonator or frequency-selective feedback element in the oscillator design to save space, although this typically results in a low-Q oscillator or "motion detector", where the frequency is sensitive to environmental disturbances. Nevertheless, these designs have been used as an expedient in many of the proof-of-concept arrays described later. A more stable and reproducible design that is well-suited to monolithic implementation has been developed at Cornell,[58] using a resonant cpw "tee" to determine the oscillation frequency. The design of high-efficiency patch oscillators has also been explored, with DC-to-RF conversion efficiencies of 44%.[59] Most recently, monolithic quasi-optical slot oscillators using the topology of figure 11b have been demonstrated at 155 GHz and 215 GHz using pseudomorphic HEMTs.[62]

The earliest quasi-optical coupled-oscillator arrays relied on the inherent (weak) mutual coupling between neighboring antennas in an array for mutual synchronization.[24,27,28] Two 4 × 4 arrays using active patch antennas were developed, one of which is shown in figure 12. The first array used Gunn diodes, and produced 415 mW of total radiated power, 22 Watts ERP, at 1% conversion efficiency. The array was designed for 10 GHz, and oscillated at 9.6 GHz. This corresponds to 26 mW per device, which was the same amount of power obtained from an isolated device, suggesting a high combining efficiency. The second array (figure 12) was very similar but used MESFETs, and produced 184 mW total power, 10 Watts ERP, and had a 26% DC-to-RF conversion efficiency. This array was designed for operation at 8.0 GHz, and oscillated at 8.2 GHz. Both arrays produced a broadside radiation pattern with a directivity of approximately 17 dB. More recently, a 2 × 4 IMPATT diode array operating at 60 GHz has been demonstrated using radiative coupling,[32] producing greater than 2 Watts.

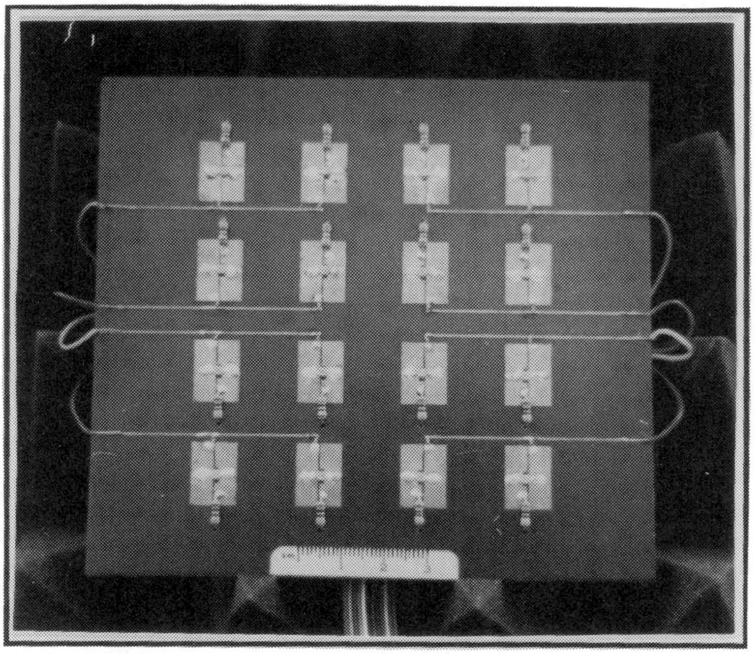

Figure 12 — Photograph of a 4 × 4 radiatively coupled oscillator array, using MESFETs and patch antennas, which operated at 8.2 GHz.[27]

The prototype array of figure 12 was especially useful as a vehicle for understanding how to establish the correct phase relationship in an array of coupled oscillators. The weak radiative coupling meant that the array elements were not strongly perturbed from the free-running state, which allowed a description of the mutual synchronization using a modified version of Adler's equation.[25,27] It was found (after much experimentation) that the phase of the mutual interaction—determined by the element separation—was crucial in determining the stable mode of operation.[28,49] Methods for characterizing this parameter were developed,[41,42] and this facilitated the design of the MESFET array in figure 12.

Unfortunately, weak radiative coupling means that the locking bandwidth is very small, so mutual synchronization can only take place when the individual array elements are very well matched. In the prototype arrays, this was accomplished by supplying individual bias to each element to allow for fine bias tuning adjustments; this would not be desirable in a large array. Even with the individual bias, it was very difficult to establish mutual synchronization without the help of external feedback reflectors.[27] For monolithic fabrication it may be possible to achieve a higher level of uniformity between the array elements, but stronger coupling mechanisms are generally preferred so that array operation is less sensitive to device variations over the array. However, there are at least two cases where weak mutual coupling between oscillators may actually be preferred: (a) arrays that are placed within a quasi-optical open resonator,[4,5] where the devices should only interact via the resonant cavity mode; (b) arrays that are injection-locked to an external source, which could be supplied quasi-optically or through an on-wafer feed network. The latter case has been demonstrated by Birkland and Itoh for a 4 × 4 FET array.[31]

Stronger inter-element coupling can be accomplished using transmission-line circuits connecting the array elements. This requires suitable modifications to the coupled-oscillator theory so that the stronger oscillator interactions can be properly controlled to insure the desired phase relationships. Such a theory has recently been described,[29] based on a modification of Stephan's work.[23,46,47] The model is shown in figure 13, where a collection of N independent sinusoidal oscillator circuits are connected via an arbitrary N-port coupling network, described by Y-parameters. Each oscillator is described by a free-running frequency, free-running amplitude, and a resonator Q-factor, all of which can be experimentally characterized. A set of differential equations describing the amplitude and phase dynamics is found, which can be used to determine steady-state phase relationships and mode stability.

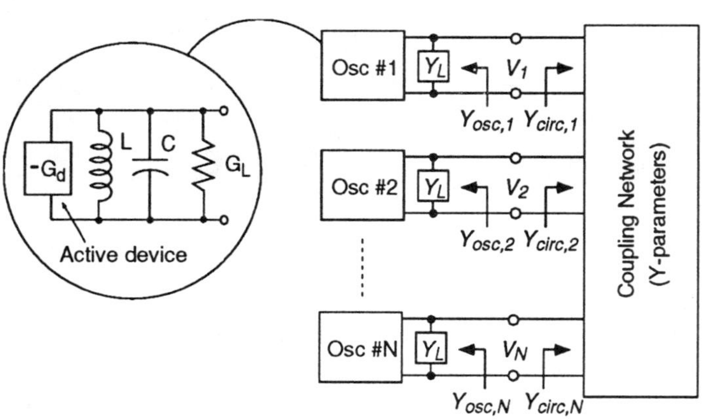

Figure 13 — General model for a coupled oscillator array. Each individual oscillator is modelled by a very simple RLC genative resistance circuit, which leads to the van der Pol equation.[30] Coupling between the oscillators is described by Y-parameters, which gives a simple set of differential equations describing the array dynamics.[29]

Most importantly, the theory describes how the coupling network parameters and frequency distributions must be configured to produce a specified phase relationship. It has been shown that there is a critical value for the coupling phase to produce a broadside beam.[28] The free-running frequency distribution (particularly the frequencies of the end elements) have a profound effect on the phase relationship. The analysis is particularly simple for coupling networks which satisfy a broad-band condition,[29] which is accomplished by introducing lossy elements. A simple array with controlled oscillator interaction has been developed and is illustrated in figure 14. The length of the coupling line sets the phase of the interaction, while the chip resistance and line impedance determine the coupling strength. Unlike the grid approach, there are size limitations on the coupled-oscillator technique. As the array gets large, the coupling strength must be increased as $\sqrt{N}$, where N is the number of array elements. For very large coupling strengths, multiple stable modes appear, and beyond this the theory breaks down since the oscillators no longer behave independently.

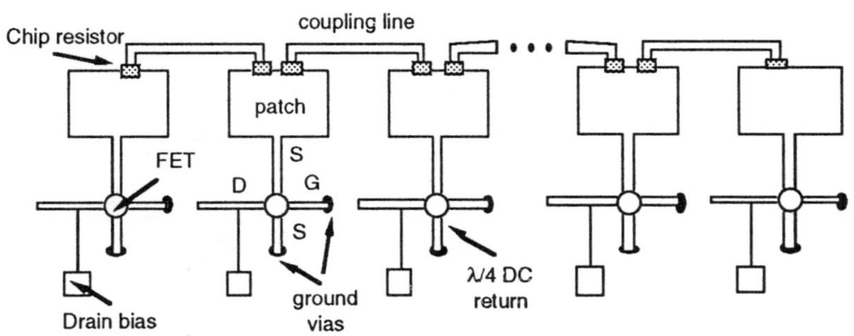

Figure 14 — An active patch array using MESFET oscillators coupled by a simple transmission-line network.[29] The chip resistor and coupling line determine the strength and phase of the oscillator coupling in a controllable fashion.

Arrays with *very* strong coupling have been described by Itoh and co-workers.[33–38] Their technique uses a long resonant transmission-line periodically loaded with active devices, as shown in figure 15. This is a multi-mode system, qualitatively similar to the Kurokawa waveguide combiner of figure 3 but with an important difference: the load is distributed along the array and separated from the coupling line, with each device coupled to its own antenna. This allows mode-suppression resistors to be placed at strategic points along the line without adversely affecting the output power. Since it is a planar approach (unlike the Kurokawa combiner), the concept extends readily to two-dimensional arrays. However there may also be limitations on the number of devices that can be combined with this technique, since unwanted modes will be difficult to suppress for long arrays with high mode density. Like the topology of figure 14, the mode resistors also complicate the DC biasing.

A modal analysis using "averaged potential" theory has been developed to explore the stability of the various possible modes in the periodic combiner technique of figure 15.[37] This theory is capable of modelling the competetive growth of the different modes, and is used to determine the proper value and position of mode-suppresion resistances to ensure the proper operating mode. A possible drawback of the analysis is the assumption of identical unit cells. The theory has been experimentally tested using linear arrays of two, four, and eight Gunn diodes in a

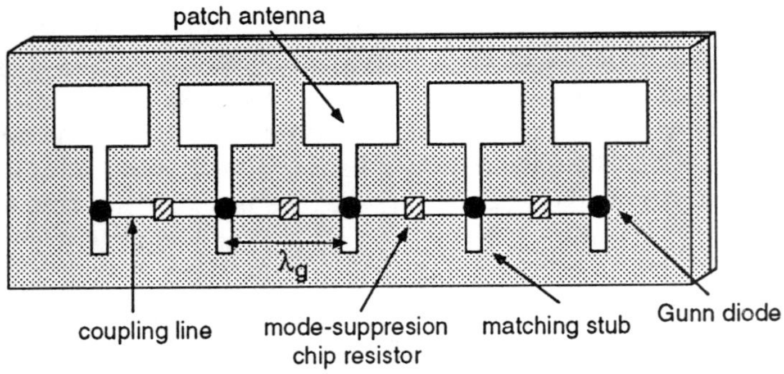

Figure 15 — Strongly coupled Gunn diode array.[37] The devices are spaced periodically (one guide wavelength) in a resonant microstrip line, loaded with a small mode suppression resistor to ensure in-phase combining.

topology similar to figure 15.37 ERPs of 19.1, 25.1, and 30.4 dBm were obtained, respectively. A 4.7 Ω chip resistance was required for mode suppression to establish the in-phase mode for power combining.

An extension of the strong coupling approach has been described by Mortazawi and DeLoach, employing an "extended resonance" technique.[39] This treats the array as a large distributed oscillator, much like the CalTech grid approach, except that the infinite array/symmetry assumptions are not made. The individual array elements are qualitatively similar to the active antennas used in previous coupled-oscillator arrays, but are not capable of self-sustained oscillation (ideally) unless connected to each other by an appropriate length of transmission line. A 3 × 3 MESFET array operating at 7.33 GHz with an ERP of 6 Watts has been demonstrated using this technique. The stablility of the desired mode has not been explored.

2.3 Beam-scanning and mode-locked arrays

Two scanning techniques have been developed using coupled oscillator arrays. In Stephan's technique, two coherent signals with a predetermined phase shift are injected into opposite ends of the array.[46] The phase shift is devided equally along the chain. This was verified experimentally at X-band.[47] The drawback of this technique is that the maximum achievable phase shift decreases as the array gets larger. The second method does not have this limitation, and was predicted using the coupled-oscillator theory.[28,48] It was found that a constant phase progression could be established by properly tuning the frequencies of the end elements in the array. This is illustrated in figure 16b and contrasted with the conventional phased-array approach of figure 16a. Both methods have a two-fold advantage over conventional phased-arrays: phase shifters are not required at each antenna, which greatly simplifies the design and manufacture of the arrays, and no RF feed network is required since the source is effectively distributed over the entire array.

When all of the oscillators are set to the same free-running frequency, the array is uniformly phased and a broadside beam results. When the frequencies of the end-elements are slightly detuned from the others (in opposite directions), a constant

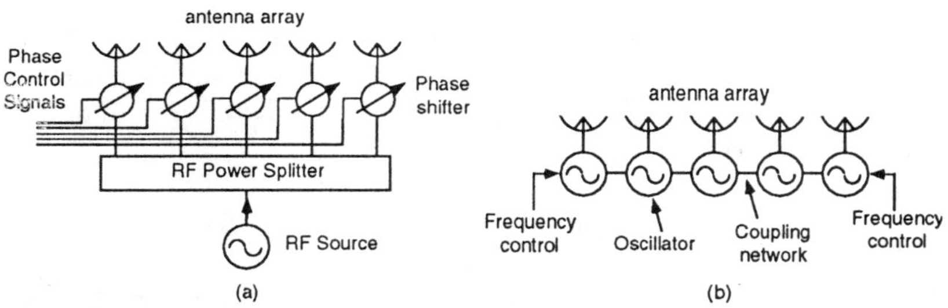

Figure 16 — Comparison of (a) conventional phased-array architecture, and (b) beam-scanning with an oscillator array.

phase progression is established along the array.[48] The magnitude of the phase shift thus incurred is proportional to the amount of frequency detuning of the end elements, and can be varied continuously over a 180° range. More importantly, the phase shift is independent of the number of oscillators. Note that the output frequency of the array will not change as the beam is scanned; only the free-running frequencies are being adjusted, while the oscillators will still lock to a common frequency in the steady-state. Thus only two VCOs are required, and these are on the array periphery where it is easier to route control signals.

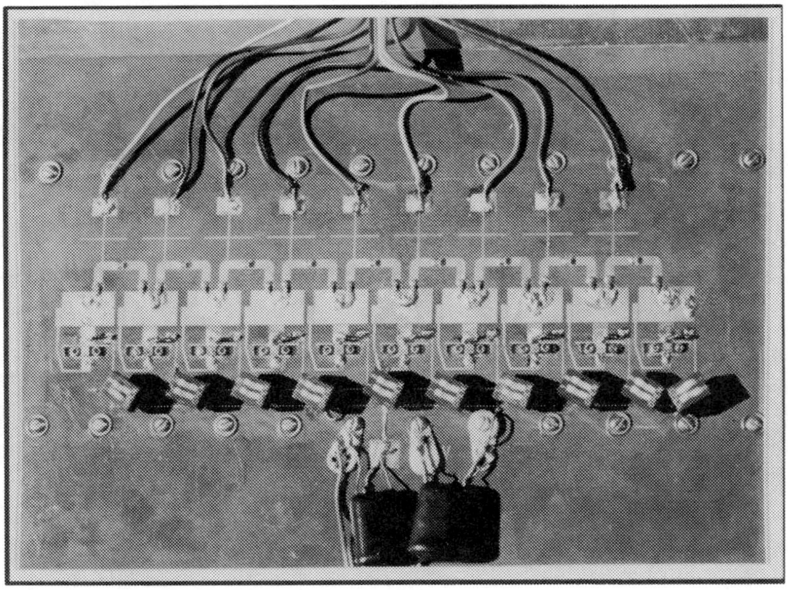

Figure 17 — A 10 × 1 VCO array for beam scanning, using a modified version of the active patch oscillators shown in figure 11. This array exhibited continuous scanning from −15° to +30° from broadside.

This scanning method has been verified using several small linear arrays. Most recently, a ten-element array was developed using a modified version of the active patch of figure 11a, which included a varactor tuning element for frequency control, and the coupling network of figure 14. This is shown in figure 17. This array used a NE900276 packaged medium power MESFET, and produced over 1 Watt total power (10 Watts ERP) at 8.4thinspace GHz with 10% conversion efficiency.[49]

By tuning only the end elements, the beam was scanned from −15° to +30° off broadside; the asymmetry was attributed to a nonuniform output power variation of the oscillators as they were tuned. The theory predicted a scanning range of −30° to +30° off broadside.

Unusual nonlinear dynamics in the arrays are also manifest in a recently observed mode-locked state.[44,45] This mode-locking is based on an analogous situation in short-pulse laser systems,[43] and is produced by adjusting the frequencies of each element to produce an evenly spaced comb spectrum. When the free-running frequencies are close to this ideal, the system will spontaneously lock into a state where the different spectral components are "in-phase", thus producing a periodic pulse train. This can be understood in a crude sense with reference to figure 18a. When two oscillators are coupled with their frequencies far enough apart to prevent mutual injection-locking, then a set of beat frequencies is produced. Each of these beat tones can be used to injection-lock other oscillators, thus forming a coherent comb spectrum. Figure 18b shows the measured output pulse power envelope for a five-element array.[44] As the number of oscillators grows, the pulse width becomes shorter and the peak power increases. This concept could be useful for producing compact pulse transmitters for millimeter-wave radar applications.

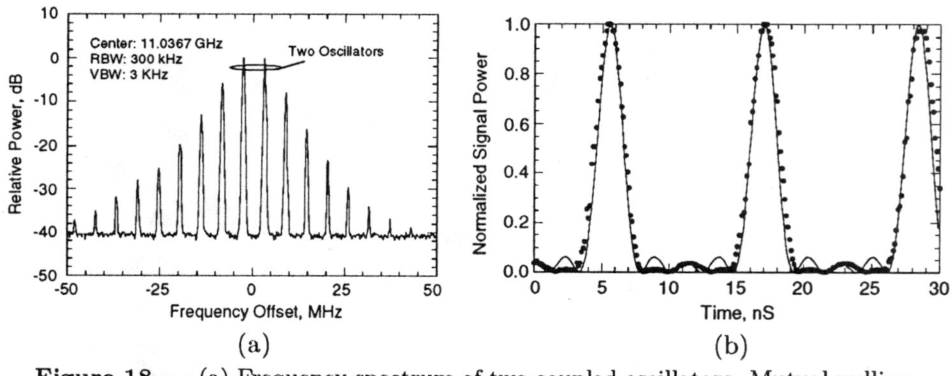

Figure 18 — (a) Frequency spectrum of two coupled oscillators. Mutual pulling effects give rise to a number of additional spectral components. These are used to injection-lock other oscillators. (b) Measured (dots) and theoretical (solid line) time dependence of the output power envelope of five mode-locked oscillators.

3. Quasi-Optical Amplifiers

Quasi-optical amplifier arrays will probably find more immediate use in mm-wave systems. Each cell of the array contains an input antenna, amplifier, and output antenna; by clever design, the input and output antennas could be coupled through a single antenna. The synchronous operation of the devices is ensured simply because all the array elements amplify the same signal. However, the input and output of the amplifier must be isolated to prevent positive feedback and possible oscillation. This is tricky because the input and output antennas are typically close together on the substrate, and can interact via mutual coupling. A convenient solution is to use orthogonal polarizations for input and output, and nearly all the arrays developed to date use this technique.

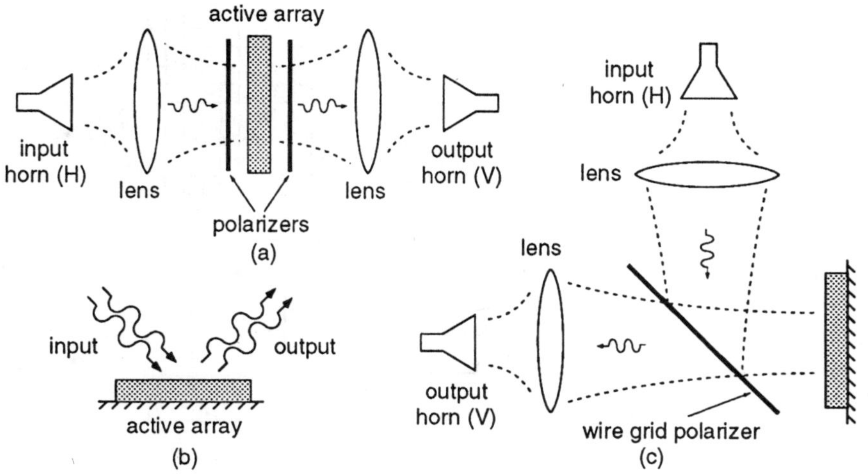

Figure 19 — Feeding arrangements for amplifier arrays. (a) Transmission amplifier with separate input and output polarizers for isolation and tuning. (b) Reflection amplifier with input/output isolation achieved via specular reflection. (c) Reflection amplifier with polarization isolation.

Both transmission amplifiers and reflection amplifiers have been developed. The feeding arrangement depends on the type of amplifier array. Transmission amplifiers are fed as shown in figure 19a, where external wire-grid polarizers are placed on opposite sides of the array, and the input and output signals are orthogonally polarized. Reflection amplifiers are somewhat more complicated, since input and output signals must be coupled from the same side of the array. Two possible techniques are shown in figure 19b-c. The input signal can be directed at the array obliquely, and the amplified output will reflect specularly; this technique would permit reflection amplifiers to be cascaded for higher gain. An alternative is to use the quasi-optical equivalent of a circulator, as shown in figure 19c. This particular example uses a tilted wire-grid polarizer, but Faraday rotation devices could also be used.[74] A single feed design using a circular horn and an orthomode transducer has also been reported.[70]

Again, both the dense grid and active array approaches have been proposed. Efficiency considerations may govern the choice of topology. As shown in figure 20a, the power saturation of a typical mm-wave amplifier leads to a critical value of input power required for maximum efficiency; this is the point at which the $P_{\text{out}} - P_{\text{in}}$ curve is a maximum. Each cell of the array must absorb this amount of power for efficient operation. In a typical high-efficiency application, the amplifier is designed to provide a specified output power and gain. This fixes the input power to the array. If this power is uniformly distributed over the array area, then the cell size will determine the amount of power absorbed by each amplifier. This is illustrated in figure 20b for two arrays, which are identical in size and are illuminated by a uniform 100 mW beam. The optimum cell size will then be determined by the amplifier characteristics.

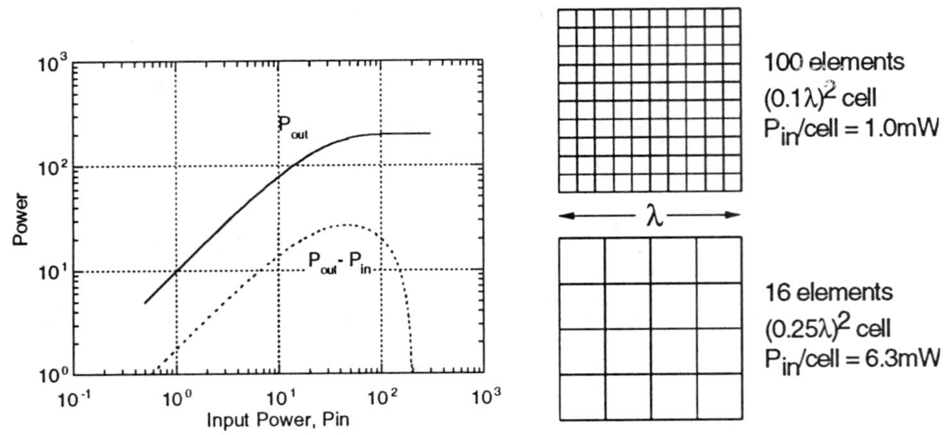

(a) (b)

Figure 20 — (a) Typical power saturation curve for mm-wave amplifiers. (b) Two arrays with identical amplifiers and incident power of 100 mW. Each array produces the same power and gain, but operate at different efficiency due to the different power incident on a unit cell.

To achieve high power amplification with high gain and good efficiency, it may be possible to cascade several arrays as shown in figure 21. The early stages would require a larger absorbing cross section for each cell, which favors the active array approach, while the final stages of the cascade would employ a large number of devices in a dense array, favoring the grid approach. This system is well suited to placement in a closed metallic waveguide. For gaussian beam systems, focusing elements can be used so that the array sizes can be made progressively larger in a cascade to accomodate more devices for increased power.

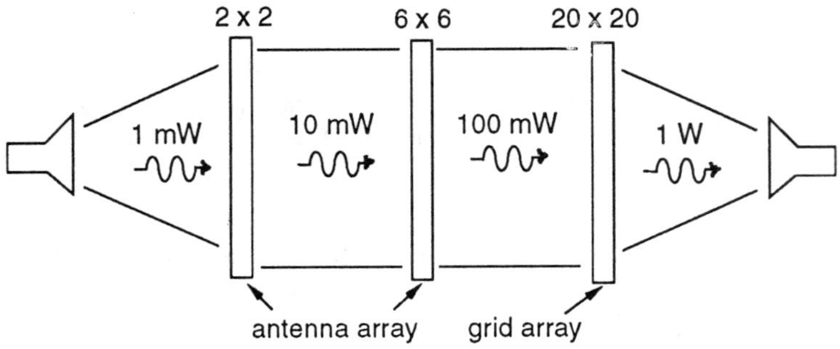

Figure 21 — For high gain, high efficiency operation, several arrays would be cascaded in a uniform beam system, with a large cell size (active antenna array approach) on the first stages and dense arrays (grid approach) in latter stages.

3.1 Active array amplifiers

A conventional phased-array is a power combiner, and provides a useful baseline for comparison to the quasi-optical active array approach. An recent example is a 21-element Ka-band amplifier array using a tapered slot antennas,[63] used as the feed for a Cassegrain antenna. The amplifiers are fed by a conventional N-way power splitter network using transmission lines, with the amplified output fed to a tapered slot antenna so that the power is combined in free space. The measured combining efficiency was only 60%, which is probably a result of using the lossy power splitter network (see figure 2b). The gain was 24 dB with a 1.5 GHz (5%) bandwidth at 32 GHz

To improve the combining efficiency, amplifier arrays that are fed quasi-optically have been developed.[64–71] With one exception[70] these designs use separate antennas for input and output signals. The challenges facing the design of a suitable amplifier cell are the limited bandwidth of common planar antenna structures, isolating input and ouput signals to prevent oscillation, and constraints imposed by the optimum cell size for efficiency. Designs based on the patch antenna,[64,69,70] integrated horns,[67] and short dipoles[68] have all produced bandwidths of less than 3%, and alternative designs are being sought. There are several possiblities for broadband antennas on high dielectric constant substrates.[88–92] The bandwidth of a patch antenna can be increased by adding parasitic resonant elements,[90,91] which is analogous to the design of broadband filter networks. Travelling-wave or complementary antennas can also be used,[88,92] but these typically occupy a large area, and in the case of the bow-tie, have an undesirable radiation pattern.

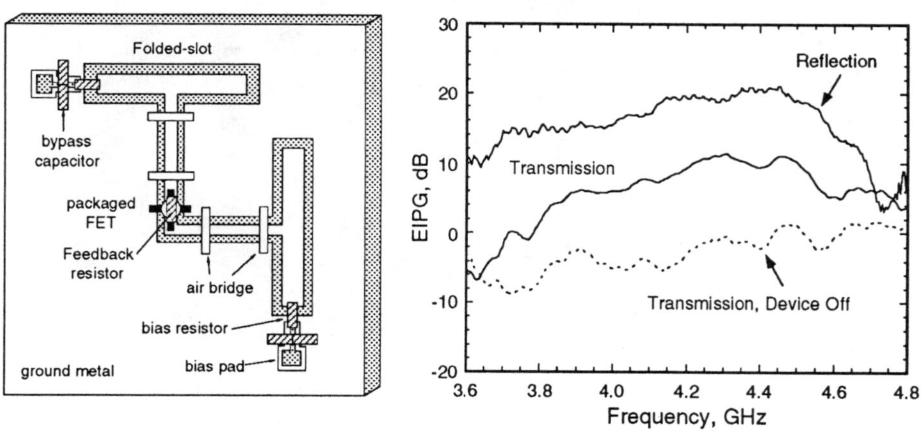

(a) (b)

Figure 22 — (a) Polarization-rotating amplifier cell using a resistive feedback FET amplifier and folded-slot antennas. (b) Measured EIPG (power gain × directive gain) for the amplifier cell, in both transmission mode and reflection mode.

Recently, an intense program has been undertaken at the Rockwell Science Center to develop a 5 Watt, 44 GHz quasi-optical amplifier using HBT technology. Amplifier cells based on wideband patch antennas and wideband slots, such as the folded-slot,[85] are being developed. The folded slot is attractive because it is easily integrated with three-terminal devices, can be fabricated in a single mask step, and is inherently wideband. A 4 GHz prototype amplifier cell using folded slots is shown in figure 22a. This circuit incorporates a simple wideband resistive feedback MESFET amplifier between the orthogonally polarized input and output antennas. The measured effective isotropic power gain (EIPG) in free space is shown in figure 22b as a function of frequency. The EIPG is written as $G_{\text{amp}}G^2_{\text{ant}}$, where G_{amp} is the amplifier power gain and G_{ant} is the gain of a single antenna. No polarizers were used in this measurement, and since the folded slot radiates both above and below the substrate, the measured gain in transmission is 3 dB lower than is achievable with polarizers. For use as a reflection amplifier, the circuit was placed in front of a mirror, which significantly increases the directivity and hence EIPG of each antenna. The measured gain in transmission with the device unbiased is also shown. The 3 dB bandwidth of this circuit was approximately 12% on a high dielectric constant substrate (Rogers Duroid 6010, $\epsilon_r = 10.8$). It is believed that bandwidths of 15–20% can be achieved with this topology by optimizing the folded-slot design and improving the amplifier matching network.

Figure 23 — A 4×4 amplifier array on Duroid 6010 ($\epsilon_r = 10.8$) using the folded-slot cell of figure 22, with packaged NE32184A devices.[65] This array operated at 4 GHz with a 10% 3 dB bandwidth.

A 4×4 array of the folded slot cells of figure 22 was fabricated and tested to explore suitable biasing schemes, and ensure that neighboring cells would not strongly couple and lead to parasitic oscillation. A photograph of this array is shown in figure 23. The element spacing is approximately a half-wavelength (in air) in both directions. The measured frequency response for this array was essentially the same as for a single element, but with a larger directivity. As expected, the output power was 16 times that of the single element for the same incident power density, but the 3 dB bandwidth of 9% was less than the single element. This is probably a result of poor etching tolerance in the fabrication process, which led to nonuniform antenna dimensions across the array. When biased for class-A operation, no oscillation was observed, but parasitic oscillation did occur for a small range of gate bias below the class-A point, which indicates either excessive coupling between neighboring elements (which could be occuring through substrate modes), or an improper matching network to a conditionally stable device.

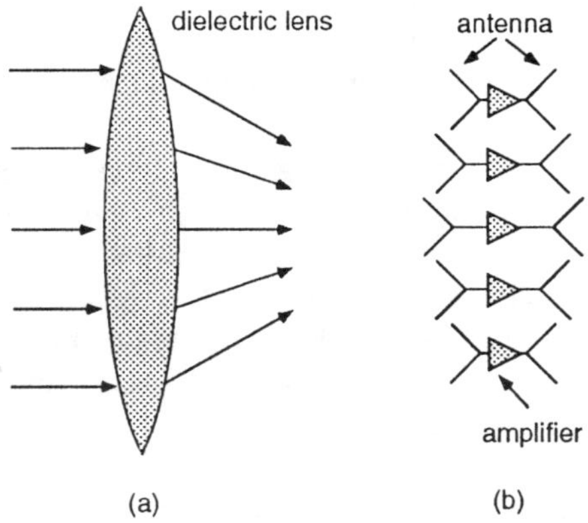

Figure 24 — The action of a dielectric lens (a) could possibly be reproduced in an active amplifier array (b) by properly adjusting the phase delay between input an output antennas.

A qualitatively similar array to figure 23 has been reported using patch antennas,[69] but differs in two respects: (a) the antennas are on opposite sides of the substrate and are shielded from each other by the respective ground planes, and (b) this allows the antennas to be oriented for the same polarization for input and output signals. This array had a measured bandwidth of approximately 4% at 10 GHz, which could be improved by using a modified patch antenna structure. A large number of via holes were required to maintain a good ground plane and to couple signals between antennas, but this could possibly be avoided by using other antennas and/or vertical-current devices like the HBT.

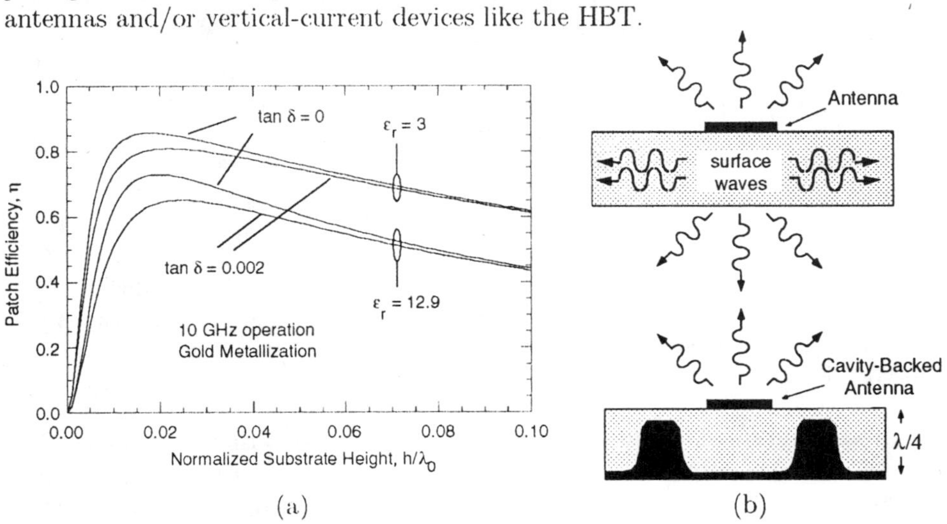

Figure 25 — (a) Theoretical radiation efficiency of a patch antenna on two common substrate materials. For thin substrates, skin-effect losses limit the efficiency, while surface wave excitation limits the efficiency for large thickness. (b) Surface wave excitation can be controlled by altering the substrate.

In these types of active array amplifiers, the phase delay between input and output antennas can be independently controlled within each array cell, simply by increasing the path length. This raises the possibility of creating an "active lens" array, as suggested in figure 24. This concept could be used to compensate for the phase errors associated with feeding the array with a waveguide horn, for example, and eliminate the need for external focusing elements such as shown in figure 19.

An additional consideration for obtaining high efficiency in active arrays is the radiation efficiency of the antenna structures. On electrically thick, high dielectric constant substrates, antennas suffer poor radiation efficiency as a result of surface-wave excitation. This is illustrated in figure 25a, showing the theoretical efficiency of a patch antenna on two different substrate materials as a function of substrate thickness. As the thickness and dielectric constant increase, surface waves are coupled to more strongly. This is doubly unfortunate because the bandwidth increases with increasing thickness and decreasing dielectric constant.[94,95] The substrate cannot be made too thin, since skin-effect losses become large and also limit efficiency. Even at the point of optimum efficiency, a significant amount of energy is lost to surface waves. In an array, the problem is not quite as bad since the relative phasing and spacing of the array elements usually limit the amount of coupling to the substrate modes. However, a simple solution is to simply alter the substrate in such a way as to inhibit the surface waves at the frequency of interest. This could be done by selectively thinning certain areas, or by etching and plating pits in the substrate to create a cavity-backed antenna, shown in figure 25b.

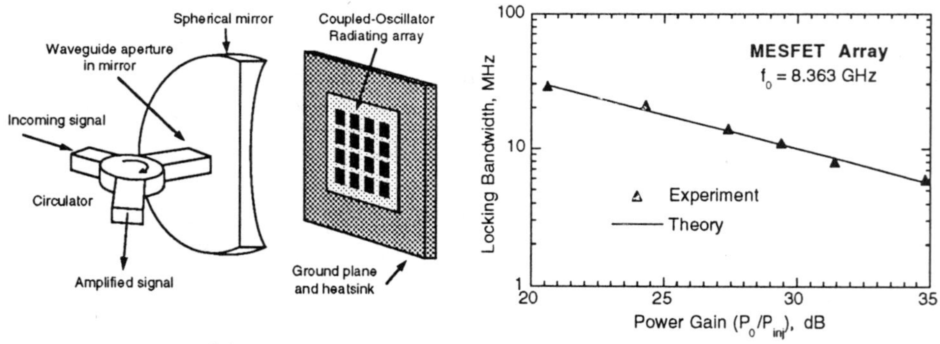

(a) (b)

Figure 26 — (a) An injection-locked amplifier for high gain and low bandwidths, using a waveguide circulator to isolate input and output signals. (b) Measured and theoretical modulation bandwidths for an experimental 16-element MESFET injection-locked amplifier.

For high gain and small bandwidths, an injection-locked reflection amplifier can be made from the oscillator arrays previously described. In that case the oscillators lock onto an externally injected signal, which can be frequency modulated. This signal can be in the form of an incident plane wave from a distant source, or a signal coupled into the cavity via a circulator as shown in figure 26a. The analysis and design of such amplifiers is not significantly different than the oscillator arrays previously described. This concept has been demonstrated using the 16-element

MESFET array of figure 12, with the input signal injected from a distant horn antenna. The measured locking bandwidth is shown in figure 26b, and indicates that very high gains can be achieved.

A reflection amplifier array using a dual-mode stacked patch antenna has also been reported.[70] This array is still in the developmental stages, but is noteworthy because it is operated entirely within an oversize metallic circular waveguide. Modified sidewalls using longitudinal ridges or dielectric loading are being considered for creating a nearly uniform field amplitude over the surface of the array for improved efficiency. The array is fed from a standard circular guide, using an orthomode transducer to isolate input and output signals.

3.2 Grid amplifiers

The grid approach is attractive for obtaining maximum power per unit area of substrate. Grid amplifiers are being explored as part of the aforementioned Rockwell amplifier project, and a 100-element HBT amplifier grid has been reported.[71] This array was designed for operation at 10 GHz, measured approximately three wavelengths on a side, and was fabricated on a low dielectric constant Duroid substrate ($\epsilon_r = 2.2$), with Rockwell differential-pair HBT chips bonded into the unit cell. A simplified illustration of the unit cell is shown in figure 27. Horizontal grid lines act as an input dipole feeding the base junctions of the differential pair chip, with the output connected to the vertical grid leads.

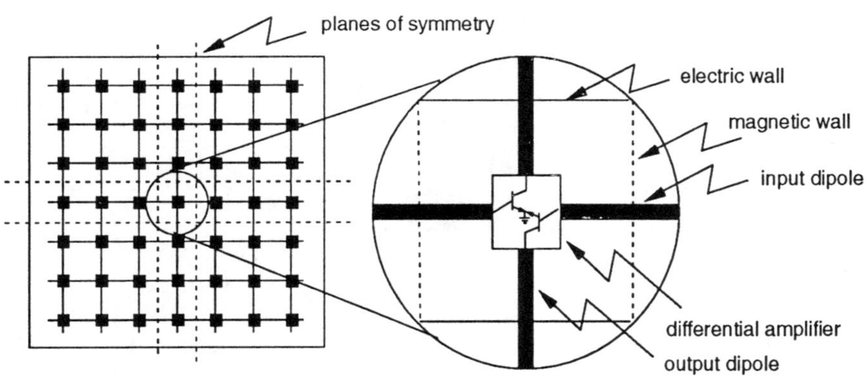

Figure 27 — Illustration of the grid amplifier developed at CalTech.[66,71] Input and output dipoles are orthogonally polarized for input/output isolation, and are coupled by a Rockwell HBT differential amplifier pair.

Both far-field measurements and lens-focused measurements (using the setup of figure 19a, described by Gagnon[73]) have been made, and gave identical results.[71] The peak gain was 10 dB at 10 GHz, with a 3 dB bandwidth of 1 GHz, or 10%. A parasitic oscillation was observed, but was empirically eliminated by bonding a sheet of high dielectric constant Duroid material ($\epsilon_r = 10.5$) to the backside of the grid. External wire-grid polarizers were used on both sides of the grid, and the relative positions of the polarizers were adjusted for optimum gain. Power saturation characteristics were also made possible by the lens-focused measurement system, with the grid placed at the beam waist. The maximum power-added efficiency was found to be 5% at an output power level of 330 mW; this efficiency was less than expected based on the measured characteristics of the naked HBT chip. Measurements of noise figure, input and output match, modulation, failure tolerance, and amplification of beams at oblique incidence have also been made, and are described in detail.[71]

By providing positive feedback to an amplifier array, an energy source is created that is a close analog of the laser. A simple mirror will not work with the arrays described so far, since they rotate the polarization during the amplification. A clever technique has recently been described[72] using a "twist reflector". This is illustrated in figure 28, and has the property that an incident linearly polarized beam is reflected with the orthogonal polarization, provided that the twist reflector is oriented properly with respect to the incident beam. Using this topology, a tunable source was created over the range of 8.2–11.0 GHz by adjusting the position of the twist reflector. The tuning range is determined by the amplifier bandwidth. This technique could provide much wider tuning ranges than is possible with an oscillator array, and could be merged with the varactor-tuning grid of figure 9 to create an wideband electronically tuned source.

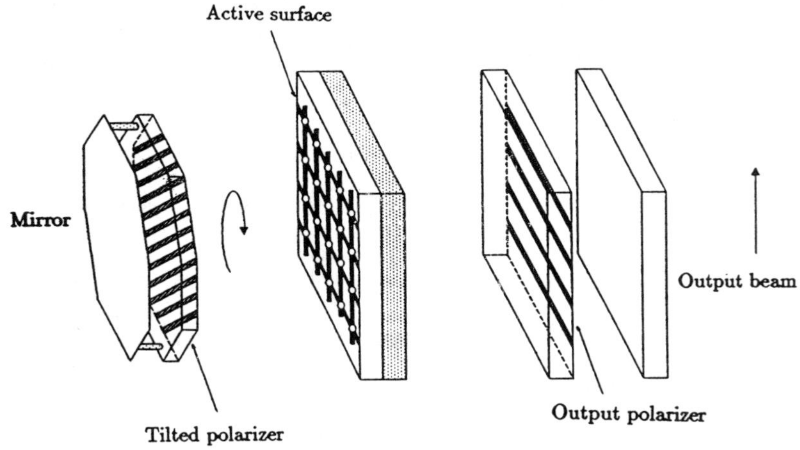

Figure 28 — A feedback oscillator using the CalTech 100-element HBT amplifier grid and a "twist reflector", which reflects the correct polarization back to the amplifier (After Kim *et al.*[72]). This arragement was tunable from 8.2 to 11 GHz by moving the reflector, nearly twice the 3 dB bandwidth of the amplifier.

4. Frequency Conversion and Beam Control

While oscillators and amplifiers are crucial components for mm-wave systems, a complete quasi-optical receiver or transmitter will require other active elements for mixing or down-conversion, modulation, filtering, beam-switching, frequency multiplication or up-conversion, and possibly phase-shifting. Some work has been done to develop these components, but the field is still wide open. Preliminary work using the grid approach has been carried out on mixers,[75–78] beam switches,[86,87] phase shifters,[79] and frequency multipliers.[80–85]

A 100 element planar Schottky diode grid for mixing or detecting quasi-optical signals has been developed at CalTech.[75] This grid is illustrated in figure 29, and was constructed using a bow-tie mesh for broader bandwidth. Both the RF and local oscillator (LO) signals are provided quasi-optically, with the same (vertical) polarization. The IF signal can be removed via a coplanar waveguide feed as shown in the figure. The advantage of such a grid is an improved dynamic range as compared to single diode mixers or detectors. However, the required LO power is correspondingly greater than single device mixers, but this could been achieved using a quasi-optical oscillator arrays as the LO source. In addition, the conversion loss and noise figure of a square grid is the same as a single diode mixer in an equivalent embedding impedance.

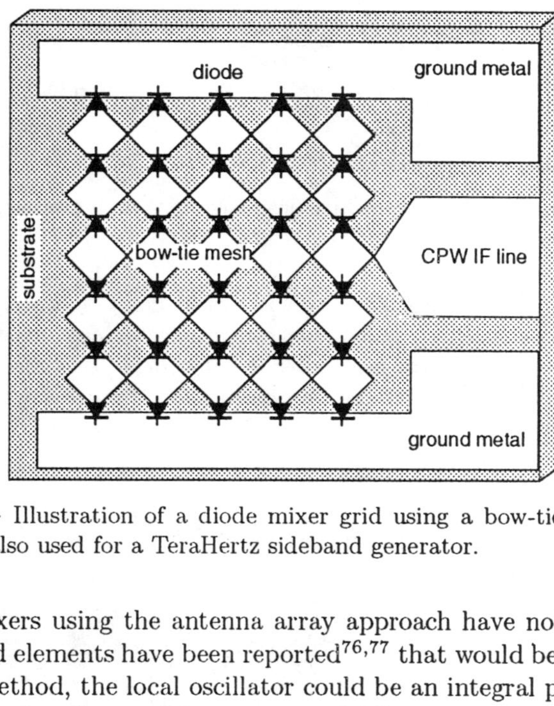

Figure 29 — Illustration of a diode mixer grid using a bow-tie mesh.[75] This topology was also used for a TeraHertz sideband generator.

Quasi-optical mixers using the antenna array approach have not yet been developed, but isolated elements have been reported[76,77] that would be suitable for such arrays. In this method, the local oscillator could be an integral part of each array element,[78] eliminating the need for an external LO array.

Grids of varactor diodes have also been proposed and developed for use as phase-shifters.[79,80] Such grids could be used for reflective beam-scanning, as illustrated in figure 30, or for electronically tuned oscillators as previously discussed.[20] In the beam-scanning application, a beam reflects of a pair of diode grids backed by a metal mirror. A progressive phase-shift is developed across the aperture that can be varied by changing the bias on the varactor diodes. A 1600 element grid using Schottky-barrier varactors demonstrated a 70° phase shift at 94 GHz. A 7 dB loss was measured, but this was primarily due to devices with a large series resistance.

Using similar diode grids, frequency multipliers have been investigated.[80,81] These arrays can provide much more power than single device multipliers. A 760 device grid produced 500 mW at 66 GHz (pumped at 33 GHz) with a second-harmonic conversion efficiency of 10%. Again, better conversion efficiency would result from the use of more advanced devices with lower series resistance.

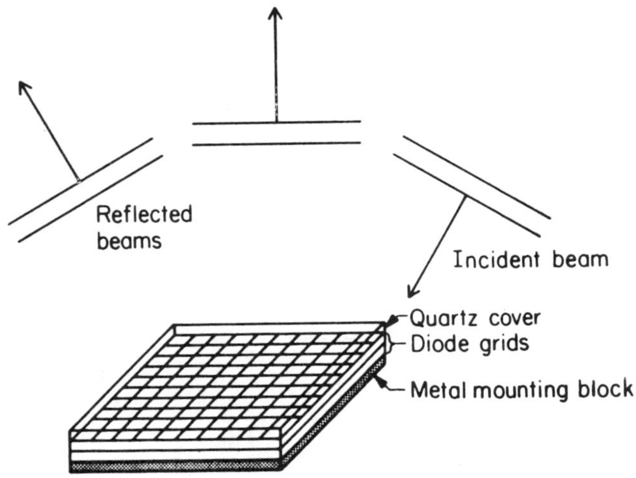

Figure 30 — Illustration of reflective beam-scanning with a varactor diode phase-shifting grid (after Lam *et al.* [79,80])

Quasi-optical multiplers have also been reported using the active antenna array approach.[82,83] More recently, parametric effects have been observed using diode-loaded antennas,[84] and arrays are under development to explore low-noise parametric amplification and parametric up-conversion.[85] Such arrays require a very strong pump signal, which could be provided quasi-optically or by an on-wafer coupled oscillator array.

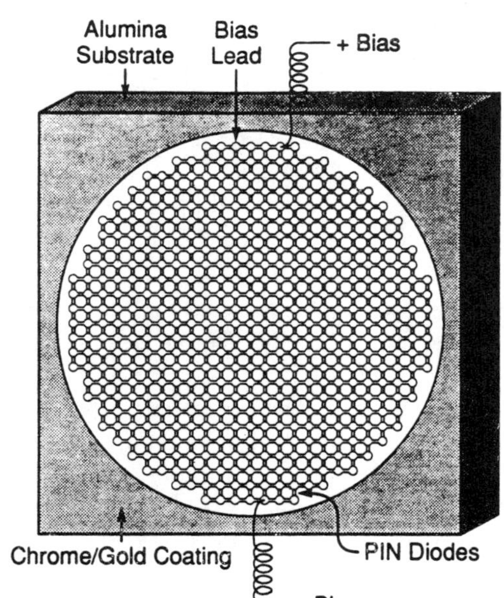

Figure 31 — Illustration of a prototype 94 GHz quasi-optical beam switch using PIN diodes (after Stephan *et al.*[87])

Diode arrays can also be used for beam-switching.[86,87] An example is shown in figure 31, which illustrates a PIN array developed as part of a Dicke-switch radiometric imaging system.[87] Such arrays can result in an enormous cost savings when compared to an array of low-loss waveguide switches. The example shown in figure 31 was fabricated both monolithically and with hybrid assembly, with similar results of better than 20 dB isolation and 1 dB insertion loss at 94 GHz. Beam switches could also be used with oscillators and amplifiers for digital signal modulation. Similar switching grids could also provide a programmable diffraction grating for beam scanning.

5. Conclusion

Quasi-optical electronics is a convenient marriage of integrated circuit techniques and optics that can solve many problems at mm-wave frequencies. Passive quasi-optical systems are a well developed technology at millimeter-wave frequencies,[8] and introduction of active array components for quasi-optical power-combining appears very attractive. An intrepid designer will find that the progression to beam guiding systems affords new, simpler, and possibly cheaper methods for achieving common circuit functions—a wire grid beam-splitter replaces a waveguide ferrite circulator, for example. An extra degree of freedom is acquired in the process, as compared to conventional integrated electronics: the mirrors and lenses can be *mechanically* adjusted for tuning or performance optimization to compensate for statistical deviations in the IC fabrication.

In what frequency range will this technology be most useful? In principle, QOPC arrays can be made from very low frequencies to near optical frequencies in the Terahertz range. At the low frequency end, the cutoff is economic; the arrays become too big, and classical combining techniques (circuit and chip level) are adequate. At the upper frequency end, electronic devices give way to optical devices. Most likely, the 30 GHz to 1 THz region will benefit the most from this technology.

As mentioned repeatedly in this paper, the active array work to date has been primarily proof-of-concept. The next generation of research in this area can take several directions: demonstrating the concepts more conclusively at very high frequencies (above 35 GHz); demonstrating much larger arrays with up to 1000 elements; developing more advanced array components that incorporate modulation functions or frequency agility; developing multiple array modules for transmitter and receiver applications. For application oriented work, several engineering issues must be addressed, including waste heat removal and device density limitations, array biasing and control signal routing, and effects of processing yield on array performance. Currently several industrial partners (Rockwell International, Hughes Aircraft Co., and Martin Marietta, among others) are working with the academic community on some of these problems. In the short term, new commercial applications for mm-waves such as automotive radar and satellite broadcasting will most likely make use of this emerging technology.

6. References

1. K.J. Sleger, R.H. Abrams, and R.K. Parker, "Trends in solid-state microwave and millimeter-wave technology", *IEEE MTT-S Newsletter*, pp. 11–15, Fall 1990.
2. K. Kurokawa, "The single-cavity multiple-device oscillator", *IEEE Trans. Microwave Theory Tech.*, vol. MTT-19, pp. 793–801, Oct 1971.
3. K. Chang and C. Sun, "Millimeter-wave power-combining techniques", *IEEE Trans. Microwave Theory Tech.*, vol. MTT-31, pp. 91–107, Feb 1983.
4. L. Wandinger and V. Nalbandian, "Millimeter-wave power-combiner using quasi-optical techniques", *IEEE Trans. Microwave Theory Tech.*, vol. MTT-31, pp. 189–193, Feb 1983.
5. J.W. Mink, "Quasi-optical power combining of solid-state millimeter-wave sources", *IEEE Trans. Microwave Theory Tech.*, vol. MTT-34, pp. 273–279, Feb 1986.
6. J.C. Wiltse and J.W. Mink, "Quasi-optical power-combining of solid-state sources", *Microwave Journal*, vol. 35, pp. 144–156, Feb 1992.
7. Special Issue on Quasi-Optical Techniques, J.W. Mink and D.B. Rutledge, eds., *IEEE Trans. Microwave Theory Tech.*, vol. MTT-41, October 1993.
8. P. F. Goldsmith, "Quasi-optical techniques at millimeter and sub-millimeter wavelengths", in *Infrared and Millimeter Waves*, vol. 6, chap. 5, pp. 277–343, K.J. Button, Ed., New York: Academic Press, 1982.

Grid Oscillators

9. H.M. Harris *et al.*, "Quasi-optical power combining of solid state sources in Ka-band", *IEEE MTT-S International Symp. Digest* (Atlanta), vol. 1, pp. 159–162, 1993.
10. Z.B. Popovic, M. Kim, and D.B. Rutledge, "Grid Oscillators", *Int. J. Infrared and Millimeter Waves*, vol. 9, pp. 647–654, July 1988.
11. Z.B. Popovic, R.M. Weikle, M. Kim, K.A. Potter, and D.B. Rutledge, "Bar-grid oscillators", *IEEE Trans. Microwave Theory Tech.*, vol. MTT-38, pp. 225-230, March 1990.
12. Z.B. Popovic, R.M. Weikle, M. Kim, and D.B. Rutledge, "A 100-MESFET planar grid oscillator", *IEEE Trans. Microwave Theory Tech.*, vol. MTT-39, pp. 193–200, Feb 1991.
13. M. Nakayama, M. Heida, T. Tanaka, K. Mizuno, "Millimeter and submillimeter wave quasi-optical oscillator with multi-elements", *IEEE MTT-S International Symp. Digest* (Dallas), vol. 3, pp. 1209–1212, 1990.
14. H. Kondo *et al.*, "Millimeter and submillimeter wave quasi-optical oscillator with multi-elements", *IEEE Trans. Microwave Theory Tech.*, vol. MTT-40, pp. 857–863, May 1992.
15. A. Pance and M.J. Wengler, 'Microwave modelling of 2-D active grid antenna arrays", *IEEE Trans. Microwave Theory Tech.*, vol. MTT-41, pp. 20–28, Jan 1993.
16. R.M. Weikle, M. Kim, J.B. Hacker, M.P. Delisio, and D.B. Rutledge, "Planar MESFET grid oscillators using gate feedback", *IEEE Trans. Microwave Theory Tech.*, vol. MTT-40, pp. 1997–2003, Nov 1992.

17. R.M. Weikle, "Quasi-optical planar grids for microwave and millimeter-wave power combining", Ph.D Thesis, California Institute of Technology, 1992.
18. D.B. Rutledge, personal communication.
19. M. Kim *et al.*, "A 35 GHz HBT monolithic grid oscillator", *17th Int. Conf. on Infrared and Millimeter Waves*, Proc. SPIE 1929, pp. 402–403, 1992.
20. T. Mader, S. Bundy, and Z. B. Popovic, "Quasi-optical VCO's", *IEEE Trans. Microwave Theory Tech., Special Issue on Quasi-Optical Techniques*, vol. MTT-41, pp. 1775–1781, October 1993.
21. J.B. Hacker *et al.*, "A 10-Watt X-band grid oscillator" (preprint), submitted to *IEEE MTT-S International Microwave Symp. Digest* (San Diego), 1994.

Coupled Oscillator Arrays

22. T. Itoh, presented at *1993 Workshop on Millimeter-Wave Power Generation and Beam Control*, (Huntsville, Alabama), sponsored by U.S. Army Research Office, Sept 1993.
23. K.D. Stephan,"Inter-injection-locked oscillators for power combining and phased arrays", *IEEE Trans. Microwave Theory Tech.*, vol. MTT-34, pp. 1017–1025, Oct 1986.
24. R.J. Dinger, D.J. White, and D.R. Bowling, "10 GHz space power-combiner with parasitic injection-locking", *Electron. Lett.*, vol. 23, pp. 397–398, 9 April 1987.
25. R. Adler, "A study of locking phenomena in oscillators", *Proc. IRE*, vol. 34, pp. 351–357, June 1946; also reprinted in *Proc. IEEE*, vol. 61, pp. 1380-1385, Oct 1973.
26. K. Kurokawa, "Injection-locking of solid-state microwave oscillators", *Proc. IEEE*, vol. 61, pp. 1386-1409, Oct 1973.
27. R.A. York and R.C. Compton, "Quasi-optical power-combining using mutually synchronized oscillator arrays", *IEEE Trans. Microwave Theory Tech.*, vol. MTT-39, pp. 1000–1009, June 1991.
28. R.A. York, "Nonlinear analysis of phase relationships in quasi-optical oscillator arrays", *IEEE Trans. Microwave Theory Tech., Special Issue on Quasi-Optical Techniques*, vol. MTT-41, pp. 1799–1809, October 1993.
29. R.A. York, P. Liao, J.J. Lynch, "Oscillator array dynamics with broadband N-port coupling networks", *IEEE Trans. Microwave Theory Tech.*, accepted for publication.
30. B. Van der Pol, "The nonlinear theory of electric oscillations", *Proc. IRE*, vol. 22, pp. 1051–1085, Sept 1934.
31. J. Birkeland and T. Itoh, "A 16 element quasi-optical FET oscillator power combining array with external injection locking", *IEEE Trans. Microwave Theory Tech.*, vol. MTT-40, pp. 475–481, March 1992.
32. A.C. Davidson, F.W. Wise, and R.C. Compton, "A 60 GHz IMPATT oscillator array with pulsed operation", *IEEE Trans. Microwave Theory Tech., Special Issue on Quasi-Optical Techniques*, vol. MTT-41, pp. 1845–1850, Oct 1993.
33. A. Mortazawi and T. Itoh, "A periodic planar Gunn diode power combining oscillator" *IEEE Trans. Microwave Theory Tech.*, vol. MTT-38, pp. 86–89, Jan 1990.

34. A. Balasubramaniyan and A. Mortazawi, "Two-dimensional MESFET-based spatial power combiners", *IEEE Microwave Guided-Wave Lett.*, vol. 1, pp. 215–218, Aug 1991.
35. A. Mortazawi, H.D. Foltz, and T. Itoh, "A periodic second harmonic spatial power combining oscillator", *IEEE Trans. Microwave Theory Tech.*, vol. MTT-40, pp. 851–852, May 1992.
36. J. Heinbockel, A. Mortazawi, and T. Itoh, "A periodic spatial power combining MESFET oscillator", *IEEE MTT-S International Microwave Symp. Digest* (Albuquerque), pp. 545–548, June 1992.
37. S. Nogi, J. Lin, and T. Itoh, "Mode analysis and stabilization of a spatial power-combining array with strongly coupled oscillators", *IEEE Trans. Microwave Theory Tech., Special Issue on Quasi-Optical Techniques*, vol. MTT-41, pp. 1827–1837, October 1993.
38. S. Kawasaki and T. Itoh, "Quasi-optical planar arrays with FET's and slots", *IEEE Trans. Microwave Theory Tech., Special Issue on Quasi-Optical Techniques*, vol. MTT-41, pp. 1838–1844, October 1993.
39. A. Mortazawi and B.C. DeLoach, "A nine-MESFET two-dimensional power-combining array employing an extended resonance technique", *IEEE Microwave Guided-Wave Lett.*, vol. 3, pp. 214–216, July 1993.
40. K. Chang, K.A. Hummer, and J.L. Klein, "Experiments on injection-locking of active antenna elements for active phased arrays and spatial power combiners", *IEEE Trans. Microwave Theory Tech.*, vol. MTT-37, pp. 1078–1084, July 1989.
41. W.P. Shillue and K.D. Stephan,"A technique for the measurement of mutual impedance of monolithic solid-state quasi-optical oscillators", *Microwave and Optical Tech. Lett.*, Dec 1990.
42. R.A. York and R.C. Compton, "Measurement and modelling of radiative coupling in oscillator arrays", *IEEE Trans. Microwave Theory Tech.*, vol. MTT-41, pp. 438–444, March 1993.
43. A.E. Siegman, *Lasers*, University Science Books, California, 1986.
44. R.A. York and R.C. Compton, "Mode-locked oscillator arrays", *IEEE Microwave Guided-Wave Lett.*, vol. 1, pp. 215–218, Aug 1991.
45. R.A. York and R.C. Compton, "Experimental observation and simulation of mode-locking in coupled-oscillator arrays", *J. Appl. Phys.*, vol. 71, no. 6, pp. 2959–2965, 15 March 1992.
46. K.D. Stephan and W.A. Morgan,"Analysis of inter-injection-locked oscillators for integrated phased arrays", *IEEE Trans. Antennas Propagat.*, vol. AP-35, pp. 771–781, July 1987.
47. W.A. Morgan and K.D. Stephan,"An X-band experimental model of a millimeter-wave interinjection-locked phased array system", *IEEE Trans. Antennas Propagat.*, vol. AP-36, pp. 1641–1645, Nov 1988.
48. P. Liao and R. A. York, "A new phase-shifterless beam-scanning technique using arrays of coupled oscillators", *IEEE Trans. Microwave Theory Tech.*, vol. MTT-41, pp. 1810–1815, October 1993.
49. P. Liao and R. A. York, "A 1 Watt X-band power combining array using coupled VCOs", to be presented at *IEEE MTT-S International Microwave Symp.* (San Diego), June 1994.

Active Integrated Antenna Circuits

50. N. Wang and S.E. Schwarz, "Monolithically integrated Gunn oscillator at 35 GHz", *Electron. Lett.*, vol. 20, pp. 603–604, 5 July 1984.
51. H.J. Thomas, D.L. Fudge, and G. Morris, "Gunn source integrated with microstrip patch", *Microwaves & RF*, pp. 87–89, Feb. 1985.
52. N. Camilleri and B. Bayraktaroglu, "Monolithic millimeter-wave IMPATT oscillator and active antenna", *IEEE Trans. Microwave Theory Tech.*, vol. MTT-36, pp. 1670–1676, Dec 1988.
53. T.O. Perkins, "Active microstrip circular patch antenna", *Microwave Journal*, pp. 110–117, 1987.
54. R.A. York and R.C. Compton, "Dual device active patch antenna with improved radiation characteristics", *Electronics Lett.*, vol. 28, pp. 1019–1021, May 1992.
55. K. Chang, K.A. Hummer, and G.K. Gopalakrishnan, "Active radiating element using FET source integrated with microstrip patch antenna", *IEEE Trans. Microwave Theory Tech.*, vol. MTT-31, pp. 91–92, Sept 1988.
56. R.A. York, R.M. Martinez and R.C. Compton, "Hybrid transistor and patch antenna element for array applications", *Electron. Lett.*, vol. 26, pp. 494–495, March 1990.
57. J. Birkeland and T. Itoh, "Two-port FET oscillators with applications to arrays", *IEEE Microwave Guided-Wave Lett.*, vol. 1, pp. 112–113, May 1991.
58. M. Vaughan and R.C. Compton, "Resonant-tee cpw oscillator and the application of the design to a monolithic array of MESFETs", *Electronics Lett.*, vol. 29, no. 16, pp. 1477–1479, Aug 1993.
59. R.D. Martinez and R.C. Compton, "High efficiency FET/microstrip patch oscillators", (preprint), submitted to *IEEE Antennas and Propagation Magazine*, Dec 1993.
60. J.A. Navarro, Y-H. Shu, and K. Chang, "Broadband electronically tunable planar active radiating elements and spatial power combiners using notch antennas", *IEEE Trans. Microwave Theory Tech.*, vol. MTT-40, pp. 323–328, Feb 1992.
61. H. P. Moyer and R. A. York, "Active cavity-backed slot antenna using MESFETs", *IEEE Microwave Guided Wave Lett.*, vol. 3, pp. 95–97, April 1993.
62. B.K. Kormanyos, S.E. Rosenbaum, L.P. Katehi, and G.M. Rebeiz, "Monolithic 155 GHz and 215 GHz quasi-optical slot oscillators", (preprint), submitted to *IEEE MTT-S Int. Microwave Symp.* (San Diego), June 1994.

Amplifier Arrays

63. J. Chang *et al.*, "Ka-band power-combining MMIC array", *15th International Conference on Infrared and Millimeter Waves*, (Orlando, FL), pp. 532–534, Dec 1990.
64. H.-S. Tsai and R. A. York, "Polarization-rotating quasi-optical reflection amplifier cell", *Electronics Lett.*, vol. 29, pp. 2125–2127, Nov 1993.
65. H.-S. Tsai, M.J.W. Rodwell, and R.A. York, "Planar amplifier array with improved bandwidth using folded-slots", to appear in *IEEE Microwave and Guided Wave Lett.*, 1994.

66. M. Kim *et al.*, "A grid amplifier", *IEEE Microwave Guided Wave Lett.*, vol. 1, pp. 322–324, Nov 1991.
67. C-Y. Yu and G.M. Rebeiz, "A quasi-optical amplifier", *IEEE Microwave Guided Wave Lett.*, vol. 3, pp. 164–166, June 1993.
68. N.J. Kolias and R.C. Compton, "A microstrip-based unit cell for quasi-optical amplifier arrays", *IEEE Microwave Guided Wave Lett.*, vol. 3, pp. 330–332, Sept 1993.
69. T. Mader, J. Shoenberg, L. Harmon, and Z.B. Popovic, "Planar MESFET transmission wave amplifier", *Electronics Lett.*, vol. 29, pp. 1600–1603, Sept 1993.
70. J.A. Benet, A.R. Perkons, and S.H. Wong, "Spatial power combining for millimeter-wave solid state amplifiers", *IEEE MTT-S International Microwave Symp. Digest* (Atlanta), pp. 619–622, June 1993.
71. M. Kim *et al.*, "A 100-element HBT grid amplifier", *IEEE Trans. Microwave Theory Tech., Special Issue on Quasi-Optical Techniques,* vol. MTT-41, pp. 1762–1771, October 1993.
72. M. Kim *et al.*, "A 6.5GHz—11.5GHz source using a grid amplifier with a twist reflector", *IEEE Trans. Microwave Theory Tech., Special Issue on Quasi-Optical Techniques,* vol. MTT-41, pp. 1772–1774, October 1993.
73. D. R. Gagnon, "Highly sensitive measurements with a lens-focused reflectometer", *IEEE Trans. Microwave Theory Tech.*, vol. MTT-39, pp. 2237–2240, Dec 1991.
74. S.J. Peters, "A compact faraday rotator for beam waveguides", *15th Int. Conf. on Infrared and Millimeter Waves,* (Orlando, FL), pp. 699–701, Dec 1990.

Frequency Conversion and Control Arrays

75. J.B. Hacker *et al.*, "A 100-element planar schottky diode grid mixer", *IEEE Trans. Microwave Theory Tech.*, vol. MTT-40, pp. 557–562, March 1992.
76. K.D. Stephan, N. Camilleri, and T. Itoh, "A quasi-optical polarization-duplexed balanced mixer for millimeter-wave applications", *IEEE Trans. Microwave Theory Tech.*, vol. MTT-31, pp. 164–170, Feb 1983.
77. K.D. Stephan and T. Itoh, "A planar quasi-optical sub-harmonically pumped mixer characterized by isotropic conversion loss", *IEEE Trans. Microwave Theory Tech.*, vol. MTT-32, pp. 97–102, Jan 1984.
78. V.D. Hwang and T. Itoh, "Quasi-optical HEMT and MESFET self-oscillating mixers", *IEEE Trans. Microwave Theory Tech.*, vol. MTT-36, pp. 1701–1705, Dec 1988.
79. W.W. Lam *et al.*, "Millimeter-wave diode-grid phase shifters", *IEEE Trans. Microwave Theory Tech.*, vol. MTT-36, pp. 902–907, May 1988.
80. W.W. Lam *et al.*, "Diode grids for electronic beam steering and frequency multiplication", *Int. J. Infrared and Millimeter Waves,* vol. 7, pp. 27–41, Jan 1986.
81. C.F. Jou *et al.*, "Millimeter-wave diode-grid frequency doubler", *IEEE Trans. Microwave Theory Tech.*, vol. MTT-36, pp. 1507–1514, Nov 1988.
82. N. Camilleri and T. Itoh, "A quasi-optical multiplying slot array", *IEEE Trans. Microwave Theory Tech.*, vol. MTT-33, pp. 189–1195, Nov 1985.

83. S. Nam, T. Uvano, and T. Itoh, "Microstrip-fed planar frequency-multiplying space combiner", *IEEE Trans. Microwave Theory Tech.*, vol. MTT-35, pp. 1271–1276, Dec 1987.
84. R.J. Ram and R.A. York, "Parametric oscillation in nonlinear dipole arrays", to appear in *IEEE Trans. Antennas & Prop.*, March 1994.
85. R.A. York and R.J. Ram, "Quasi-optical frequency conversion arrays", *1993 Fourteenth biennial IEEE/Cornell University conference on Advanced Concepts in High-Speed Semiconductor Devices and Circuits*, (Ithaca, NY), August 1993.
86. L.B. Sjogren *et al.*, "A monolithic diode array millimeter-wave beam transmittance controller", *IEEE Trans. Microwave Theory Tech., Special Issue on Quasi-Optical Techniques*, vol. MTT-41, pp. 1782–1790, October 1993.
87. K.D. Stephan, P.H. Spooner, and P.F. Goldsmith, "Quasi-optical millimeter-wave hybrid and monolithic PIN diode switches", *IEEE Trans. Microwave Theory Tech., Special Issue on Quasi-Optical Techniques*, vol. MTT-41, pp. 1782–1790, October 1993.

Planar Antennas for Arrays

88. D.B. Rutledge, D.P. Neikirk, and D.P. Kasilingam, "Integrated Circuit Antennas", in *Infrared and Millimeter Waves*, vol. 10, chap. 1, pp. 1–90, K.J. Button, Ed., New York: Academic Press, 1983.
89. G.M. Rebeiz, "Millimeter-Wave and TeraHertz Integrated Circuit Antennas", *Proc. IEEE*, vol. 80, pp. 1748–1770, Nov 1992.
90. D.M. Pozar, "Microstrip Antennas", *Proc. IEEE, Special Issue on Antennas*, vol. 80, pp. 79–91, Jan 1992.
91. J.R. James and P.S. Hall, eds., *Handbook of Microstrip Antennas* (two volumes), Peter Peregrinus Ltd., London, 1989.
92. K.S. Yngvesson *et al.*, "The tapered slot antenna — a new integrated element for millimeter-wave applications", *IEEE Trans. Microwave Theory Tech.*, vol. MTT-37, pp. 365–374, Feb 1989.
93. D.M. Pozar,"Considerations for millimeter-wave printed antennas", *IEEE Trans. Antennas Propagat.*, vol. AP-31, pp. 740–747, Sept 1983.
94. N.G. Alexopoulos, P.B. Katehi, and D.B. Rutledge, "Substrate optimization for integrated circuit antennas", *IEEE Trans. Microwave Theory Tech.*, vol. MTT-31, pp. 550–557, July 1983.
95. D.H. Schaubert, D.M. Pozar, A. Adrian, "Effect of microstrip antenna substrate thickness and permittivity: comparison of theories with experiment", *IEEE Trans. Antennas Propagat.*, vol. AP-37, pp. 677–682, June 1989.
96. D. H. Schaubert and K. S. Yngvesson, "Experimental study of a microstrip array on high permittivity substrate", *IEEE Trans. Antennas Propagat.*, vol. AP-34, pp. 92-97, Jan 1986.

A Review of Millimeter Wave Tubes

Man Wong

Litton Electron Devices
San Carlos, California 94070-4194
USA

1. INTRODUCTION

Ever since microwave tubes were first invented in the 1930's [1], they have been integral parts of all systems requiring high power in the microwave range (from approximately 1 GHz to 300 GHz). Applications of microwave tubes include radars, electronic countermeasures, communications, materials processing, heating, particle accelerators, and energy transfer. With a few exceptions, the development of millimeter wave (MM-wave) tubes operating from 30 to 300 GHz has in general been sequential to the development of their lower frequency counterparts. Therefore, it is quite common for MM-wave tubes to be scaled versions of the lower frequency tubes. Although a variety of MM-wave tubes have been invented, and developed throughout the electron vacuum tube history, only a few basic tube types have received continued interest in the industry due to technical and economical factors such as fabrication complexity, reliability, cost and market demand.

The common commercially available MM-wave tubes which are currently being manufactured, with output power exceeding tens of watts, include[2-15],

a. Klystrons
b. Traveling Wave Tubes (TWTs)
c. Magnetrons
d. Gyrotrons

Other MM-wave tubes such as free electron lasers, and Cerenkov devices are being developed especially for high power applications. These devices are still in the research and development domain, and will not be covered in this article.

The emphasis of this review article is on commonly available MM-wave tubes which not only can provide relatively high peak power but also high average power. The commercial tubes cited herein are representative ones only and are not intended to be exhaustive. Since the design concepts and operating principles for these tubes in the microwave region are well-documented[1,16-25], no attempt will be made to include the detailed theories. However, common terms or parameters which are necessary to illustrate the design considerations unique to MM-wave frequencies will be introduced.

2. ELECTRON-ELECTROMAGNETIC WAVE INTERACTION

Irrespective of the tube types, MM-wave tubes generate or amplify signal by extracting energy from an electron beam. The energy extraction is achieved by passing the electron beam near an interaction structure or circuit along which MM-wave propagates. The appropriate electric field component of the wave interacts with and reduces the rotational or translational velocities of the electrons converting the electron energy into electromagnetic wave energy. The interaction can be continuous or snapshot in nature. In most cases, some form of synchronism between the wave and the beam is required to build up the interaction process. This section describes, in simple terms, the basic physics involved in the energy conversion process.

2.1 Lorentz Force

An electron interacts with the electric field, **E**, and magnetic field, **B**, of an electromagnetic wave (EM) through the Lorentz force

$$\mathbf{F} = -e(\mathbf{E} + \mathbf{v} \times \mathbf{B}) \tag{1}$$

where e is the absolute electron charge, and **v** is the velocity of the electron. The electric field will either accelerate or decelerate the electron in the direction of the field and changes the energy of the electron. The magnetic field can only change the direction of motion of the electron but not its energy. To convert the electron energy into EM wave energy, the following time integral,

$$e \sum_i \int \mathbf{v}_i \bullet \mathbf{E} \, dt \tag{2}$$

where $\mathbf{v}_i$ is the velocity of the ith electron, should be a positive quantity over the period of interaction in one or many RF cycles.

2.2 Requirements for Electron Energy Extraction

From equation (2), to be able to extract energies from the electrons, the following general requirements can be stated:

1. There should be a component of electric field along a velocity component of the electrons.
2. A number of electrons should participate in the process.
3. The phase between the modulation on the velocities of the electrons and the electric field should remain favorable over the period of interaction.

2.3 Electron bunching in Linear-beam Tubes

In linear-beam tubes such as klystrons and TWTs, the axial component of the circuit electric field interacts with the collinear electron beam as illustrated in Figure 1. Starting with an electron beam with uniform velocity, as the electrons interact with the axial electric field, some electrons will be accelerated and some will be decelerated. Electron bunches will be formed. When these bunches see a retarding field at the output cavity or near the output portion of the circuit, a substantial amount of the electron energy will be converted into EM wave energy.

2.4 Electron Bunching in Gyrotrons

In gyrotrons, the transverse component of the cavity electric field interacts with rotational motions of the electrons in the same plane. The angular velocity of the electrons is given by the cyclotron frequency,

$$\omega_c = eB_{dc}/\gamma\, m_o \qquad (3)$$

where γ is the relativistic mass factor, m_o is the electron rest mass and B_{dc} is the applied dc magnetic field. Starting with an uniform distribution of electrons in a given cyclotron orbit, electrons accelerated by the transverse electric field will have a larger relativistic mass and a lower cyclotron frequency, and electrons decelerated by the field will have a lower relativistic mass and a higher cyclotron frequency. If the operating frequency, ω, in radians per second, is synchronized with the cyclotron frequency, then electrons will continue to see approximately the same field at successively half EM wave cycles. Electron bunching will occur in the cyclotron orbit as illustrated in Figure 2.

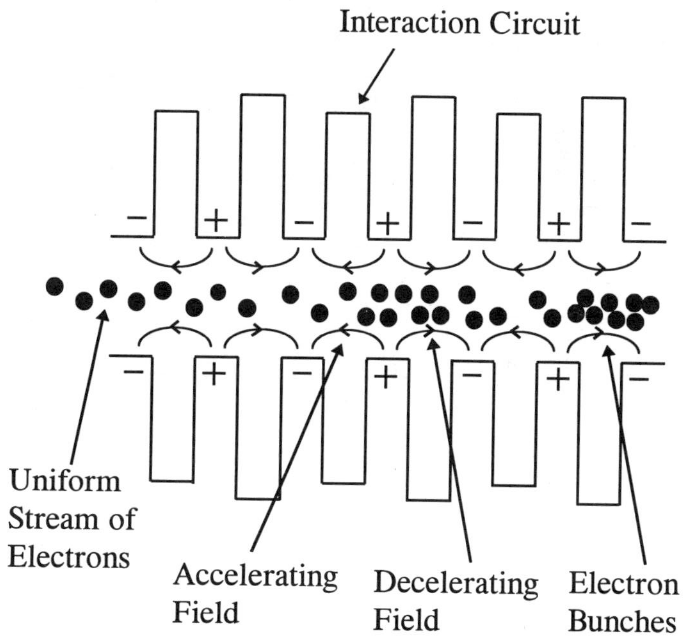

Figure 1. Electron bunching Process in a Coupled Cavity Traveling Wave Tube.

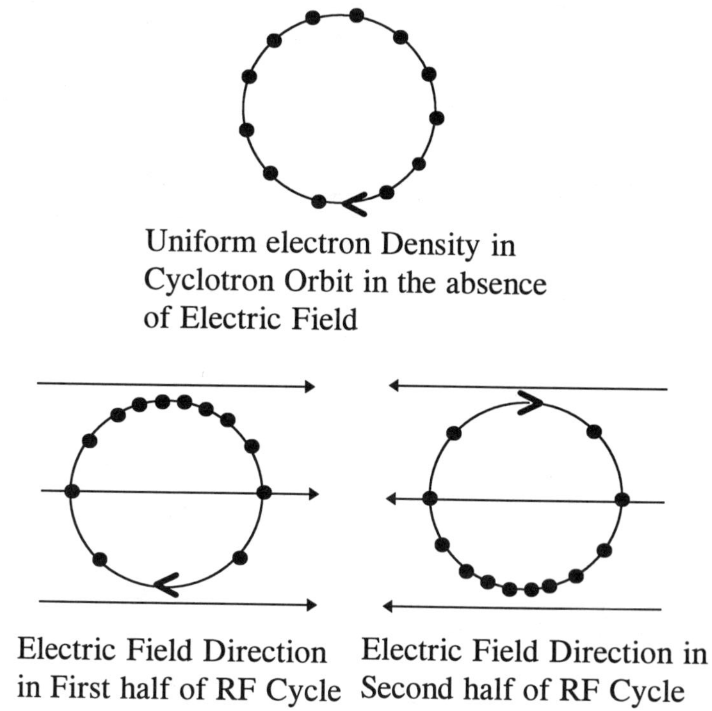

Figure 2. Electron bunching Process in a Gyrotron.

2.5 Phasing between Electron Bunches and Electric Field

The electric field of a circuit wave can be represented in the general form

$$\mathbf{E} = \mathbf{E}\, e^{j(\omega t - \kappa_{11} z)} \qquad (4)$$

where κ_{11} is the propagation constant in the z-axis. The phase velocity of the wave is given by

$$v_p = \omega / \kappa_{11} \qquad (5)$$

For slow-wave devices such as a TWT, the phase velocity is less than the velocity of light, that is,

$$v_p < c \qquad \text{or} \qquad \kappa_{11} > \kappa_c = \omega / c \qquad (6)$$

The transverse propagation constant is given by

$$\kappa_{\perp} = j\,(\kappa_{11}^{2} - \kappa_c^{2})^{1/2} \qquad (7)$$

For a linear-beam velocity modulated by the circuit wave, the signal is carried by the beam at the dc beam velocity, u. In the non-relativistic regime,

$$u = 0.0626\, c\; V_o^{1/2} \qquad (8)$$

where V_o is the beam voltage in kV. The rate of change (reduction) of energy of the electrons is proportional to the time integral of the velocity and electric field dot product, equation (2). The energy exchange (or time integral) is favorable when

$$\omega = \kappa_{11}\, u \qquad (9)$$

(For simplicity in illustrating the phasing concept, space charge effects, which are essential in describing the detail interaction process, have been omitted.)

For a coupled cavity traveling wave tube (CCTWT), the phase velocity is related to the period, L, of the circuit. For operation in the forward wave of the first space harmonic as shown in Figure 3, synchronization between the beam and the circuit wave occurs between 1.0 π to 2.0 π phase change per period. For stability and bandwidth reasons, typically

$$\kappa_{11} L \sim 1.3\pi \tag{10}$$

From equations (8), (9) and (10), the circuit period L is proportional to the square root of the voltage,

$$L \sim (1.3) \times 0.0313\, \lambda\, V_o^{1/2} = 0.0407\, \lambda\, V_o^{1/2} \tag{11}$$

where λ is the free space wavelength. For example, at 95 GHz, and 20 kV, L ~ 0.023" or 0.58 mm. The circuit period is proportional to the wavelength and to the square root of the operating voltage.

For fast-wave devices such as a gyrotron,

$$v_p > c \qquad \text{or} \qquad \kappa_{11} < \kappa_c = \omega / c \tag{12}$$

The transverse propagation constant is given by

$$\kappa_\perp = (\kappa_c^2 - \kappa_{11}^2)^{1/2} \tag{13}$$

The electron bunches rotate at the cyclotron frequency. The time integral, equation (2), is enhanced when the following synchronism condition is satisfied,

$$\omega = N\omega_c + \kappa_{11}\, v_{11} \tag{14}$$

where N is the harmonic number, and v_{11} is the axial component of the beam velocity. A typical dispersion diagram of a gyrotron operating in the fundamental mode is shown in Figure 4. The transverse velocity component, $v_\perp$, is usually larger than v_{11} . The velocity ratio,

$$\alpha = (v_\perp / v_{11}) \tag{15}$$

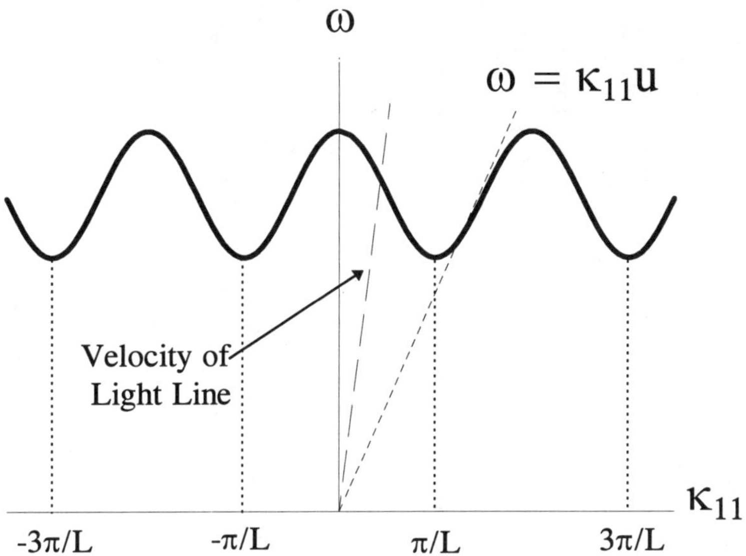

Figure 3. Dispersion Diagram of a Coupled Cavity Traveling Tube Operating in the first Space Harmonic.

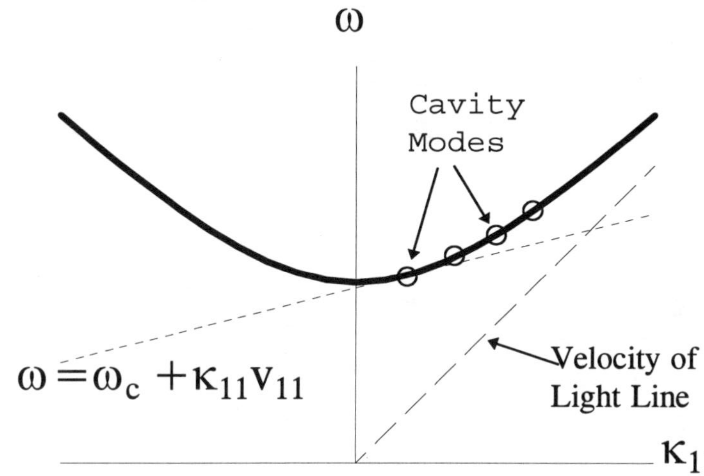

Figure 4. Dispersion Diagram of a Gyrotron Oscillator Operating in the Fundamental Mode.

is typically around 1.5 to 2. For an 80 kV beam and for $\alpha = 2$, the axial energy is 20% of the total energy. The axial velocity is given by equation (8) using a voltage of 16 kV.

2.6 Electric Field Distribution

For cylindrical linear-beam circuits, the axial electric field is described by the modified Bessel function $I_0(\kappa_\perp r)$. From equations (6), (7) and (9),

$$|\kappa_\perp| = \kappa_{11} \, (1 - u^2/c^2)^{1/2} \tag{16}$$

For beam velocity up to about 43% of the speed of light, $|\kappa_\perp|$ is less than κ_{11} by no more than 10%. Therefore,

$$|\kappa_\perp| \cong \kappa_{11} \tag{17}$$

Since $\kappa_{11} = \omega/ v_p = \omega/ u = (\omega/c)/(u/c) = \kappa_c/(u/c)$, κ_{11} as well as $|\kappa_\perp|$ is larger than the free space wavenumber by the ratio of the velocity of light to the phase velocity of the circuit or the beam velocity. At 28% of the velocity of light (corresponding to $V_o = 20$ kV), $\kappa_\perp$ is approximately equal to 22.4 cm^{-1} at 30 GHz and 224 cm^{-1} at 300 GHz. The large values of $|\kappa_\perp|$ implies that the field will drop-off rapidly from the circuit with distance. The electric field distribution of a CCTWT is illustrated in Figure 5. To maintain the on-axis field to about 60% of the field at the edge of the beam tunnel, the beam tunnel radius, a_L , is usually designed to be

$$a_L \sim 1.5/ |\kappa_\perp| \tag{18}$$

To minimize circuit interception, the beam radius, b, is typically about 50% of the beam tunnel radius for MM-wave linear-beam tubes. Therefore,

$$\begin{aligned} b \sim 0.75/|\kappa_\perp| &\cong 0.75/ \kappa_{11} \\ &= 0.75(u/c) / (\omega/c) \\ &= 0.00747 \, \lambda \, V_o^{1/2} \end{aligned} \tag{19}$$

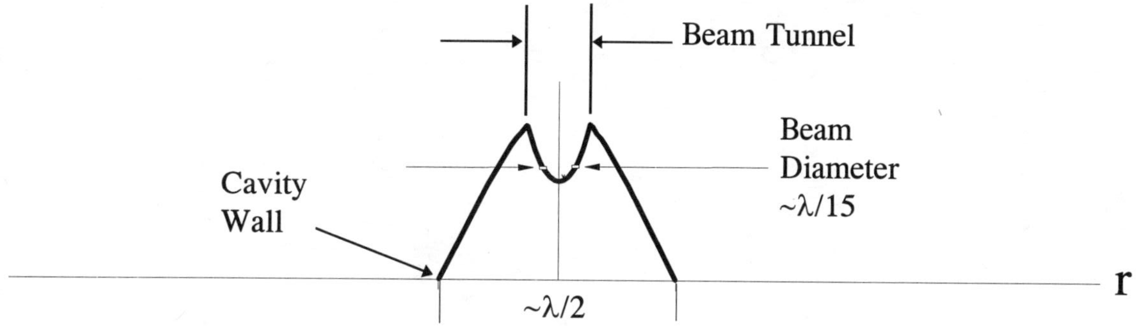

Figure 5. Electric Field Distribution in a Coupled Cavity Traveling Wave Tube.

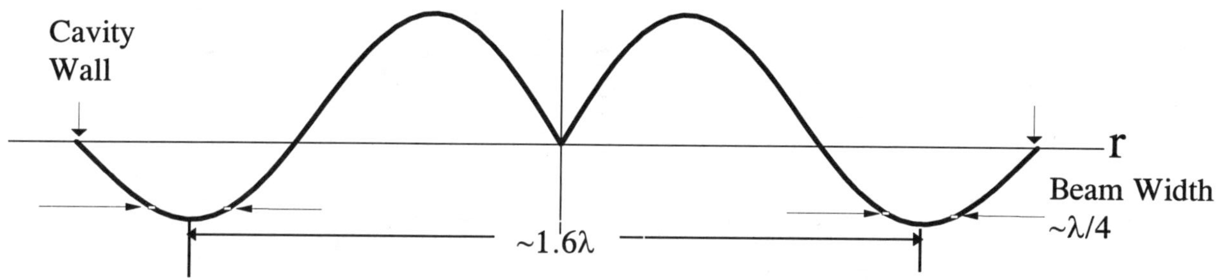

Figure 6. Electric Field Distribution in a Gyrotron Operating in a TE_{02} Mode.

The beam radius is then approximately proportional to the square root of beam voltage and the wavelength. The electron-EM wave interaction cross-sectional area for a linear-beam tube is approximately given by

$$A_L \sim 0.000175\,\lambda^2\,V_o \qquad (20)$$

At 30 GHz, and V_o = 20 kV, $A_L \sim 0.0035\ \text{cm}^2$.

For a gyrotron operating in a TE_{on} mode, the transverse electric field varies as Bessel function $J_1(\kappa_\perp r)$. For a TE_{021} mode, the cavity wall is placed at

$$\kappa_\perp = 7.016/a_G \qquad (21)$$

The radius of the cavity, a_G, and the length of the cavity, d, are adjusted to resonate at a given operating frequency. Choosing the length of the cavity to be (qλ),

$$\kappa_{11} = \pi/\,(q\lambda) \qquad (22)$$

Substituting equations (21) and (22) into equation (13),

$$a_G = 2.23 \times \{4 - (1/q)^2\}^{-1/2}\ \lambda \qquad (23)$$

a_G is independent of the operating voltage. By placing the beam around the field maximum at r ~ 0.8 a_G as shown in Figure 6, the tolerable variation in radius around the maximum is about (λ/4). Therefore, the electron-EM wave interaction cross-sectional area is

$$A_G \sim 2\pi\,(0.8\,a_G)\,(\lambda/4) \qquad (24)$$

for a TE_{02} mode. Choosing a cavity length of 2λ, that is, q=2,

$$a_G = 1.15\ \lambda \qquad (25)$$

The interaction area is

$$A_G \sim 1.45\,\lambda^2 \qquad (26)$$

At 30 GHz, $A_G \sim 1.45\ \text{cm}^2$.

From equations (20) and (26), the interaction cross-sectional areas of a CCTWT and a gyrotron can differ by as much as two orders of magnitude or more. The difference in the interaction cross-sectional areas accounts for differences in output power, in beam current density and in cathode loading between linear-beam tubes and gyrotrons.

3. LINEAR-BEAM TUBES

3.1 Beam Generation, Focusing and Collection

Linear electron beams can be generated in various forms, such as sheet beams, pencil beams, and hollow beams. Solid cylindrical pencil beams are commonly used on linear-beam tubes. With the exception of reflex klystrons, the generation, focusing and collection of the electron beam is illustrated in Figure 7 using a helix TWT. Although the three functions are accomplished normally in separate portions of the tube, their correlation must be considered in optimizing the tube performance.

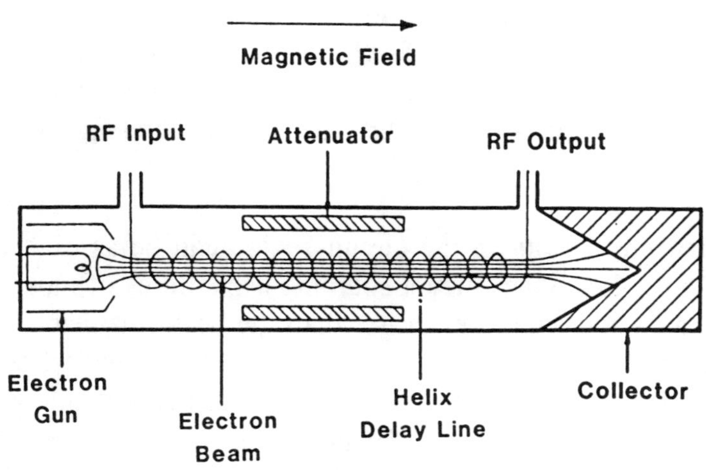

Figure 7. Schematic of traveling wave tube.[18]

Briefly, electrons emitted from the cathode are focused electrostatically into a narrow cylindrical beam. The beam flows essentially in a rectilinear path through a circuit, such as a resonant cavity, a helix circuit, a ladder circuit or a folded-waveguide circuit , which provides the necessary interaction space for the conversion of electron energy into EM wave energy. An axial static magnetic field is usually applied to control the diameter of the beam within the interaction circuit to minimize electron beam interceptions. After losing some of its energy, the beam is gathered by the collector. The residual energy of the beam is converted into heat at the collector.

3.1.1 Electron Gun Perveance and Cathode-to-Beam Area Compression Ratio

To reduce the cathode emission sensitivity to the exact temperature of the cathode as well as the temperature gradient across the cathode, the cathode is usually operated in the space-charge limited domain. Under the space-charge limited emission condition, the beam current, I_o (in Ampere units) and the beam voltage, V_o (in kV units), are related by the equation

$$I_o = 0.0316\, \mu_P\, V_o^{3/2} \tag{27}$$

where μ_P is the perveance in units of 1×10^{-6} A/ $V^{3/2}$. For example, when V_o = 20 kV and I_o = 0.1 A, the micro-perveance μ_P is 0.035. The perveance is a geometric factor and, as long as all dimensions are scaled in a proportional manner, it is independent of physical dimensions.

The beam current density in the circuit region can be as high as a few hundred A/cm^2. Since the current sourcing capabilities of most cathodes are limited, a cathode diameter larger than the beam diameter is used to lower the cathode current density. For dispenser cathodes, which are often used in linear-beam tubes, the maximum cathode loading is about 2 A/cm^2 for CW and 10 A/cm^2 for pulsed operation. The cathode-to-beam area compression ratio, $C_{A,}$ ranges from 10 to 300 at MM-wave frequencies.

3.1.2 Beam Control Electrodes

Although the electron beam can be modulated by adjusting the cathode voltage, the voltage swing required is rather high. To reduce the beam modulation voltage, suitable beam control electrodes can be added to the gun such as modulation anode (mod-anode), control focus electrode or aperture grid, intercepting grid (control grid only) , and non-intercepting grid (control grid and shadow grid).

Mod-anode and control focus electrode are commonly used in MM-wave tubes because of the ease of fabrication. Intercepting grid and non-intercepting grids are more difficult to implement for MM-wave tubes due to the small size. However, gridded tubes have been built at 35 GHz.

3.1.3 Beam Focusing

As the millimeter circuit is small and has rather low thermal capacity, it is important to minimize beam circuit interception. Without focusing, the electron beam spreads due to repulsive space charge forces. The effect of space charge forces is exacerbated by electron bunching in the electron-EM wave interaction process. To maintain the diameter of the beam throughout the circuit, a magnetic field coaxial with the beam is often applied. The magnetic field can be generated by a solenoid (SOL), by permanent magnets (PM) or by periodic permanent magnets (PPM).

3.1.4 Brillouin Flow

Assume that the magnetic field is applied abruptly at the beam minimum with no magnetic flux at the cathode or before the beam minimum, and that the electron beam at the beam minimum has uniform density and uniform axial velocity without radial or azimuthal velocity components. The beam acquires rotational velocity when it crosses the radial magnetic flux lines (the $\mathbf{v} \times \mathbf{B}$ effect) at the beam minimum. As the beam rotates, the Lorentz force acts radially to oppose the centrifugal and space charge forces. It can be shown that there is a minimum magnetic field, called the Brillouin field[1] , which allow the three forces to be exactly balanced. Under this conditions, the electron beam diameter remains unchanged and the smooth rotating beam is called the Brillouin flow.

The Brillouin field, in Gauss, is given by the equation,

$$B_B = 58.1\, I_o^{1/2}/bV_o^{1/4}$$

$$= 10.3\, (\mu_P\, V_o)^{1/2} / b \qquad (28)$$

where the beam radius, b, is in inches.

3.1.5 Practical Field Required for Brillouin Flow

In practice, it is necessary to use a higher focusing field than the Brillouin field due to factors such as thermal velocity effects, grid scattering (in gridded guns), non-uniform current density, the finite magnetic field transition length and the uncertainties in the position of the beam minimum. It is possible to determine the practical rms level of magnetic field, B_P, necessary to obtain Brillouin flow from the following equation,[25]

$$B_P = 1.28\, \{B_B^2 + 9.05 \times 10^{-4}\, C_A\, T_C / b^2\}^{1/2} \qquad (29)$$

where T_C is the temperature of the cathode in Kelvin. This relationship provides the field at the outer envelope of the beam necessary to hold the non-laminar beam in equilibrium at that radius. From equation (29), for non-thermal beams, the field is about 28% higher than the Brillouin field. For a highly thermal beam, the required field can be as high as 2 to 3 times the Brillouin field.

As an example, for a 95 GHz linear-beam tube operating at a beam voltage of 20 kV and a beam current of 0.1 A, the beam radius is approximately 0.004" from equation (19) and the Brillouin field is 2172 Gauss from equation (28). The beam current density is ~ 308 A/cm^2. To maintain the average cathode loading to 2 A/cm^2 maximum, the cathode-to-beam area compression ratio, C_A, should be at least 154. Using a cathode temperature of 1373 ^{o}K, the practical level of magnetic field for Brillouin flow is derived from equation (29) to be ~5248 Gauss which is about 2.4 times the Brillouin field.

3.1.6 Space-Charge Balanced Flow

In Brillouin flow, there is no magnetic flux linkage to the electron beam before the beam minimum. All of the magnetic flux in the electron beam is introduced at or after the beam minimum. Although the field required is near minimum, the beam is rather sensitive to perturbations such as electron bunching due to electron-EM wave interactions in the circuits. If the magnetic flux is adjusted such that a significant portion of the magnetic flux within the beam also links the cathode, a stiffer space-charge balanced flow is obtained. Assuming that the magnetic field at the cathode is B_C and the flux lines follow the electron trajectories, the magnetic field at the beam minimum is then $C_A B_C$. At the beam minimum, the balance of the total magnetic flux is introduced. The total magnetic field B_o required for space charge balanced flow is estimated to be

$$B_o = \{ B_P^2 + (C_A B_C)^2 \}^{1/2} \tag{30}$$

3.1.7 Electron Beam Collection

The collector design is related to the velocity profile of the spent beam - the beam after passing through the circuit. The spent beam will have speed distributed from a fraction of the original beam voltage V_o to a beam voltage somewhat higher than the beam voltage. Some electrons will be traveling parallel to the axis but many will be moving at an angle with respect to the axis as they enter the collector.

In Brillouin flow, the beam's rotational energy acquired when the beam entered the circuit from the gun side will be converted into forward translation energy in a magnetic field free collector. In space-charge balanced flow, the beam will have little rotational energy while the beam is still in the magnetic field but it will acquire reverse rotational energy as the beam crosses the magnetic flux lines into the collector. Therefore, the beam will expand faster in a magnetic field free collector in space-charged balanced flow than that in Brillouin flow.

By collecting the spent beam at potentials depressed closer to the cathode potential (than the ground potential), some of the unused beam energy can be recovered. For MM-wave linear-beam tubes, the electronic efficiency is usually from a few % to about 20%. The efficiency of the tube is enhanced approximately by the ratio of the cathode voltage to the collector voltage (measured with respect to the cathode). Therefore, the efficiency of a MM-wave tube can be quite high even enough the intrinsic electronic efficiency is low compared with the lower frequency microwave tubes.

The fact that the collector can be depressed close to the cathode potential also helps to reduce the beam power density as the beam intercepts the collector. In addition, due to collector depression the spent beam expands faster and the collector can be made shorter.

3.2 Klystrons

In a conventional klystron, which is illustrated in Figure 8, the interaction circuit consists of one or more resonant cavities. The electron beam interacts with the electric field component of the stored EM energy at the gap of the cavity. The electron beam provides the coupling between cavities and carries the EM signal from cavity to cavity. Due to the small size, it is difficult to fabricate a conventional klystron at MM-wave frequencies. However, MM-wave conventional klystrons are commercially available up to about 45 GHz.[8,12]

Some of the fabrication difficulties associated with conventional klystrons can be overcome by using the extended interaction technique. In an extended interaction klystron (EIK), which is illustrated in Figure 9, the axial interaction region of a conventional klystron cavity is extended by the use of a slow-wave structure enclosed in a cavity. The effective interaction impedance is increased roughly proportional to the number of gaps used. However, since a slow-wave circuit is used, the beam velocity must be synchronized with the circuit velocity to take advantage of the increase in impedance. A commercial version of the extended interaction klystron uses a biperiodic ladder circuit.[13, 26-29]

The biperiodic ladder circuit, which is shown in Figures 1 and 9, is independently fabricated and sandwiched between the cavity halves to form a complete cavity. The biperiodic ladder circuit is machined by

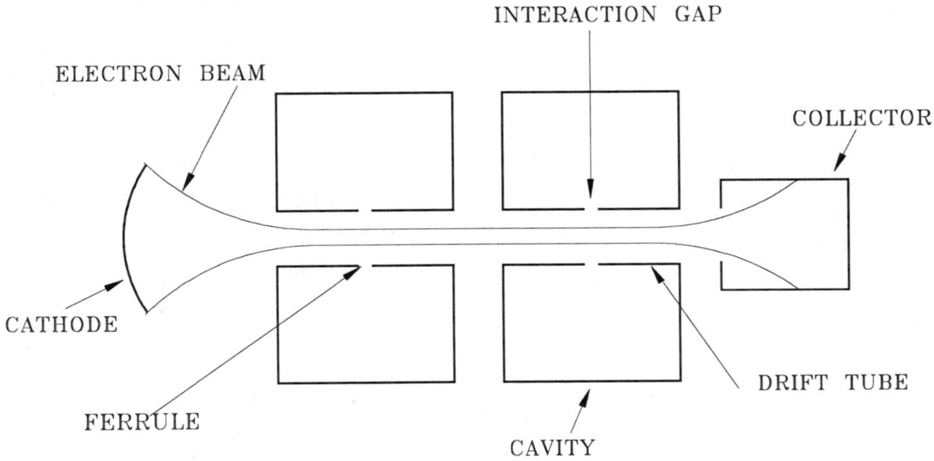

Figure 8. Schematic Diagram of a Conventional Klystron.

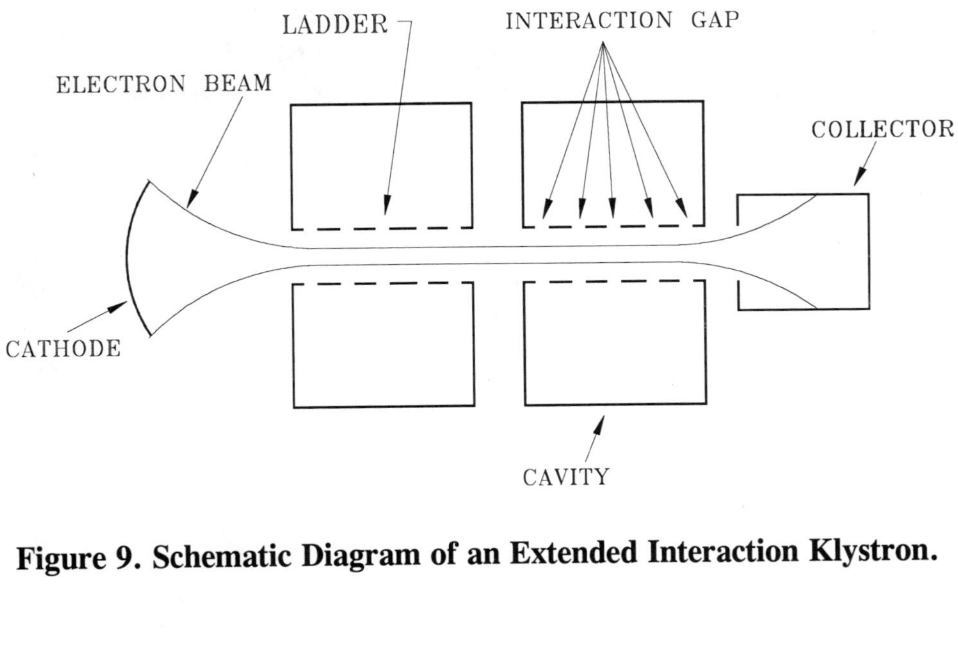

Figure 9. Schematic Diagram of an Extended Interaction Klystron.

electric discharge machining. These EIKs have been fabricated up to 280 GHz. EIKs can be built as oscillators (EIO's) or amplifiers (EIA's). Oscillators are usually designed with only one cavity and are typically tunable over a 5% bandwidth by physically changing the dimensions of the cavity. EIAs are multi-cavity klystrons amplifiers. The cavity frequencies in an EIA are normally set during fabrication in the factory to optimize the performance in power, gain and bandwidth at specific frequencies. EIAs are narrow-banded amplifiers, usually less than 1%. At 95 GHz, a 3-dB bandwidth of about 500 MHz or more has been reported.[27]

Some typical MM-wave conventional klystrons and EIKs are included in Table 1. An 1.0 kW peak 95 GHz EIA is illustrated in Figure 10.

Table 1. Typical MM-wave Klystrons [8,12,13]

CW Power (W)	Frequency (GHz)	Manufacturer
1000	30	Varian VKQ2445 (Osc., PM)*
450	30	NEC LD4406 (Amp., PM)
100	95	Varian VKB2468 (Amp., PM)*
1	220	Varian VKY2441 (Osc., PM)*

Pulsed Power (W)	Duty (%)	Frequency (GHz)	Manufacturer
2,500	10	35	Varian VKA7852 (Amp., PM)
1,000	1	95	Varian VKB2469 (Amp., PM)*
200	.5	140	Varian VKT2419 (Osc., PM)*
60	.5	220	Varian VKY2429 (Osc., PM)*

*Extended Interaction

Figure 10. An 1.0 kW Peak 95 GHz Varian Extended Interaction Klystron Amplifier (EIA).

Figure 11. An 120 W CW 27.5 to 30.0 GHz Varian Helix Traveling Wave Tube.

3.3 Helix Traveling Wave Tubes

In a helix traveling wave tube (HTWT), the circuit is helical as shown in Figure 7. The average diameter of the helix is proportional to the wavelength of operation. The helix must be wound tighter and tighter as frequency increases. At 95 GHz, the diameter could be as low as 0.012". The size of the dielectric rods required to support the helix inside the vacuum envelope also has to be reduced. The reduction in size decreases the thermal conductance and the difficulties associated with fabricating the helix assembly have limited the HTWTs to rather low power levels and to the lower end of the MM-wave band. Despite these limitations, the frequency and power performance of HTWTs have been increasing in recent years. Some of the typical HTWTs currently being manufactured are listed in Table 2. An 120 W CW 27.5 to 30.0 GHz HTWT, an 80 W CW 42.0 to 45.0 GHz HTWT, and an 800 W peak pulsed 30.0 to 35.0 GHz HTWT are illustrated in Figures 11, 12 and 13 respectively.

Table 2. Typical MM-Wave Helix TWTs [2,4,5,9,12]

CW Power (W)	Frequency (GHz)	Manufacturer
120	27.5 to 30.0	Varian VTA 6193
80	26.0 to 34.0	Thomson TH 3875
80	42.0 to 45.0	Hughes 8905HA
13	59.0 to 64.0	AEG TL 62010

Pulsed Power (W)	Duty (%)	Frequency (GHz)	Manufacturer
800	5	30.0 to 35.0	Litton L-5837
1000	20	34.5 to 35.5	Hughes 8909H

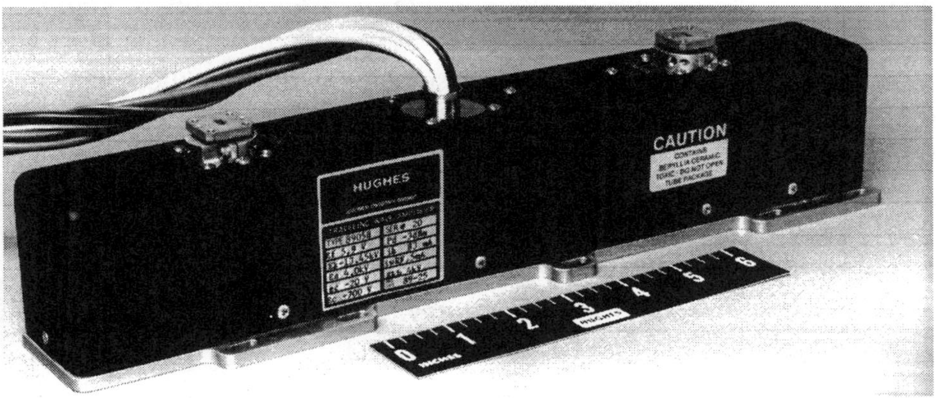

Figure 12. **An 80 W CW 42.0 to 45.0 GHz Hughes Helix Traveling Wave Tube.**

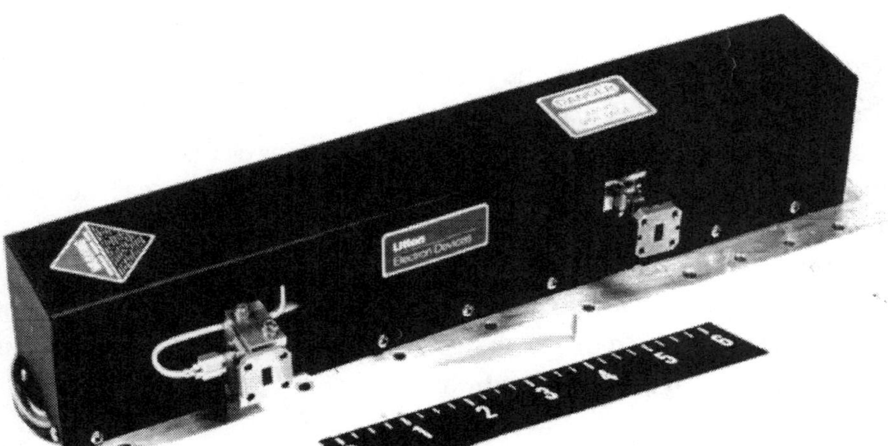

Figure 13. **An 800 W Peak Pulsed 30.0 to 35.0 GHz Litton Helix Traveling Wave Tube.**

3.4 Coupled Cavity Traveling Wave Tubes

A conventional coupled cavity circuit is shown in Figure 14(a). The cylindrical cavities are electrically coupled by placing irises or coupling slots between the common wall of the cavities. If the coupling slots are aligned from cavity to cavity, the resulting circuit is called a in-line circuit. If the coupling slots are rotated between successive slots, the circuit is called a staggered circuit. Depending on the circuit type and the relative frequency of the coupling slots and the cavity, the dispersion of the circuit changes. It is quite common for a CCTWT to operate in the forward wave of the first space harmonic with the coupling slot resonant frequency located above the cavity frequency. The dispersion diagram of such a circuit is shown in Figure 3 .

It is quite common to use re-entrant cavities in a conventional CCTWT especially at lower frequencies. The ferrule forms the re-entrancy and it concentrates the electric field across the interaction gap. The ferrule area is usually minimized to reduce the stored energy in the circuit. The minimization of stored energy helps to improve the interaction impedance of the circuit.

In a CCTWT, the period and the gap per cavity must be set separately. Hundreds of cavity plates and spacers may need to be stacked for a tube with reasonable gain and efficiency. It is increasingly difficult to curtail accumulation of errors on the circuit period and to fabricate cavities with frequencies which are sufficiently close to each other to avoid mismatches in the circuit. Despite these difficulties, conventional CCTWTs have been manufactured up to 95 GHz.[4] The fabrication complexity is slightly reduced by going to a cavity design without ferrules trading off against a reduction in efficiency.

Instead of cavities with cylindrical geometry, cavities with square or rectangular topology are gaining acceptance at MM-wave frequencies. The prominent ones include ladders[30-32] and folded-waveguides.[33,34] These circuits are normally machined by milling and/or electric discharge machining with respect to a fixed reference along the axis such that the errors on the period do not accumulate. In addition, the cavity frequencies are determined by common walls which enclose all the cavities. Deviations of cavity frequencies are also reduced.

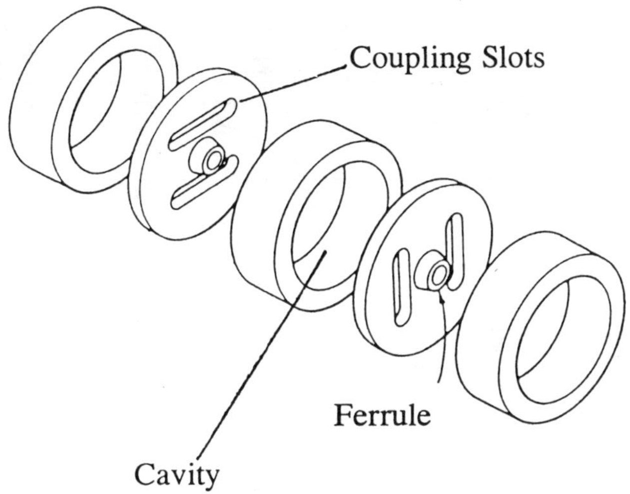

(a). A Conventional Coupled Cavity Circuit

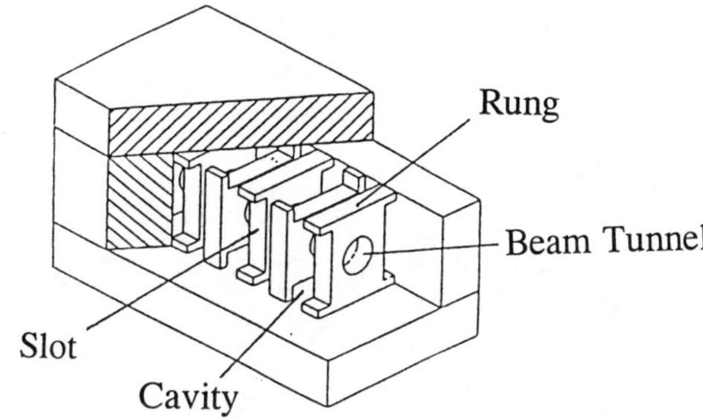

(b). A Staggered Ladder Circuit.

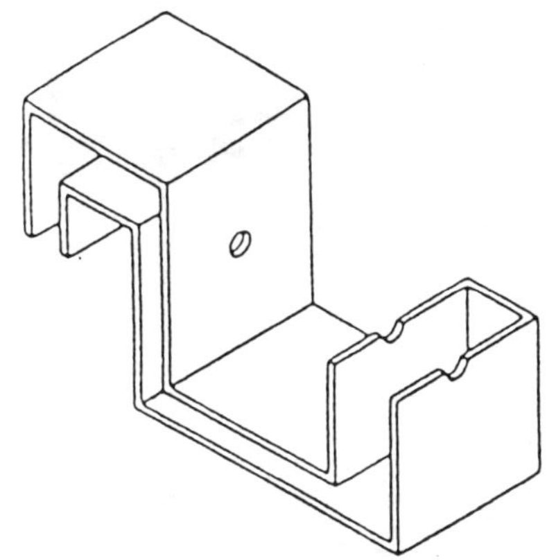

(c). A Folded-Waveguide Circuit.

Figure 14. Illustrations of Some of the Typical Coupled Cavity Circuits.

A double staggered ladder circuit is shown in Figure 14(b)[31]. This circuit has two coupling slots on each cavity rung. The coupling slots are rotated by 90 degrees on successive rungs. The circuit has been used in a 95 GHz TWT with up to 100 W CW output.[31] An in-line ladder circuit [32] is more dispersive than a double staggered ladder circuit but is also more efficient. These ladder circuits have been used in some of the most powerful MM-wave solenoid focused CCTWTs at 35 GHz and at 95 GHz.[12,32,35]

The biperiodic ladder circuit can also be used in CCTWTs. However, the circuit is more dispersive and less thermally rugged compared with staggered ladders and with folded-waveguides.

The folded-waveguide circuit has been revived recently in the commercial world separately by Litton and Northrop.[33,34] A folded-waveguide circuit is shown in Figure 14(c).[33] The folded-waveguide circuit has potential in offering broadband performance and lower manufacturing cost.

Some of the typical MM-wave CCTWTs are tabulated in Table 3. A 250 W CW 42-45 GHz CCTWT is illustrated in Figure 15.

Table 3. Typical MM-wave CCTWTs [4,12]

CW Power (W)	Frequency (GHz)	Manufacturer
400	29.0 to 30.0	Hughes 933H (PPM)
1,500	30.0 to 31.0	Varian VTA6630 (Sol)
1,000	43.5 to 45.5	Hughes 948H (Sol)
200	59.0 to 61.0	Hughes 964H (PPM)
40	93.0 to 97.0	Varian VTW-6495 (PPM)

Pulsed Power (W)	Duty (%)	Frequency (GHz)	Manufacturer
4,000	5	34.0 to 36.0	Hughes 8910H (PPM)
50,000	20	34.5 to 35.5	Varian VTA5700 (Sol)
8,000	10	94.5 to 95.5	Varian VTA5795 (Sol)

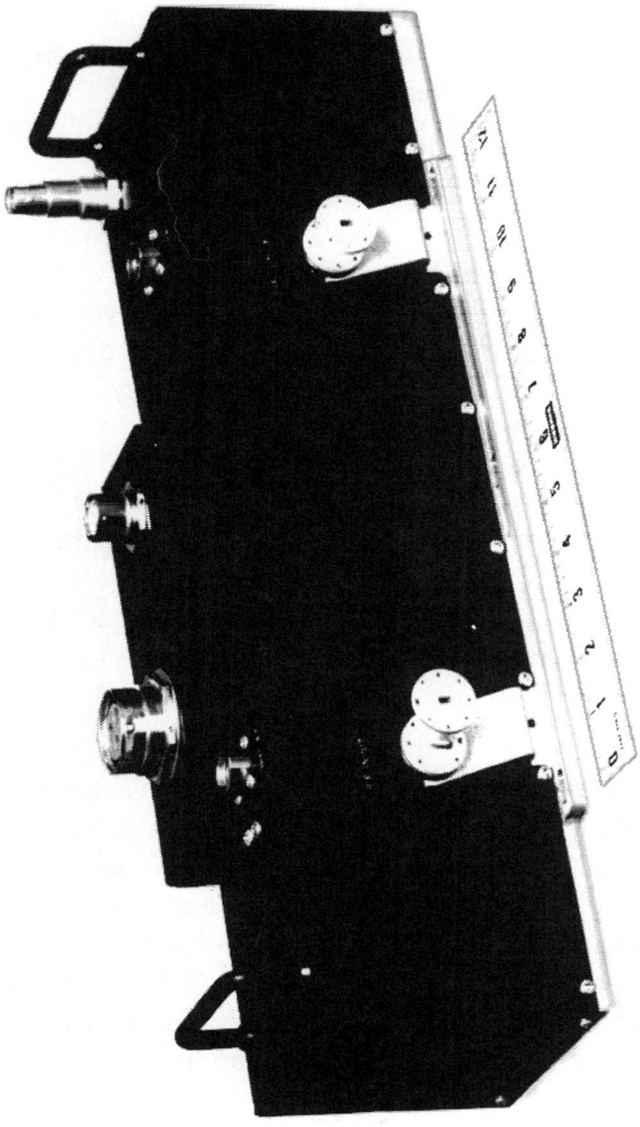

Figure 15. A 250 W CW 43.5 to 45.5 GHz Hughes Coupled Cavity Traveling Wave Tube.

3.5 Linear-beam Tube Design Considerations

MM-wave linear tubes invariably use lower beam currents and higher beam voltages compared with their lower frequency counterparts. There are a number of reasons for these characteristics.

3.5.1 Circuit Dimensions

Circuit dimensions are proportional to the wavelength. Therefore, holding other parameters constant, a circuit at 95 GHz will be ten times smaller than one at 9.5 GHz. As the circuit size gets smaller many fabrication difficulties start to surface. Some of these difficulties include machining, brazing, alignment of parts and assembly, and maintaining tolerance proportional to wavelength. For example, for a cavity circuit, the transverse dimension is about half a wavelength. At 95 GHz, the transverse dimension is about 0.0621". To maintain the circuit to a spread of 95 GHz +/- 100 MHz, the tolerance on the dimension is +/- 0.000065". Tolerance control is especially challenging for narrower banded circuits such as klystrons. Consequently, MM-wave klystrons usually contain tuners to correct for fabrication errors as well as to provide capability to tune to different frequencies.

The gap, period or pitch as well as the tunnel diameter of a circuit increases by a factor proportional to the square root of the beam voltage. See for example, equations (11) and (19). Consequently, circuit fabrication difficulties will be alleviated to some degree if the beam voltage is higher. For example, at a beam voltage of 10 kV and at 95 GHz, a cavity gap spacing is about 0.0080". By designing the same tube at 25 kV, the gap spacing is increased by a factor of 1.58 to 0.0126". The absolute tolerance on the gap is relaxed by the same factor.

3.5.2 Thermal Considerations

When the longitudinal circuit dimensions are larger, the circuit is thermally more robust because the cavity and drift tubes have more surface area. This is especially critical for MM-wave tubes because thermal constraints often limit the output power.

3.5.3 Magnetic Field Requirements

The minimum magnetic field required to focus a beam is inversely proportional to the beam radius, b as given in equation (28). When the magnetic field becomes a constraint in the design, the perveance will have to be reduced to offset the effect of small beam size. From equations (19) and (28), for a given operating voltage, the perveance will be proportional to the square of wavelength. Therefore, MM-wave beam tubes generally use guns with low microperveance ranging from 0.01 to 0.5. For linear-beam tubes, the practical limit for the magnetic field is about 10,000 Gauss for solenoid or permanent magnet focusing systems and it is about 5,000 Gauss rms for periodic permanent magnetic focusing systems, due to saturation effects in pole-pieces. The higher beam voltage will help to circumvent to some degree the magnetic field constraint.

3.5.4 Current Density

For a given beam current, the beam current density is inversely proportional to the square of the beam radius. Therefore, the beam current density increases rather rapidly with frequency. Beam current density of hundreds of A/cm^2 is not uncommon for MM-wave tubes. This high level of current density is usually provided by taking the cathode current density (loading) to its highest level consistent with life requirements and by using high convergence guns with large cathode area to beam area compression ratios. The area compression ratio can be as high as 300 at low perveance. This high ratio causes problems in focusing as is shown earlier that higher magnetic field is required to confine thermal electrons. The future availability of long life high emission cathodes will help to reduce the area compression ratio or reduce the magnetic field required for focusing.

4. 0 MAGNETRONS

Magnetrons are widely used in many diversified applications, especially those which do not require a high degree of phase coherency. However, the applications of MM-wave magnetrons have been limited by their low average power performance, and relatively short lifetimes.

Figure 16 shows a schematic representation of the geometry of a magnetron cavity employing a rising sun circuit - one of the common circuits used at MM-wave frequencies. Another popular one is the coaxial circuit.[6] The rising sun circuit is also the anode and it surrounds a central cathode. Radial electric field is established between the circuit and the cathode by applying a positive voltage to the anode. Electrons emitted from the cathode are drawn towards the circuit. An axial dc magnetic field, which is perpendicular to the electric field lines, causes the electrons to move in cycloidal trajectories and to drift at a speed in the azimuthal direction determined by the ratio of the dc electric field and the dc magnetic field.

The rising sun circuit in the cavity supports traveling waves with phase velocity V_p which is less than the speed of light. The electron drift velocity and wave phase velocity are synchronized to enhance interaction similar to linear-beam traveling wave tubes. Electrons accelerated by the RF electric field will be deflected back towards the cathode. Electrons decelerated by the RF electric field give up some of their energy and continue to move towards the anode. Due the actions of the RF/DC electric field and the dc magnetic field, the latter electrons are trapped by the RF wave and form spoke-like patterns rotating in synchronism with the RF wave. As the electrons flow from the cathode to the anode through the spokes, they gain kinetic energy in the radial direction due to dc electric potential. However, due to the action of the dc magnetic field, the radial kinetic energy is converted to azimuthal kinetic energy in the direction of the flow of the electrons. Since these electrons are being decelerated by the RF wave, they give up energy to the wave. In other words, the potential energy acquired by the electrons as they move towards the anode is continuously being converted into RF energy until they strike the anode.

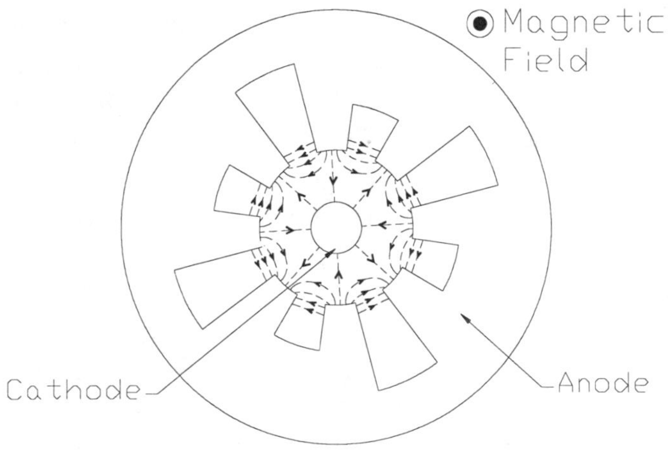

Figure 16. Cross Section of a Magnetron Using a Rising Sun Circuit.

Figure 17. An 150 kW Peak 35 GHz Litton Magnetron.

Magnetrons are quite distinct from other tubes in that beam generation and beam collection are in proximity with beam interaction. Electrons encountering accelerating RF fields as they emit from the cathode are driven back into the surface of the cathode. This causes secondary emission and cathode heating. The amount of heating is related to the total anode current. For some magnetrons, the filament voltage has to be proportionally reduced as the anode current increases to maintain constant cathode temperature and to prolong the life of the cathode.

The anode structure, besides functions as the circuit, also collects the electrons. Therefore, the circuit has to dissipate the residual power of the electrons. As much as 20 or 30% of the beam power may have to be dissipated by the anode. It is increasingly difficult to remove the heat from the anode as wavelength decreases and as the circuit cross-section decreases proportional to the square of the wavelength. This is in sharp contrast to linear-beam tubes or gyrotrons where the collectors can in principle be designed to reduce the power density more or less independent of wavelength.

Due to thermal or other considerations, the development of MM-wave cross-field tubes has not received as much attention as the linear-beam tubes or the fast-wave devices.[36] Some of the typical performance of pulse magnetrons manufactured are given in Table 4. Although these magnetrons have fairly high peak power, the duty cycle for these devices is rather low. An 150 kW peak 35 GHz Magnetron is illustrated in Figure 17.

Table 4. Typical MM-wave Magnetrons [3,6,13]

Pulsed Power (W)	Duty (%)	Frequency (GHz)	Manufacturer
150,000	.05	35	Litton L-4524
10,000	.1	70	Varian BL-221
2,500	.02	95	EEV MG5294

5. GYROTRONS

In linear-beam tubes, the EM wave propagating on the circuit provides a longitudinal electric field component collinear with the axial velocity of the electron beam. Some of the tubes such as TWTs employ slow-wave circuits to synchronize the axial wave velocity with the beam velocity to enhance interaction throughout the circuit. The strength of the longitudinal electric field falls off at a distance from the circuit. Therefore, the beam has to be close to the circuit for efficient interaction. For high power and high frequency, it is increasingly difficult to place the high energy beam close to a circuit with rather low thermal capacity.

In contrast to the classical MM-wave tubes, a gyrotron utilizes the interaction of rotating electrons and the electric field of the transverse component of the EM wave in a cavity. The electrons are injected into a strong axial magnetic field in the shape of a thin hollow cylinder. It is not required, in principle, to place the beam as close to the cavity wall as the electric field does not decrease rapidly from the cavity wall with distance compared with linear-beam tubes as discussed in section 2.6.

For a gyrotron operating in the fundamental mode, the beam is placed in field maximums. For a TE_{02} mode, the electric field pattern has been depicted in Figure 6. The interaction cross-section can be increased by going to a higher order mode, such as a whispering gallery mode, $TE_{m,p}$, where $m >> p$. The higher order mode allows the use of cavities which are much bigger in size. This will reduce the RF power density on the wall of the cavity for a given power. In addition, the use of whispering gallery mode provides better mode selectivity than the lower order TE_{0n} modes. Due to these reasons, high average power gyrotrons generally use whispering gallery modes of a high order such as $TE_{22,2}$.[37] An 110 GHz 500 kW peak gyrotron is illustrated in Figure 18.

It is also possible to operate a gyrotron at an harmonic of the cyclotron frequency, the magnetic field required for a Nth harmonic gyrotron is given by

$$B_{dc} = \gamma\omega_c/(e/m_o)N \qquad (31)$$

Figure 18. An 110 GHz Varian Gyrotron with Beam/RF Separation - 500 kW peak, 2.5 sec; 1 MW peak, 1 ms.

The advantage of operating in the harmonic mode is that the magnetic field is greatly reduced at some expense of interaction efficiency. By keeping the magnetic field to values achievable with non-superconducting magnets or with permanent magnets, the tube will be more attractive for more compact and light weight system applications. Gyrotron amplifiers are also being developed in the form of gyro-klystrons, and gyro-TWTs.[38,39] These gyro-amplifiers are still in the developmental stage.

Some of the typical commercial or industrial gyrotrons are summarized in Table 5. It appears that gyrotrons with industrial quality operating with 1 MW CW output at frequencies at 110 GHz or beyond may be a reality in the next few years. However, a number of major obstacles will have to be overcome. They include developing an output window which will withstand the high average power, reducing the collector heat load by splitting off the RF from the electron beam, and selectively exciting the desired cavity modes in an overmoded cavity.

Table 5. Typical MM-Wave Gyrotrons [10,11,22,37]

Pulsed Power (kW)	Pulse Length (ms)	Frequency (GHz)	Manufacturer
200	≥1000	28	Varian VGA-8000
200	≥1000	53	Varian VGE-8005
500	≥1000	83	Russia
200	100	100	Thomson 1503
500	500	110	Russia
500	≥1000	110	Varian VGT-8011
600	≥1000	140	Russia
900	300	140	Russia
500	700	167	Russia
25	250	250	Varian VGT-8025

6. SUMMARY AND CONCLUSIONS

A list of the current MM-wave tube activities in the tube industry is provided in Table 6. Developmental work are being carried out on both slow-wave and fast-wave devices. A summary of the performances of some of the existing commercial devices is given in Figure 19.

Driven by the requirements of high resolution radars, development work on lower power millimeter tubes in the tube industry have shown resurgence in recent years. It appears that most of the development work are concentrated at the key radar bands such as 35 GHz , 95 GHz and 140 GHz. There are more competition at 35 GHz among klystrons, EIKs, HTWTs and CCTWTs. At 95 GHz or beyond, for average power level at 100 W or more and bandwidth exceeding 2 GHz, CCTWTs using ladder or folded-waveguide circuits will likely be prevalent.

Gyro-amplifiers are being developed for airborne radars with peak powers in the high tens of kilo-watts at 95 GHz. These amplifiers will likely be harmonic gyro-devices focused with permanent magnets to minimize size and weight. The quest for gyrotrons capable of generating average power exceeding one million watts will likely bring about new technologies in the development of suitable window materials, more efficient heat transfer mechanism and the separation of MM-wave power from electron beam power in the collector.

Table 6. Summary of Some of the MM-Wave Tube Development Activities in the Industry (02/94)

- High Peak and Average Power PPM focused TWTs at Ka-band (All)
- 95 GHz Helix TWT (Hughes)[40]
- 95 GHz Double Staggered Ladder CCTWTs (Varian)[31]
- MM-Wave Folded-Waveguide CCTWTs (Litton, Northrop)[33,34]
- 140 GHz Biperiodic Ladder CCTWT (Varian Canada)[41]
- 95 GHz Harmonic Gyro-devices (Litton, Varian, Russia)[22,38,39]
- 110 GHz (or higher) 1 MW CW Gyrotron (Varian, Russia)[22,37]

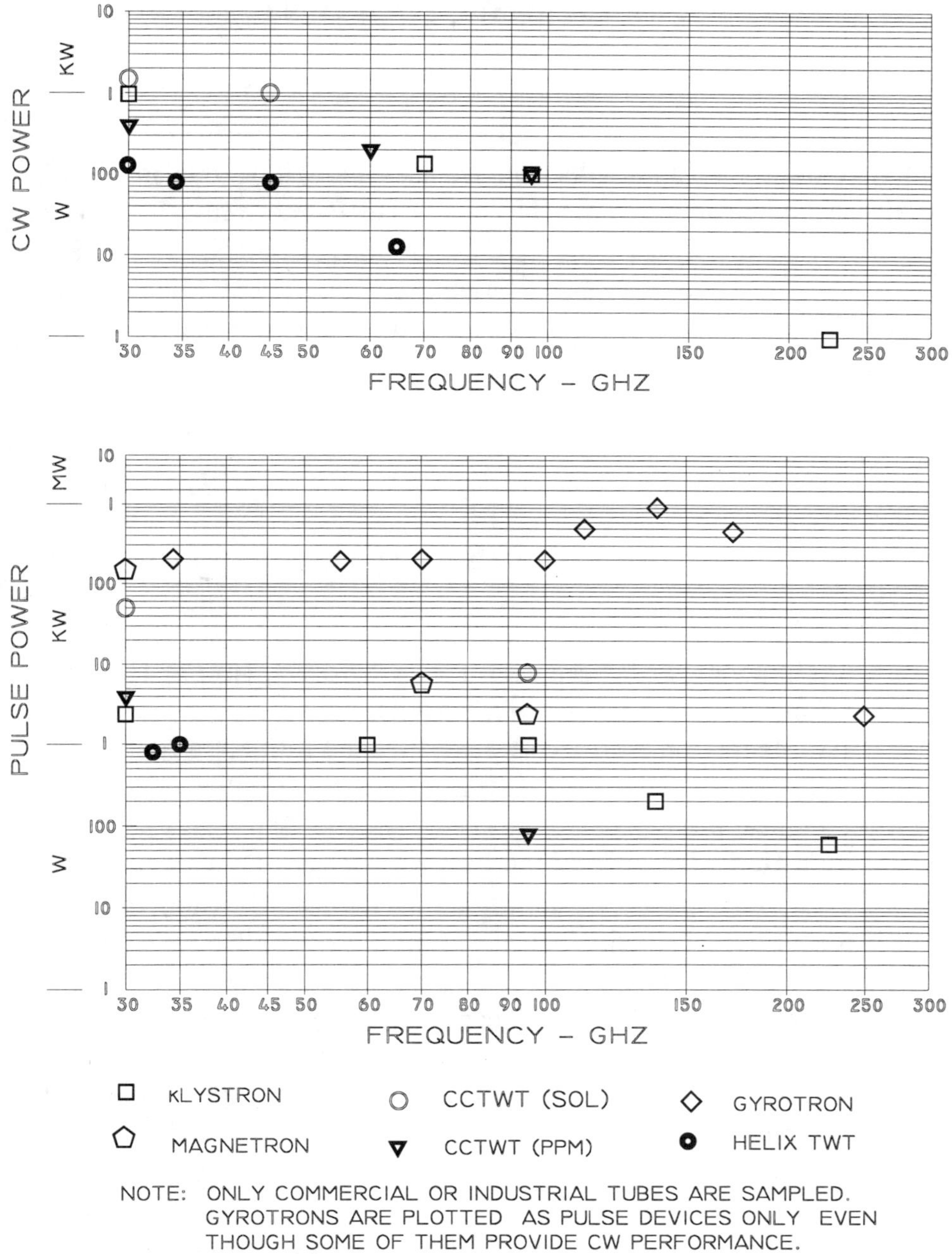

Figure 19. Performance of Some of the Commercial MM-Wave Tubes (02/94).

7. ACKNOWLEDGMENT

The author acknowledges the guidance and encouragement from Dr. Robert Symons of Litton Electron Devices in writing this manuscript. He also acknowledges the assistance of Alan Patmore and Dr. Richard True of Litton Electron Devices, Mike Cascone, Brian Steer, Dr. Kevin Felch, and Dr. Hugo Huey of Varian Associates, and Mike Smith of Hughes Aircraft Company in providing product pictures and/or technical information for the article.

8. REFERENCES

1. A.S. Gilmour, Jr.,"Microwave Tubes," Artech House, 1986.
2. "Microwave Tubes from AEG," AEG.
3. "EEV Magnetron Abridged Data," Second Edition, EEV.
4. "Hughes TWT and TWTA Handbook, " Hughes, and private communication.
5. "Litton Helix Tubes," Litton Electron Devices Division.
6. "Litton Magnetrons," Litton Electron Devices Division, 1984.
7. "Litton Coupled Cavity Tubes," Litton Electron Devices Division.
8. "Klystron Catalog," NEC.
9. "Microwave Tubes for Telecommunications," Thomson Tubes Electroniques, 1991.
10. "Microwave Tubes for Scientific Applications," Thomsom Tubes Electroniques, 1993.
11. "Introduction to Gyro Devices," Varian Microwave Power Tube Products, 1993.
12. "Millimeter Waves - Amplifiers, Sources, Components, and Subsystems," Varian, 1988.
13. "Cross-Field Amplifier, Klystron, and Magnetron Catalog," Varian Beverly.
14. "Varian Extended Interaction Klystron Selection Guide," Varian.
15. "Coupled Cavity Tubes," Varian, 1988.

16. G.B. Collins, "Microwave Magnetrons," New York: McGraw-Hill, 1984.
17. J.F. Gittins, "Power Travelling Wave Tubes," American Elsevier, Inc., 1965.
18. J.T. Mendel, "Helix and Coupled-Cavity Traveling Wave Tubes," Proc. IEEE, vol. 61, No.3, 1973.
19. J.D. Pierce, "Travelling Wave Tubes," Princeton, N.J.:Van Nostrand, 1950.
20. J.D. Pierce, "Theory and Design of Electron Beams," 2nd Edition, New York: Van Nostrand, 1954.
21. A. Staprans, E. W. McCune, J. Ruetz, "High-Power Linear-Beam Tubes," Proceedings of the IEEE, No. 3, March 1973.
22. V.A. Flyagin, A.L. Goldenberg, V.E. Zapevalov, "State of the Art of Gyrotron Investigation in Russia," 18th International Millimeter Waves, James R. Birch, T.J. Parker, Editors, Proc. SPIE , 581-584, 1993.
23. V. Granatstein and I. Alexeff, Editors, "High-Power Microwave Sources," Artech House, Inc. , 1987.
24. R.S. Symons and H. Jory, "Cyclotron Resonance Devices," Advances in Electronics and Electron Physics, Vol. 55, Academic Press, 1981.
25. R. True, "Electron Beam Formation, Focussing and Collection in Microwave Tubes,", Chapter 17 of Handbook of Microwave Technology, Academic Press, in press, 1994.
26. C. Nilsen, M. Viant and M. Wong, "Low Noise MM-Wave Transmitter Tube Oscillators," Microwave Journal, July, 1988.
27 C. Nilsen, B. Steer, M. Viant and M. Wong, "95 GHz 1 kW Peak High Performance Radar Transmitter," Proceedings of Military Microwave Conference, July, 1990.
28 C. Nilsen, M. Viant and M. Wong, " A Versatile High Power 95 GHz Transmitter," 3rd Annual TRIO/CRL Radar Symposium, October, 1990.
29 M. Viant and M. Wong, "Advances in Millimeter Wave Extended Interaction Klystrons and Transmitter Subsystems," 15th International Conference on Infrared and Millimeter Waves, December, 1990.
30 M. Wong, C. Nilsen and D. Yake, "Development of a K_a-Band High Power Amplifier for Olympus Earth Stations," Olympus Utilization Conference, Vienna, Austria, 1989.

31. B. James, and M. Wong, "A 95 GHz Broadband Millitron™ Traveling Wave Tube," Workshop on Millimeter Wave Beam Power Generation and Beam Control, The University of Alabama, Huntsville, September, 1993.
32. B. James, "Coupled-Cavity TWT Designed for Future MM-Wave Systems," MSN & Communications Technology, Dec., 1986.
33. A.J. Theiss, D.B. Lyon and Y. Hiramatsu, "An Integral-Pole-piece Folded-Waveguide Slow-Wave Circuit for High Power Millimeter-Wave TWTs," Technical Digest, IEDM Meeting, 1993.
34. G. Dohler, D. Gallagher, and J. Richards, "Millimeter Wave Folded Waveguide TWTs," Vacuum Electronics, Annual Review, 1993.
35. B. James, "A High Power Coupled-Cavity TWT at 35 GHz," IEDM 1980.
36. V. Naumendo, "Investigations of Millimeter-Wave Magnetrons," Digest, International Conference on Millimeter and Submillimeter Waves and Applications, January, 1994.
37. K. Felch, T.S. Chu, J. Feinstein, H. Huey, H. Jory, J. Neilson and R. Schumacher, "Long Pulse Operation of a Gyrotron with Beam/RF Separation," 17th International Conference on Infrared and Millimeter Waves, Proc. SPIE, pp. 184-185 (1992).
38. W. DeHope, G. Hu, M. Mizuhara, J. Neilson, R. Schumacher, "Design and Development of a Third Harmonic, 95 GHz GyroTWT," IEDM digest, 1992.
39. T. A. Hargreaves, G.P. Scheitrum, T. Bemis and L. Higgins, "Design and Construction of a 95 GHz, 4th Harmonic Oscillator Experiment," Technical Digest, IEDM Meeting, 1993.
40. H.C. Limburg, O. Sauseng, R.R. Hull, J.A. Davis, C. Goodman, and G. Lee, "Development of a 10 Watt 95 GHz Helix TWT," Vacuum Electronics, Annual Review, 1993.
41. M. Viant, and C. Nilsen, "Millimeter Wave Amplifiers Using a Ladder Circuit with Alternating Long and Short Slots," Vacuum Electronics, Annual Review, 1993.

ULTRAFAST OPTICS AND MILLIMETER-WAVE TECHNOLOGY

Chi H. Lee
Department of Electrical Engineering
University of Maryland
College Park, MD 20742

ABSTRACT

Progress in the application of ultrafast optics to microwave and millimeter-wave technology using picosecond photoconductors and photoresistors is reviewed. Generation, control and characterization of both pulsed and CW high frequency waves are presented. An optical-microwave inter-mixing technique has been applied to phase lock a free running microwave oscillator to the picosecond laser pulses. A time-domain network analyzer which uses optoelectronic techniques for on-wafer monolithic microwave and millimeter-wave integrated circuit measurements is described.

1. Introduction

Rapid advances in both microwave and optical technologies make it possible to merge them together to form a new class of devices that can perform high speed/high frequency electronic functions. With the advent of femtosecond lasers we envision such devices to work in the picosecond time domain. We refer to the new technology as picosecond optical electronics, which combines the fields of optics and electronics. The link for these two fields is picosecond photoconductivity and/or picosecond photoresistivity.

Optical technique for controlling the microwave devices offers unique advantages in: (1) near perfect isolation between controlling and controlled devices, (2) elimination of RF feed in a large array system, (3) immunity from electromagnetic interference, (4) light weight and compact size, (5) extremely fast response, (6) high power handling capability, and (7)possibility for monolithic integration.

In general there are two approaches in generating microwaves from optical sources: (1) CW light source and (2) picosecond pulse train. In the CW approach the microwave generation is achieved by beating two optical sinusoidal oscillations. For example, optical generation and control of IMPATT, [1-2] TRAPPATT, [3], MESFET [4] and HEMT [5] are in the CW approach category. The other approach is using a picosecond or femtosecond optical pulse train from CW mode-locked laser. An electrical pulse train is instantaneously generated via picosecond photoconductivity effect when the optical pulse train is illuminated on a photoconductor. Through this effect one strips off the optical carrier frequency while retaining the ultrawide bandwidth of the optical pulses. Since the electrical pulses thus generated have picosecond time duration, it is a natural match to millimeter-waves whose periods are in the tens of picosecond range.

Picosecond photoconductors can now be engineered to generate electric pulse as short as 1 ps [6] with jitter free switching. All subsequent electronic events can be slaved to this initial pulse. Picosecond optical techniques can be utilized again to generate other pulses that can be precisely time controlled. Using picosecond optical electronic techniques, we are able to generate microwave and millimeter-wave signals that are in complete time synchronization with the exciting optical pulses. RF waveform generation, ranging from one monocycle [7] to continuous wave, with peak power up to a few megawatts in the pulsed case and frequency up to 20 GHz in the cw case, [8] have been achieved. Phase control of the microwave signals can be accomplished by simply controlling the arrival time of the exciting optical pulses, potential for a true time delay phased array radiation.

Picosecond photoresistors are another type of device that can be used to generate ultrashort electrical pulses. The discovery of high T_c superconductors (HTS) has attracted wide interest in picosecond "opening" switches and modulators based on the transition of superconductoring state of the HTS to the normal state. Thus these types of devices are based on photoresistivity effect. Since this state transition is fully reversible, the switch can be operated repetitively. A high current opening switch can be constructed from HTS material which is limited only by the critical current density, J_c, of the superconductor. Usually this value is about $\sim 10^3 A/cm^2$ for bulk HTS materials and $\sim 10^7 A/cm^2$ for HTS thin films. On the other hand, the state-switching of the HTS could be carried out by irradiating on the HTS materials with various kinds of lasers because all HTS materials are highly absorbing at visible and infrared lights. HTS devices could be operated with a rise time of nanoseconds or picoseconds because of their fast response to the laser pulses [9-15].

The merging of ultrafast optoelectronics with high temperature superconductors represents an opportunity for a new class of devices with unique properties. Not only HTS switches for generating electrical pulses with high power gain and pulse compression but also HTS microwave devices, such as generators, modulators, resonators, filters, antennas, phase shifters or transmission lines, could be controlled by picosecond laser pulses or combined with picosecond optical semiconductor devices [16]. The system bandwidth can be further improved by using optics. The laser beam can, in some cases, be regarded as a third control terminal which adds a degree of freedom for the device design and permits novel device concepts.

I will review the progress in the application of ultrafast optics to microwave and millimeter-wave technology through picosecond optoelectronics. The term "picosecond optoelectronics" is defined as electronic function performed by picosecond optical pulses via photoconductivity or photoresistivity effect. In addition to the millimeter-wave generation and control, picosecond optical pulses can also be used to characterize the monolithic millimeter-wave integrated circuits and to phase lock free running microwave oscillator to the optical pulses.

2. Picosecond Photoconductors

A photoconductor exhibiting picosecond response time is referred to as a picosecond photoconductor. This phenomena is referred to as picosecond photoconductivity effect which was first reported in 1972 [17]. Picosecond photoconductors used as ultrafast switching and gating devices have been demonstrated. In this application, the photoconductor is dc biased at the input terminal and a fast rise electrical output signal is obtained instantaneously when it is irradiated by a picosecond optical pulse.

There are a large number of devices based on the picosecond photoconductivity effect. They include switches, gates, samplers, electronic impulse function correlators, A/D converters, optical detectors, dc to rf converters and coherent microwave generators. The principle of operation for some of these devices will be discussed in this paper. They are mostly based on the interaction of electromagnetic waves with photoinduced electron-hole pairs in a bulk semiconductor. Recently a number of photoconductive devices using semiconductor films have been reported. [18, 19]

Most of the applications of picosecond photoconductors are conductive mode devices based on photoconductive switches. The actual design of the switch depends on the specific application. The switch is a bulk device which can be scaled in size. For high frequency and high speed application the output voltage requirement is low and thus the photoconductive gap length is typically of the order of a few micrometers. For high power switches, however, the device size is usually in millimeters or even centimeters. Fig. 1 shows some of the typical designs. To achieve extreme speed one has to use the coplanar microstrip design to reduce the circuit capacitance to a minimum. Electric pulses of 350 femtosecond duration have been generated with this new design [6].

In order to preserve the inherent bandwidth capability of the picosecond photoconductive device, the photoconductor is used as a switch in a charged transmission line pulser arrangement. In this arrangement the charging segment of the transmission line is attached to one electrode of the photoconductive switch at one end of the segment and is connected to the charging voltage source on the other end. The output end of the switch is connected to the output line which is terminated by a matched load. In order to achieve good isolation between the input and output, the photoconductive materials chosen always have high dark resistivity. Table 1 shows some of the picosecond photoconductors. Note the high dark resistivity of all materials. Among them silicon has the lowest dark resistivity, resulting in serious thermal runaway problem.

Picosecond photoconductors have many applications based on their unique characteristics, e.g., extremely high speed, large dynamic range, scalability, and jitter-free response. There are two categories of applications: low voltage and high voltage. In the former case the most important characteristics of the device is its high speed. The switching efficiency is not crucial. One needs a material with carrier lifetime as short as possible even at the expense of sacrificing mobility. In the latter case, the device is required to deliver as much power to the load as possible. The switch transfer efficiency should be near 100%. Speed is not the primary concern. The mobility of the switching material should maintain a reasonable value. One basic function of all these devices is electronic switching and gating. Other applications include exampling, waveform generation, microwave modulation, generation of microwave pulse, pulsed power and particle detection.

3. Picosecond Photoresistors

A photoresistor exhibiting picosecond response time is refered to as a picosecond photoresistor. In this work we use high temperature superconductors as the photoresistors. When the HTS film is illuminated with a laser pulse, Cooper pairs are broken by the absorption of photons, resulting in a superconducting to normal state transition. This is an intrinsically fast process which results in an immediate resistivity jump. Thus it is essentially an opening switch. [9]

In HTS optoelectronic devices, the optical response, both in sensitivity and speed, is directly related to the quasi-particle nonequilibrium dynamics. In the laser activated high temperature superconductors, a study is neeed to understand the equilibrium and nonequilibrium dynamics of quasi-particles and Cooper pairs, in particular, the effect of phonons on the response of Cooper pairs and quais-particles to external stimuli. Pair breaking and quasi-particle recombination, influenced by phonon dynamics, occur over various time scales ranging from femtoseconds to nanoseconds.

Opening switches using three different HTS materials were tested. The first two types of opening switches were prepared using T1PbSrCaCu0 and T1BaCaCuO HTS films of thickness 0.7-0.8 μm, respectively. The $Tl_2Ba_2CaCu_2O_8$ and $Tl_{0.5}Pb_{0.5}$
$Sr_2CaCu_2O_7$ films were epitaxially oriented, with their c-axes normal to the surface and their a-axes aligned with the $< 100 >$ directions of the underlying substrate. The T_cs of the $Tl_2Ba_2CaCu_2O_8$ and $Tl_{0.5}Pb_{0.5}Sr_2CaCu_2O_7$ films were measured inductively to be 105 K and 83 K respectively. The switches, as shown in Fig. 2, were 50Ω coplanar waveguides with center line of 30 μm wide. The devices had frequency response up to 20 GHz. The third type HTS opening switch was a simple microstripline of YBCO film with a thickness of 0.65 μm and a width of 100 μm grown on a substrate of 0.5 mm thick $LaA1O_3$ crystal.

Fig. 3 shows the experimental setup for the testing of the HTS opening switches. The HTS switch was mounted on a cold finger in a vacuum cryogenic chamber (VCC). The switch was biased up to 40mA by a dc current source, I_o; the signal switched-out was monitored by a Tektronix 11802 digital sampling oscilloscope with a temporal resolution of 20 picosecond. The laser used in this experiment was a Nd: glass oscillator-regenerative amplifier laser system [20] producing 10 ps, 10 μJ laser pulses at a repetition rate of 500 Hz. The lase was focused onto the HTS switch by a 10 cm cylindrical lens.

We have investigated the switching risetime, decay time, and switching efficiency for the three different types of HTS opening switches. A risetime as short as 50 ps, shown in Fig. 4, was observed for the first two types of switches, while a risetime about 100 ps for the YBCO switch. We believe the slower risetime was caused by a narrower bandwidth of the YBCO switch. The decay time could be from subnanosecond to a few microseconds depending strongly on the current, laser pulse energy, and the thickness of the HTS films, which is apparently related to the heat escape from the film to the substrate. The switching efficiency, defined as the ratio of the switched-out current to the charging current, depends on how large the switch off-impedance is, compared to the load impedance. In order to obtain high switching efficiency, not only a high energy laser pulse, but also a high impedance of the HTS switch at normal state is needed. Under our experimental condition, a switching efficiency of 80% has been achieved.

4. On-Wafer Photoconductive Sampling of MMICs

Photoconductive (PC) sampling has been demonstrated as a useful technique for the characterization of microwave/millimeter-wave integrated circuits [21-23]. This technique is needed to test devices with a frequency bandwidth larger than that of conventional electronic network analyzers. One of the main advantages of this measurement technique is that the electrical pulse signal containing the microwave spectrum is generated on-wafer, alleviating the need for microwave/millimeter-wave probe contacts. Only low-frequency sampled signals must be extracted from the device under test (DUT). In addition, this technique allows broadband characterization from a single measurement. Subpicosecond electrical pulse generation by a PC switch has been reported [6]; this pulse width corresponds to Terahertz measurement bandwidth.

On-wafer PC sampling refers to PC sampling on the DUT before it is diced from the wafer and packaged. With this constraint added to conventional PC sampling, two requirements must be met. First, fabrication of the PC switches must be compatible with the gallium arsenide (GaAs) monolithic microwave integrated circuit (MMIC) production process. As previously demonstrated for the hybrid case, PC sampling can be performed with switches fabricated to these specifications [24]. This compatibility allows monolithic integration of the PC switches with the MMIC. Second, only dc probes should be used to generate and extract the signal from the wafer. This rules out the need for complicated microwave/millimeter-wave probes and their associated problems such as reproducibility and invasiveness. In addition to the reduced cost of MMIC production, on-wafer characterization also alleviates the possibility of accidental oscillation of the DUT due to improper transition from the microstrip line to other guiding structures.

On-wafer one-port testing of passive elements has been reported [25]. This is the first report of monolithic integration of optical test structures to MMIC amplifiers, and on-wafer two-port PC sampling.

4.1 Hybrid Testing Configuration

Hybrid testing configuration means that the GaAs optical test structures are wire-bonded to the MMIC which were fabricated using an O^+-implant process rather than proton bombardment. This configuration is tried first to prove the validity of the PC switches.

As mentioned earlier, the first requirement for achieving on-wafer PC sampling is the compatibility of PC switch fabrication to that of the MMIC. O^+ is one of the typical ion species used for the isolation process by implantation during MMIC fabrication. The same energy and dose as used in the isolation process were used to implant the PC switches. Therefore, these switches can be fabricated with the MMIC, without adding a new process step.

The O^+ implantation was performed at two energies, 140 keV with a dose of $5 \times 10^{11} cm^{-2}$ and 360 keV with a dose of $1.3 \times 10^{12} cm^{-2}$, which are the same as the regular oxygen isolation process for the TRW MMICs. It was then annealed at 410degC for 10 minutes after implantation to reduce defects generated by the ion implantation. The O^+ is midgap hole trap that serves as an efficient recombination center. All of this was done before the formation of the metal contact to allow on-wafer optical probing to be included in the TRW MMIC fabrication process. [24]

The laser system is a Coherent CW Mode-locked ND:YLF laser which generates 50-ps pulses at 1.054 μm with a 76-MHz repetition rate. The output was compressed to approximately 3 to 5 ps by a fiber pulse compressor then frequency doubled to 0.527 μm by a potassium titanyl phosphate (KTP) crystal. The data are then collected using the standard pump-probe PC sampling technique. Typical electrical auto correlation full width at half maximum of the O^+ switch is about 19.5 ps. The DUT is a three-stage 12-GHz automatic gain control (AGC)(MMIC)amplifier with a 7-GHz bandwidth.

The output sampling time is deliberately extended to 2 ns to study the effect of time windowing. Due to dark current, photovoltaic effect, and difference of responsivity inherent to the switches, calibration of the acquired data is necessary. Both input and output switches show good linear response at low bias voltage. The responsivity gives a 17-percent difference for the two sampling switches, which is equivalent to a 1.34 dB in $|S_{21}|$ calibration.

S_{21} parameter measured by network analyzer and PC sampling with different time window length is shown in Fig. 5. Good agreement with the network analyzer measurement is obtained up to a 40-dB DR in the case of the 2-ns time window length. Two distinct features of the PC sampling measurements are evident. First, the separation of the peaks of the ripples is approximately equal to the inverse of the length of the time window. Second, the DR increases as the time window increases. At 7 GHz from the center frequency, the DRs are 31 and 39 dB for the 1.15- and 2-ns time window lengths, respectively. The calculated DRs of 35 and 40 dB agree well with the observed results. The conclusion is that the DR is limited primarily by the time window length. The 40-dB DR is the highest DR experimentally measured by optoelectronic techniques.

4.2 Monolithic Testing Configuration

With the success of the characterization of an MMIC by the hybrid testing configuration using O^+ implanted PC switches, the optical test structures were monolithically integrated with the MMIC. This is the first report of a PC testing structure having been fabricated monolithically with the MMIC. Different testing structure designs were implemented in an effort to reduce the real estate required for the optical test structure. [24]

The optical system is the same as that used in the hybrid case. To facilitate on-wafer testing, a probe station using only dc tungsten probe needles was developed. The gold plated probes were used to bias the MMIC, and the generation gap switch, as well as to extract the cross-correlated signal from the sampling ports. A high-resistivity probe is required to bias the generation switch gap. The dc probes provide the only electrical contacts to the MMIC. A microscope was also implemented to assist in placing the dc tungsten probes at the 100-μm^2 pads. The schematic diagram of the on-wafer sampling system is shown in Fig. 6. With the exception of this probe station, the standard optical generating and sampling method was used for the measurement.

One important consideration limiting the commercial viability of on-wafer optical measurement techniques is the size of the GaAs real estate required for the built-in optical test structure. Two different methods were used in an attempt to reduce the size of these structures. One utilized a bent microstrip line, while the other used a test structure whose size is comparable to that used in a conventional coplanar waveguide (CPW) probe, referred to here as the de-embedding structure. De-embedding is a computational procedure devised to extract S-parameters from a DUT in the presence of other scattering sites.

Accurate characterization of the MMIC using the PC sampling techniques requires careful calibration of the PC switches. A dc signal is usually applied to the microstrip, and the on-state and off-state resistances are measured. In the case of the monolithic test structure, the microstrip is terminated in a 50-Ω integrated resistor.

The typical input and output waveforms from the built-in optical test structure configurations are shown in Fig. 7. The multiple peaks of the input waveforms are a result of reflections from the end of the biasing side line, which is a shunt microstrip line designed to minimize the discontinuity seen by the electric pulse on the main microstrip line. Reasonable results have been obtained compared to the conventional scalar network analyzer measurement, as shown in Fig. 8. Some discrepancies can be attributed to the nonuniform nature of the specific circuits tested.

Since the reflection a the end of biasing side line depends on the exact position of the probe on the contact pad, different probe positioning can slightly alter the shape and timing of the reflection. This effect is more pronounced in the case of a de-embedding structure because the needle probe may not be at the corresponding position when measuring the responses of the MMIC and the through-line calibration structure. The exact position of the generation gap biasing probe cannot be duplicated from device to device, or from measurement or measurement. With the implementation of a dc probe, which has a built-in chip resistor, the mechanical constraints in positioning the probes on the contact is not critical. It was found that a 50 to 100Ω chip resistor gives the best attenuation. Larger resistance may cause more radiation coupling across the chip resistor, resulting in less attenuation. This also suggests that it is possible to use a lossy biasing side line to reduce the reflection from probe/side line contact.

5. Optically Controlled Generation and Phase Shifting Broad-Band Millimeter-Wave Signals

One important requirement in system operation of millimeter-wave devices is the ability to control phase variation of propagating wave along a transmission line [16, 26]. Semiconductor phase shifters using field-effect transistor, Schottky diode, and PIN diode technologies suffer from considerable RF insertion loss, and are limited in operation at millimeter-wave frequencies. The application of optical techniques to obtain phase-shift was first demonstrated by optically illuminating a semiconductor dielectric waveguide [27]. Similar approaches have been applied to integrated transmission lines [28]. Recently, injection locking of oscillation have been demonstrated for phase-shift up to 18 GHz [29]. Using microwave modulated optical signals and microwaves generated by beating two lasers of slightly different frequencies true-time-delay phase shift up to 50 GHz have been demonstrated [30]. Ultrafast optics can also be used to control the phase of millimeter waves by synchronizing millimeter waves to mode-locked laser pulse trains. This work demonstrates the true-time-delay phase shift of broad-band 60-GHz signals using this new scheme. Such true time-delay is necessary in a phased-array radar, whereby broad-band signals are generated, in order to achieve the required antenna pointing accuracy [31].

A block diagram of the optically controlled broad-band signal generator and phase shifter is shown in Fig. 9. A 5-ps electrical signal with a 76-MHz repetition rate is generated by focusing a 3-ps mode-locked laser pulse train on a gallium arsenide (GaAs) photoconductive (PC) switch. The PC switch has a length of 10 μm, and a substrate thickness of 4 mil [32]. The present switch was proton implanted at a dose of $10^{14} cm^{-2}$ and multiple sequence of energies to achieve an uniform damage from the surface to a sufficient depth for fast optical response.

The PC switches are connected to the input and output of a multistage V-band monolithic integrated low-noise amplifier (LNA) through wire bonds. The amplifier is fabricated using InGaAs-GaAs pseudomorphic high electron mobility transistor (HEMT) technology. It has a power gain of 20 dB (or 30 dB at high-gain bias), and a bandwidth of 8GHz. The electrical pulses, which cover a frequency spectrum from dc up to about 150 GHz, are input to the amplifier. Only signals of frequencies within the gain bandwidth of the LNA are amplified. This optical/millimeter-wave scheme results in coherent, broad-band 60-GHz signals at the output of the LNA in response to the input electrical impulses generated by the optical excitation at the PC switch. The broad-band 60-GHz signal at the output of the amplifier is then mapped out using the PC sampling techniques [24]. By time-delaying the triggering optical pulse trains, the generated broad-band 60-GHz signal can be phase-shifted.

The input waveform to the broad-band 60 GHz amplifier is shown in Fig. 10 and the corresponding output waveform is shown in Fig. 11. Fig. 12 shows the measured waveforms of the time-delayed broad-band 60-GHz signal. The timing jitter of the output waveform from the LNA is negligible as evident from the excellent signal to noise ratio of the time-domain waveform. With optical round-trip dealy of only 2.4 mm, the 60-GHz signals can be continuously shifted by one complete cycle, 360deg. The present feature of requiring only short optical delay is important for phased-array applications [33], since it allows very fast and accurate control of the phase shift of the millimeter-wave signal, which can be readily accomplished by properly designing of the optical system using piezoelectric devices.

The generation of the 60-GHz broad-band signals follows closely the delay of the optical triggering signal, with the corresponding phase-shift shown in Fig. 13. In order to demonstrate continuous phase shifting of the 60-GHz millimeter wave, nine equally spaced time-delay measurements were made. In general, the amount of phase shift, $\phi(f)$, is equal to

$$\phi(f) = \frac{-2\pi f}{c_0}\Delta L,$$

where c_0 is the velocity of millimeter-wave signal in the free space, f is the frequency, and ΔL is the single-pass optical delay length.

With the present setup, the accuracy of the phase control is better than 0.1deg. By performing Fourier transform of the time-domain waveforms, the phase information can be obtained. The phases of these millimeter-wave signals are fairly linear as a function of frequency over the broad-band spectrum.

The incorporation of optical techniques into millimeter-wave technologies has created a new class of devices that can perform electrical functions that are not easily attainable by purely electronic means. Synchronization between the ultrafast optical pulse trains and millimeter wave makes it easy to control millimeter-wave devices. In this work, the technique of optically controlled generation and phase shift of broad-band 60-GHz signals was presented. Potential applications of this technique are to control beam steering of satellite-borne communication system and phased-array radar. For satellite-borne communication, it is possible to modulate the broad-band carrier frequencies by digital or analog modulation on the bias for the PC switch. We measured the timing jitter of the mode-locked laser pulse train to be about 1 ps that could match the critical demand for radar applications.

6. Ultra-Wideband Pulse Generation Using a High T_c Superconductor Opening Switch

Opening switch is a key element in an inductive energy storage system. Common opening switches, such as explosive and plasma erosion switches, can provide high-current and high-voltage operation but lack in fast opening and high repetition rate. Photoconductive semiconductor switches (1), on the other hand, have shown their unique characteristics as opening switches: fast opening time, jitter-free response, and scalability to handle extremely high power. Recently, we have demonstrated for the first time in our group the operation of a GaAs opening switch (IESPPS), resulting in electrical pulse compression and peak power gain [34, 35]. However, the semiconductor opening switches require specially tailored laser pulses to keep the opening switches conducting during the system storage stage. Using photons to keep a material in the conducting state is an inefficient and expensive approach. Furthermore, the performance of semiconductor opening switches is limited by "lock-on" effect. The discovery of high T_c superconductors (HTS) has attracted interests in picosecond opening switching based on the transition of the superconductor from its superconducting state to normal state. Since this state transition is fully reversible, the switch can be operated repetitively. A high current opening switch can be devised using HTS materials owing to their high critical current densities, usually $\sim 10^7 A/cm^2$ for HTS films. On the other hand, the state-switching of the HTSS can be carried out by irradiating the HTS with various kinds of lasers since HTS materials are highly absorbing at visible and infrared lights, and the state-switching can be achieved with a switching time of nanoseconds or picoseconds due to their fast response to the laser pulses [11, 14, 36]. When used in an inductive storage system, the HTS opening switch becomes even more attractive since in the ideal case there will be no energy loss during storage and the energy can be stored for a very long time. In this review we report on the operation of optically controlled HTS opening switches with switching risetime as short as 50 picosecond and switching efficiency up to 80%. The HTS opening switch was employed in a prototype IESPPS with a Si switch as the closing switch. Nanosecond electrical pulses with peak power gan of 40 have been achieved [20].

The HTS switches were employed in a prototype IESPPS. As shown in Fig. 14, the IESPPS used in this work was a current charged transmission line [37] with a HTS opening switch and a Si semiconductor closing switch coupled to a 0.1 μF capacitor. The Si switch had a 50Ω microstripline design with a straight gap of 0.5 mm. Its measured photoconductive lifetime was several microseconds. When the Si switch is closed by a laser pulse, the energy initially stored in the charged capacitor is transfered in the current form to the transmission line through the HTS switch in its closing state. Since the transmission line is shorted, the charging current wave is reflected at both ends and builds up in a staircase waveform which increases one step after each round-trip. Fig. 15 shows the ideal and the experimentally observed stair-step current buildup from an IESPPS with a transmission line of 165 cm long. Once the current reaches its maximum, the HTS switch is opened by another laser pulse and the energy stored in the transmission line is delivered to an impedance matched load. Fig. 16 shows the electrical pulse switched out by the IESPPS with a 16 cm long transmission line. At a charging voltage of 1V, about 6.4V was delivered into the load. This corresponds to a voltage gain of 6.4, and a peak power gain of 40, defined as the ratio between the peak power delivered by the IESPPS and the peak power that would be delivered directly into the load from the capacitor. Higher gain is possible if the closing-state resistance of the Si switch, a few ohm, is further reduced. At higher charging voltage, thermal runaway was observed for the Si switch, and a lower power gain was obtained. The thermal runaway can be avoided by using pulse charging.

High Tc superconductor (HTS's) have been shown as efficient fast-opening laser activated switch [20]. A HTS can be transformed from its superconducting state to its normal state with a switching time of nanoseconds or picoseconds with laser pulses of various wavelengths When combined with a pulse-forming network, it is possible to use the HTS as an opening switch to generate square nanosecond electrical pulses with risetimes of less than 100 ps. These pulses serve as almost ideal step functions for excitation of an antenna. As such, they can be used to assess the time-domain performance of wideband antennas, and to produce ultra-wideband (UWB) microwave radiation [38]. Furthermore, triggering of these pulses is jitter-free with respect to the laser pulse, a desirable property for the generation of UWB pulses for impulse radar applications [39].

In this work we demonstrate a current charged transmission line (CCTL) [37] pulse forming network controlled by an HTS opening switch with wideband conical monopole transmitting and receiving antennas. The CCTL pulse forming network utilized in the experiment is illustrated in Figure 17(a). It consists of a current source charging a 50 Ω characteristic impedance transmission line (TL) through an HTS opening switch. The switch is a 50Ω coplanar waveguide fabricated from a 0.7 μm $Tl_2Ba_2CaCu_2O_8$ film on a (100) $LaAlO_3$ substrate [40, 41]. The switch has a rise time shorter than the SMA connector limit (~60 ps). Since the load is typically the radiating antenna, it is placed outside the Dewar, 1m along the TL as illustrated.

At $t = 0^-$ the HTS is in its superconducting (closed) state, and the TL has charged to the current I_o. Half of the current is contained in a forward traveling wave and half in a reverse traveling wave. The net current passing through the load is zero. The HTS is then activated by a 15 ps full width at half maximum, 20 μJ, mode-locked, Nd:Glass laser pulse at $t = 0^+$. The laser pulse transforms the HTS to a high-resistance (open) state causing the energy stored in the TL's to be delivered to the load. The energy is delivered in the form of a pulse with a 6 ns duration corresponding to the round-trip time in the 0.6m TL to the right of the load. The pulse delivered to a $Z_L = 50\Omega$ resistive load is shown in Fig. 18(a).

The resistive load is then replaced by a $Z_a = 50\Omega$ characteristic impedance solid Cu conical monopole antenna mounted above a Cu ground plane as shown in Figure 17(b). The conical antenna can be modeled to first order as a waveguide with characteristic impedance determined by the cone angle. A half angle of 47deg was chosen corresponding to $Z_a = 50\Omega$ and various cone diameters were used. A cone receiving antenna was also used, terminated in a Tektronix 11802 sampling oscilloscope which presents a matched 50Ω load. The transmitting and receiving antennas share a common ground plane and thus a common axis. The received voltage pulse was recorded for various transmitting and receiving antenna separations and diameters.

The received pulses from the radiation and receiving system is shown in Fig. 18(b). In this case identical transmitting and receiving antennas were used, each with an end diameter of 1.9 cm and a 47deg half angle; the transmitting and receiving antennas were spaced 12 cm apart, and a charging current of $I_o = 35$ mA was used.

The radiated field corresponds to the second time derivative of the driving pulse, while the received voltage corresponds to the first time derivative of the field or the third time derivative of the waveform at the input to the transmitting antenna. Since the 100 ps risetime of the driving pulse is of the same order as $2a_t/c = 87$ ps for the transmitting antenna which was used, we expect the double differentiation of the transmitting antenna to be approximate in this case. The approximation is also limited by the fact that the antenna is mismatched at its feed point due to the connection to two TL's in parallel.

Since the time domain transfer function of the transmitting antenna is proportional to the derivative of the step-function response, the frequency domain transfer function of the transmitting antenna is proportional to the Fourier transform of the received waveform, for frequencies below the reciprocal risetime of the driving pulse (1/100ps = 10 GHz). Provided that the receiving antenna is short enough to take the derivative of the radiated field, this technique can be used to assess the transfer function of a wide variety of transmitting antennas. The power spectral density of the received waveform reveals a center frequency near 3.5 GHz with a half-power bandwidth of 2.8 GHz and an 80qualifying these pulses as UWB.

While conical antennas were used in this work, the HTS pulse forming technique is adaptable to other wideband antenna designs. The system demonstrated here was limited primarily by the response time of the microwave connectors and the oscilloscope response-time. However, if an integrated antenna is used with an electro-optic sampling technique, we predict that the limiting frequency will be near 70 GHz, corresponding to the rise-time of the laser pulse. Thus it may be possible to achieve a real-time measurement of antenna response with a bandwidth exceeding that of any commercially available network analyzer.

We have demonstrated that an HTS opening switch controlling a CCTL pulse forming network and wideband conical antenna can be used to produce UWB radiation in the 3.5 GHz regime. In addition, the step-function excitation allows one to assess the antenna transfer function directly from the received pulse. This method of generation is adaptable to a wide variety of ultra-wideband antenna designs.

7. Optical Microwave Intermixing and Phase Locking

As mentioned in the previous sections, the photoconductive sampling technique has applications in microwave measurements including the on-wafer characterization of MMICs. The essential requirement for the measurement is the time sychronization of the event to be measured with the measuring optical pulses. In the method discussed earlier in this paper, the impulse technique is employed. In the pulsed technique, the test signal and the sampling pulse are derived from the same laser pulse with one appropriately delayed with respect to the other by using optical beam splitter and delay line. The sampling is done by stepping the time delay with a stepping motor driven optical delay line. The data acquisition rate is limited by the speed of the electro-mechanical stepping motor. A typical measurement takes a few minutes. There is an alternative technique which is ideal to measure the CW microwave event. It requires first the synchronization of the microwave signals with the optical pulses and then using the optical pulse to measure the microwave event by equivalent time sampling technique. Since it involves continuous sampling, it does not require stepping motor to control the time delay. The data acquisition rate is much faster. A typical sampling run takes only a few microseconds to complete. In this section, I will describe the approach to carry out the equivalent time sampling using only the mode-locked laser as the main source.

The technique involves a new phase locking scheme which can potentially lock any microwave oscillation to the optical pulses for signal with frequency up to the 100 GHz range. This can be done on-wafer before the MMIC chip is diced and packaged. The technique does not require launching high frequency signal onto the MMIC chip and thus avoids the use of dedicated high frequency probe contact.

The technique uses a broadband optical-microwave mixer which will mix the optical pulse signal with that of the microwave. The picosecond photoconductor can be used as the mixer. It has the necessary bandwidth and also has an excellent signal to noise ratio. Electrooptic mixer can also perform the task except its signal to noise ratio is not as good. The principle of optical pulse train and microwave intermixing can be understood as follow: The optical pulse train shown in Fig. 19 can be regarded as a comb signals either in the time as frequency domain generated by the mode-locked laser. The fundamental base band frequency is the same as the pulse repetition frequency of the mode-locked laser. It contains harmonics with spectral components extending to the reciprical of the pulse width. The high frequency microwave signal will mix with a certain harmonic (say N^{th}) so that the mixing condition

$$f_m = N f_L \pm f_{IF}$$

is satisfied. This fixes the intermediate frequency signal to a frequency f_{LF} determined by the above mixing condition. This IF signal preserves all the information of the high frequency RF, including its relative phase (or timing) with respect to the optical pulses in the pulse train (Fig. 20). In order to maintain the bandwidth, the laser pulse width should be much smaller than the period of the high frequency RF wave to be sampled. Since the laser repetition frequency is about 76 MHz. The IF frequency will be less than this value.

The optical system used is a continuous wave mode-locked Nd:YLF laser which generates 50-ps pulses at a 75.7-MHz repetition rate. These laser pulses are compressed to 3 ps, which are then frequency-doubled to produce green light. The green beam is then split into two beams, one for optical-microwave phasing-locking (OMPL) and the other for waveform sampling. The time coherence between the test and sampling signals is therefore maintained. Figure 21 is a schematic diagram of the optical-microwave phase-locked loop (OMPLL). The laser is directed to a GaAs PC switch whose substrate material has been deliberately ion-damaged to reduce the carrier lifetime. A 30-V DC is superimposed onto the output of a voltage-controlled oscillator (VCO) to bias the PC switch. This results in enhancing the signal to noise ratio (S/N) of the intermixed signal by more than 25 dB. The intermixed signal is then amplified through a high gain intermediate frequency (IF) amplifier. The IF phase is compared to a reference signal which is frequency divided from the laser mode-locker frequency. The resulting error signal is then delivered through a loop filter to tune the VCO. Once OMPL is achieved, the reference and IF waveforms can be displayed on a conventional oscilloscope, either one can be used as the trigger source. When the signals are locked, stable traces appear on the oscilloscope.

The schematic diagram of the waveform sampling system is shown in Figure 22. After the microwave source was phase-locked to the laser pulses, the output signal was delivered to another fast PC sampling switch for waveform display. The response of the PC switch was measured by the standard pump and probe PC sampling technique. The autocorrelated pulse width was measured to be 14.5 ps, which corresponds to an actual FWHM of 10 ps after deconvolution. The rise time of the PC response is about 3 ps which is limited by the laser pulse width. The measurement bandwidth can be calculated based on the rise time, which is more than 100 GHz [18]-[19]. Since the frequency of the microwave oscillator was selected to be about 1 GHz, signals of up to the 100th harmonic of the IF should be detected to achieve a 100-GHz sampling system.

The value of the intermediate frequency (f_{if}) is based on a tradeoff between time resolution and noise. The f_{if} was selected to be 300 kHz, which corresponds to a time resolution of 4-ps. To validate this OM sampling system, we first tested the system using a low frequency signal, which was generated by saturating the phase-locked microwave signal. The waveform was measured with both a conventional sampling scope (TeK11802/SD-24, 20-GHz bandwidth) and with the present optical system. Figure 23 shows the comparison of the waveforms obtained by these two methods, indicating that excellent agreement was achieved.

The phase-locked microwave signal was then used to drive a step recovery diode (SRD) in order to generate high frequency signals for circuit characterization. The measured waveform is compared in real-time with a sampling scope. The OM sampling system clearly shows a faster response, as shown in the inset in Figure 24.

The technology involving the interaction between ultrafast optical pulses and microwave signals offers a number of potential new applications. A real-time sampling system with 100-GHz instantaneous bandwidth has been developed using optoelectronic techniques with high sensitivity, low signal distortion, minimum electrical invasiveness, and picosecond time resolution. This optical system can be further enhanced to demonstrate real-time network analysis.

The time-domain data acquisition time less than 40 μs has been achieved. Additional data analysis to achieve the fequency-domain response only requires less than a few hundred microseconds with the latest commercially available signal processing chip [42]. For monolithic integration of the optical test structure onto the MMIC wafer, size of the optical structure can be less than 500-μm in length to converse valuable wafer real estate [43]. Further development of this real-time network analysis technique should provide potential applications in the microwave area.

8. Conclusion

Progress in the application of ultrafast optical technical using photoconductive switching to the microwave and millimeter-wave technology has been reviewed. The generation of both pulsed and cw microwave signals has been demonstrated. A time-domain network analyzer using photoconductive sampling method for onwafer MMIC characterization has also been described. A new microwave phase lock scheme which synchronizes the microwave output with picosecond laser pulses has been demonstrated. With this synchronized microwave source, it is possible to measure the microwave waveform. Since this technique allows the conversion of RF phase signal to DC error signal without requiring RF probe physically contacting (or connecting) with the microwave circuit, this makes on-wafer phase locking and characterization of a MMIC circuit possible, prior to the wafer being diced, and individual MMIC packaged.

9. References

1. H. W. Yen, M. K. Barnoski, R. G. Hansperger, and R. T. Melville, "*Switching to GaAs IMPATT diode oscillator by optical illumination*," *Appl. Phys. Lett.* **Vol. 31**, pp. 120-122, July 1977.
2. J. R. Forrest and A. J. Seeds, *"Optical control of IMPATT microwave oscillators,"* in 1978 *Intl. Electron Device Meeting Tech.* Dig. pp. 282-285, Dec. 1978.
3. R. A. Keihl, *"Behavior and dynamics of optically controlled TRAPATT oscillators," IEEE Trans. Electron Devices.* **Vol. ED-25**, pp. 703-710, June 1978.
4. A. A. A. DeSalles, *"Optical control of GaAs MESFET's IEEE Trans. Microwave Theory Tech.* **Vol. MTT-31**, pp. 812-820, Oct. 1983.
5. R. N. Simons and K. B. Bhasin, *"Analysis of optically controlled microwave/ millimeter wave device structures,"* in 1986 *IEEE MTT-S International Microwave Symposium* Dig. pp. 551-554, June 1986.
6. M. B. Ketchen, D. Grischkowsky, T. C. Chen, C-C. Chi, i. N. Duling, III, N. H. Halas, J. -M. Halbout, J. A. Kash, and G. P. Li, *"Generation of subpicosecond electrical pulses on coplanar transmission lines," Appl. Phys. Lett.* **Vol. 48**, pp. 751-753, 1986.
7. H. A. Sayadian, M. -G. Li, and Chi. H. Lee, *Generation of high-power broad-band microwave pulses by picosecond optoelectronic technique," IEEE Trans. Microwave Theory Tech.*, **Vol. MTT-37**, pp. 43-50, 1989.
8. D. Butler, E. A. Chauchard, K. J. Webb, K. A. Zaki, Chi H. Lee, P. Polak-Dingels, H.-L.A. Huang and H. C. Huang, *"A cw 20GHz optoelectronic source with phased-array applications, Microwave and Opt. Tech. Lett.* **Vol. 1**, pp. 119-123, 1988.
9. Y. Tzeng, C. Cutshaw, T. Roppel, C.W. Tanger, M. Belser, R. Williams, L. Czekala, M. Fernandez and R. Askew, "High Temperature Superconductor Opening Switch," Appl. Phys. Lett. 54, 949 (1989).
10. W.R. Donaldson, A.M. Kadin, P.H. Ballentine and R. Sobolewski, "Interaction of Picosecond Optical Pulses with High Tc Superconductor Films," Appl. Phys. Lett. 54, 2470 (1989).
11. A. Frenkel, M.A. Saifi, T. Venkatesan, P. England, X.D. Wu and A. Inam, "Optical Response of Nongranular High Tc YBCO Superconducting Thin Films," J. Appl. Phys., 67, 3054 (1990).
12. D.P. Osterman, M. Radparva and S.M. Faris, IEEE Trans. Magn. MAG-25, 1319 (1989).
13. M. Leung, P.R. Broussard, J.H. Claassen, M. Osofsky, S.A. Wolf and U. Strom, Appl. Phys. Lett. 51, 2046 (1987).
14. W.S. Brocklesby, D. Monroe, A.F.J. Levi, M. Hong, S.H. Liou, J. Kwo, C.E. Rice, P.M. Mankiewich and R.E. Howard, Appl. Phys. Lett. 54, 1175 (1989).
15. Mark Johnson, "Nonbolometric photoresponse of YBaCuO films," Appl. Phys. Lett. 59, 1371 (1991).

16. Chi H. Lee, "Picosecond Optics and Microwave Technology," IEEE Transaction on Microwave Theory and Techniques, 38(5), May, 1990.

17. S. Jayaraman and C. H. Lee, *"Observation of two-photon conductivity in GaAs with nanosecond and picosecond light pulse," Appl. Phys. Lett.* **Vol. 20**, pp. 392-395, 1972.

18. F. W. Smith, H. Q. Le, V. Diadiuk, M. A. Hollis, A. R. Calawa, S. Gupta, M. Frankel, D. R. Dykaar, G. A. Mourou and T. Y. Hsiang, *"Picosecond GaAs based photoconductive optoelectronic detectors," Appl. Phys. lett.,* **Vol. 54**, pp. 890-892, 1989.

19. M. C. Nuss, D. W. Kisker, P. R. Smith and T. E. Harvey, *"Efficient generation of 480 fs electrical pulses on transmission lines by photoconductive switching in metalorganic chemical vapor deposited CdTe," Appl. Phys. Lett.,* **Vol. 54**, pp. 57-59, 1989.

20. Wei-lou Cao, Yuan-Qun Liu, Chi H. Lee, S. N. Mao, X. X. Xi, and T. Venkatesan, "Nanosecond Inductive Energy Storage Pulsed Power System Using YBCO Superconductor Opening Switches," IEEE Photon. Tech. Lett., Vol. 5, 109, 1993.

21. M. Matloubian, S.E. Rosenbaum, H.R. Fetterman, and P.T. Greiling, "Wideband Millimeter Wave Characterization of Sub-0.2 Micrometer Gate-Length AI-InAs/GaInAs," *IEEE Microwave and Guided Wave Letters*, 1, pp. 32-34, Feb. 1991.

22. J.A. Valdmanis and G. Mourou, "Subpicosecond Electrooptic Sampling: Principles and Applications," *IEEE J. Quantum Elec.*, **QE-22**, pp. 69-78, Jan. 1986.

23. S.L. Huang, H.-L.A. Hung, and C.H. Lee, "On-Wafer Optoelectronic Techniques for Millimeter-Wave Generation, Control, and Circuit Characterization," *Ultrafast Electronics & Optoelectronics Topical Meeting*, paper ME11, San Francisco, CA, Jan. 1993.

24. S.L. Huang, E.A. Chauchard, C.H. Lee, H.-L. A. Hung, T.T. Lee, and T. Joseph, "On-Wafer Photoconductive Sampling of MMICs," *IEEE Transactions on Microwave Theory and Tech., MTT-40*, pp. 2312-2320, Dec. 1992.

25. H.-L. A. Hung, T.T. Lee, P. Polak-Dingels, E.A. Chauchard, K.J. Webb, C.H. Lee, H.C. Huang, "Characterization of GaAs Monolithic Circuits by Optical Techniques," *SPIE Proc. Optical Technology for Microwave Application*, **1102**, pp. 98-106, Mar. 1989.

26. M. Tamburrini, M. Parent, L. Goldberg, and D. Stillwell, "Optical feed for a phased array microwave antenna," *IEEE Electron. Lett.*, Vol. 23, pp. 680-681, June 1987.

27. C.H. Lee, P.S. Mak, and A.P. DeFonzo, "Optical control of millimeter-waveguides," *IEEE J. Quantum Electron*, vol. QE-16, pp. 277-288, Mar. 1980.

28. P. Cheung, D.P. Neikirk, and T. Itoh, "Optically controlled coplanar waveguide phase shifters," *IEEE Trans. Microwave Theory Tech.*, Vol. 38, pp. 586-595, May 1990.

29. D.J. Sturzebecher, X. Zhou, X. Zhang, and A.S. Daryoush, "Design of oscillators for optically fed MMW phased array," *IEEE Int. Microwave Symp. Dig.*, Albuquerque, NM, 1992, pp. 325-328.

30. G.J. Simonis and K.G. Purchase, "Optical generation, distribution, and control of microwaves using laser heterodyne," *IEEE Trans. Microwave Theory Tech.*, vol. 38, pp. 667-669, May 1990.

31. R.J. Mailloux, "Phased array theory and technology," *Proc. IEEE*, vol. 70, pp. 246-291, 1982.

32. C. Rauscher, "Optoelectronic approach to on-chip device and circuit characterization at microwave and millimeter-wave frequencies," *IEEE Trans. Microwave Theory Tech.*, vol. 39, pp. 1179-1193, July 1991.

33. J.D. Kraus, *Antennas*, 2nd ed. New York: McGraw-Hill, 1988.

34. Chi H. Lee, "Optical control of semiconductor closing and opening switches," *IEEE Trans. Electron Devices*, vol. 37, pp. 2426-2438, 1990.

35. E.E. Funk, E.A. Chauchard, M.J. Rhee, and Chi H. Lee, "80-kW inductive pulsed power system with a photoconductive semiconductor switch," *IEEE Photon. Tech. Lett.*, vol. 3, pp. 576-577, 1991.

36. H.S. Kwok, J.P. Zheng, Q.Y. Ying, and R. Rao, "Nonthermal optical response of Y-Ba-Cu-O thin films," Appl. Phys. Lett., vol. 54, pp. 2473-2475, 1989.

37. E.A. Chauchard, C.C. Kung, C.H. Lee, M.J. Rhee, and V. Diadiuk, "A new method to generate square pulses: optoelectronic switching in a current charged transmission line," *IEEE Trans. Plasma Sci.*, vol. PS-15, pp. 70-72, 1987.

38. X.-C. Zhang, Y. Jin, and K. Ware, Appl. Phys. Lett., 64, 622 (1994).

39. Arye Rosen and Fred Zutavern, Eds., *High-Power Optically Activated Solid-State Switches*, (Artech House, Boston, 1994).

40. W.L. Holstein, L.A. Parise, C. Wilker, and R.B. Flippen, Appl. Phys. Lett., **60**, 2014, (1992).

41. W.L. Holstein, C. Wilker, D.B. Laubacher, D.W. Face, P. Pang, M.S. Warrington, C.F. Carter, L.A. Parisi, J. Appl. Phys., **74**, 1426 (1993).

42. Alacron, i860 digital signal processing chip, model FT200/2-AT. The processing time for 1K FFT takes 299 μs.

43. S.L. Huang, L.P. Golob, C.H. Lee, and H.-L.A. Hung, "A Novel Approach to Miniature Photoconductive Sampling of Microwave Circuits," *Dig. Conf. Lasers and Electro-Opt.*, paper JFD2, Anaheim, CA, May 1992.

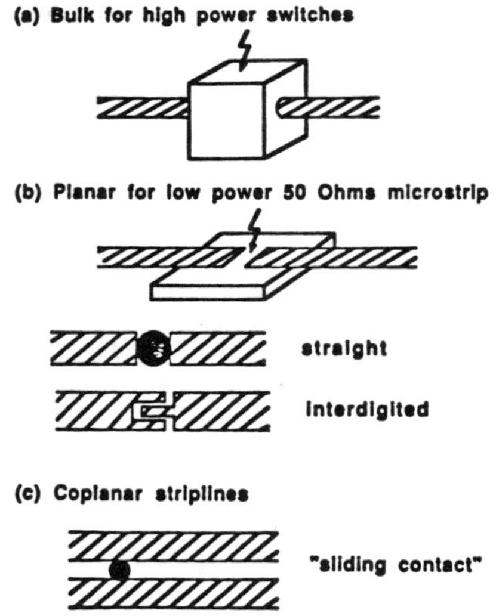

Fig. 1. Different photoconductive switch designs: (a) bulk switch for high power application, (b) 50Ω microstrip design for low power, high-speed application, the gap can be straight gap or interdigitated; and (c) coplanar striplines for very high-speed applications.

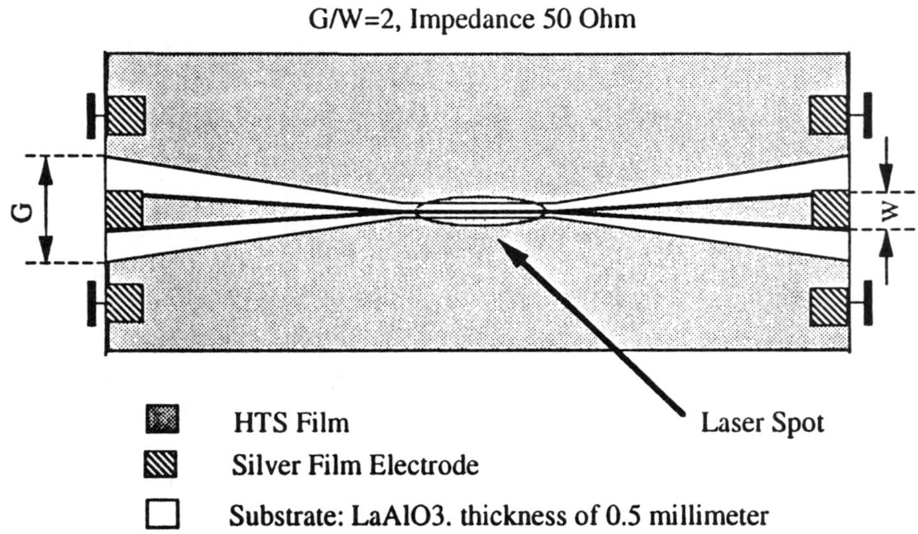

Fig. 2. High temperature superconductor coplanar waveguide switch.

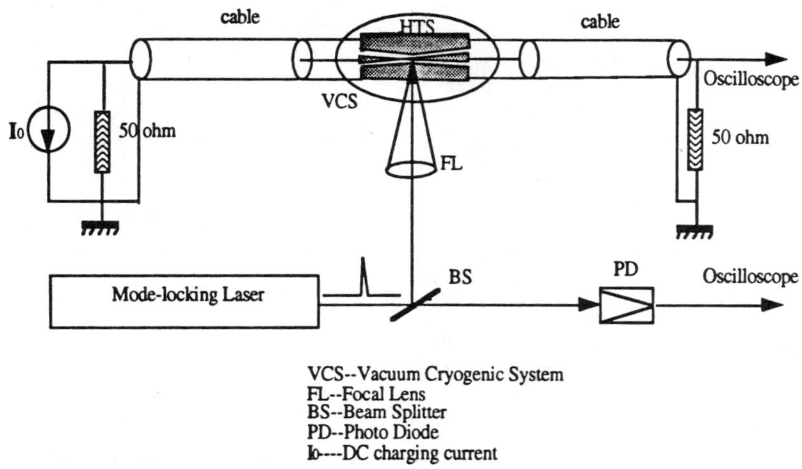

Fig. 3. Experimental setup for HTS opening switch testing.

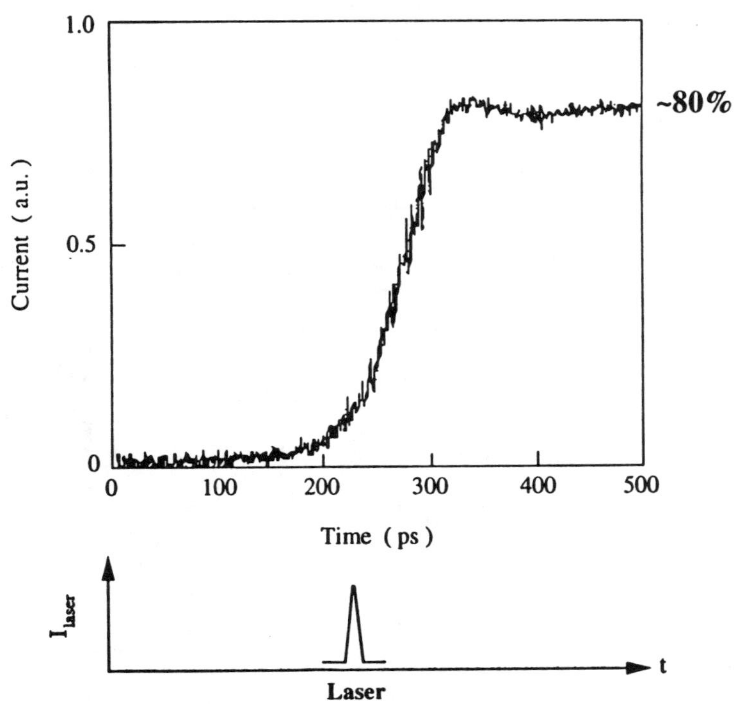

Fig. 4. Ristime of the HTS opening switch.

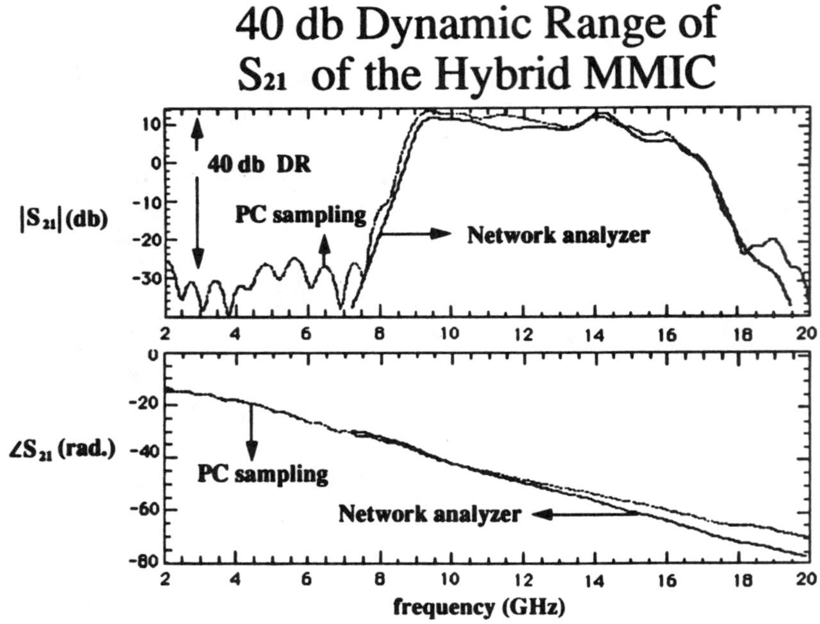

Fig. 5. S_{21} of a hybrid MMIC obtained by sampling.

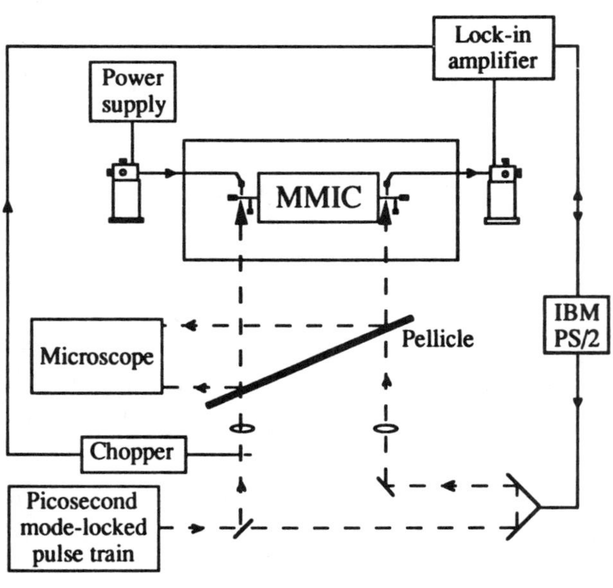

Fig. 6. Schematic of on-wafer PC sampling experiment.

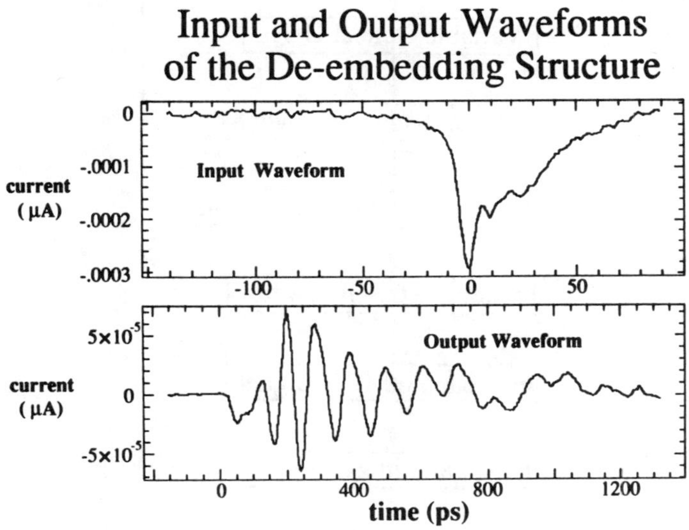

Fig. 7. Input and output waveforms of the de-embedding structure.

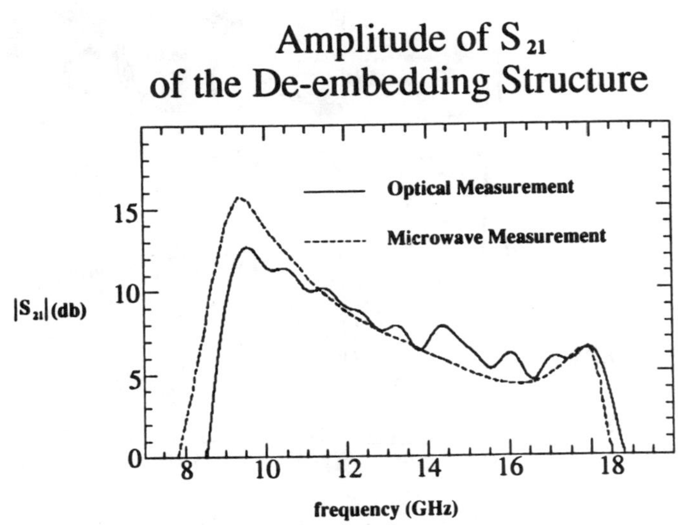

Fig. 8. S_{21} of the MMIC with the de-embedding testing structure.

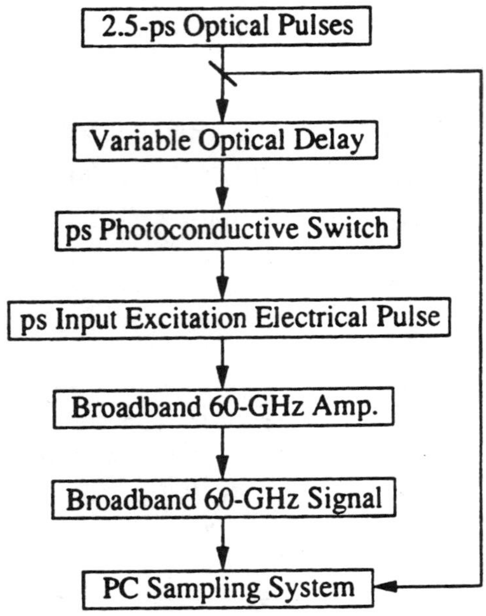

Fig. 9. Block diagram of the experimental setup for generating 60 GHz signals.

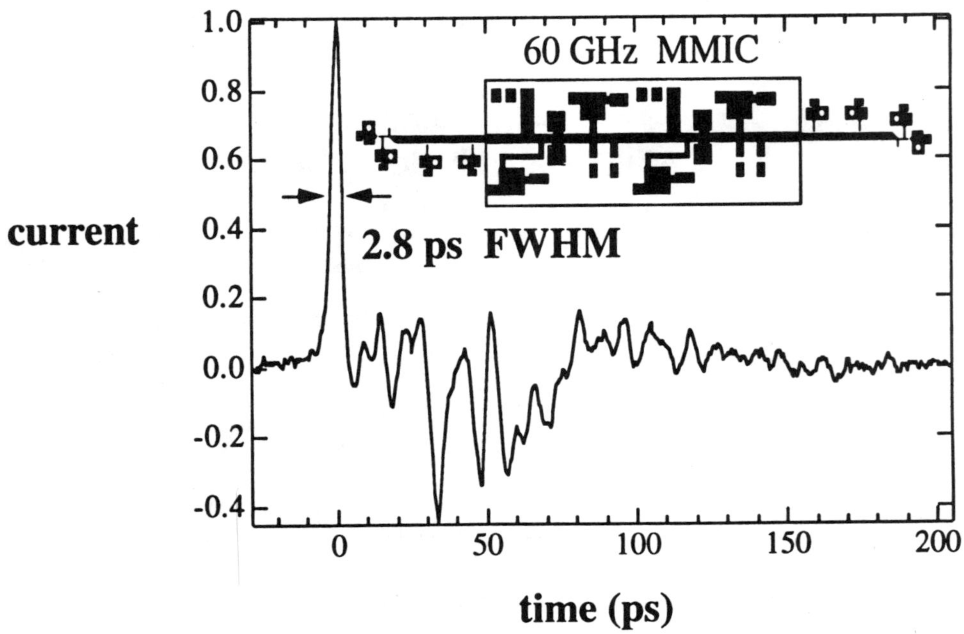

Fig. 10. Input and reflection waveform of the 60 GHz MMIC amplifier

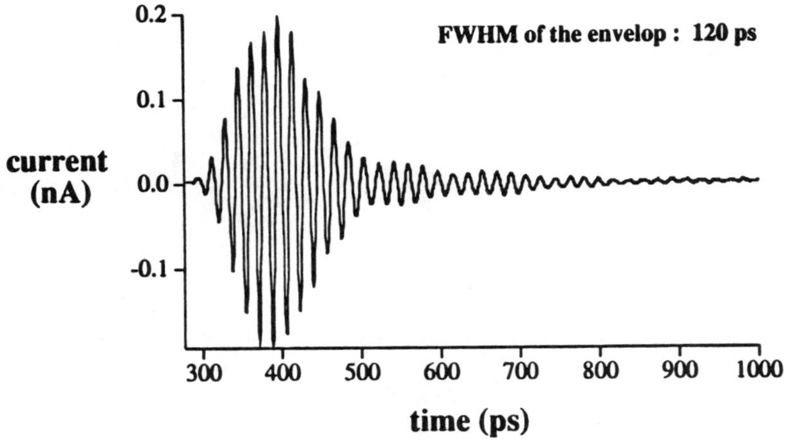

Fig. 11. The corresponding output waveform.

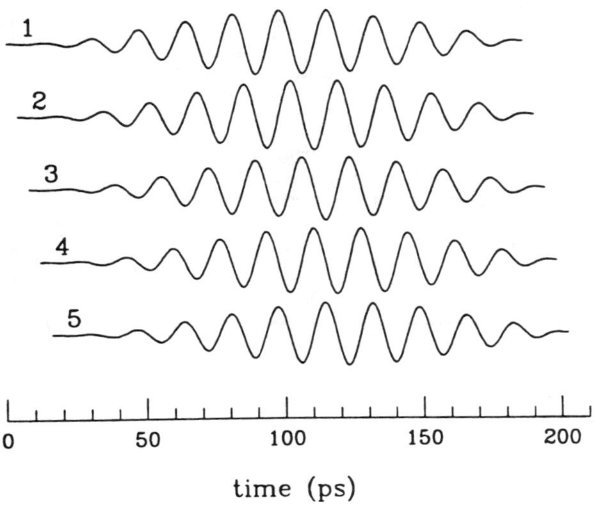

Fig. 12. True time delay of the broadband 60-GHz signals, successive time delay on the onset of 60-GHz signal is achieved by delaying the arrival time of the optical pulses.

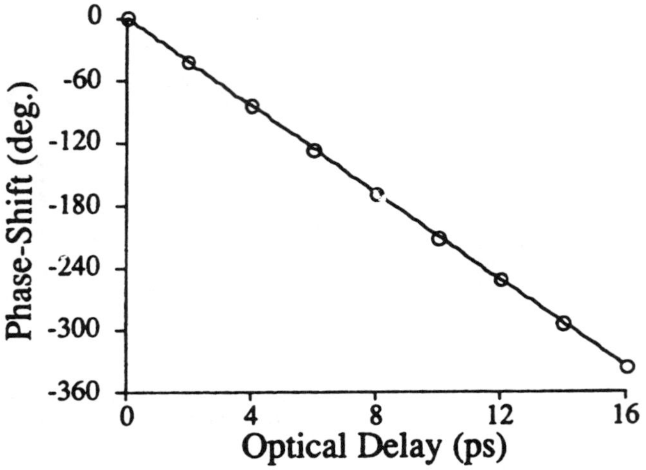

Fig. 13. Optically controlled phase shift of the millimeter-wave signals at the center frequency of 60 GHz.

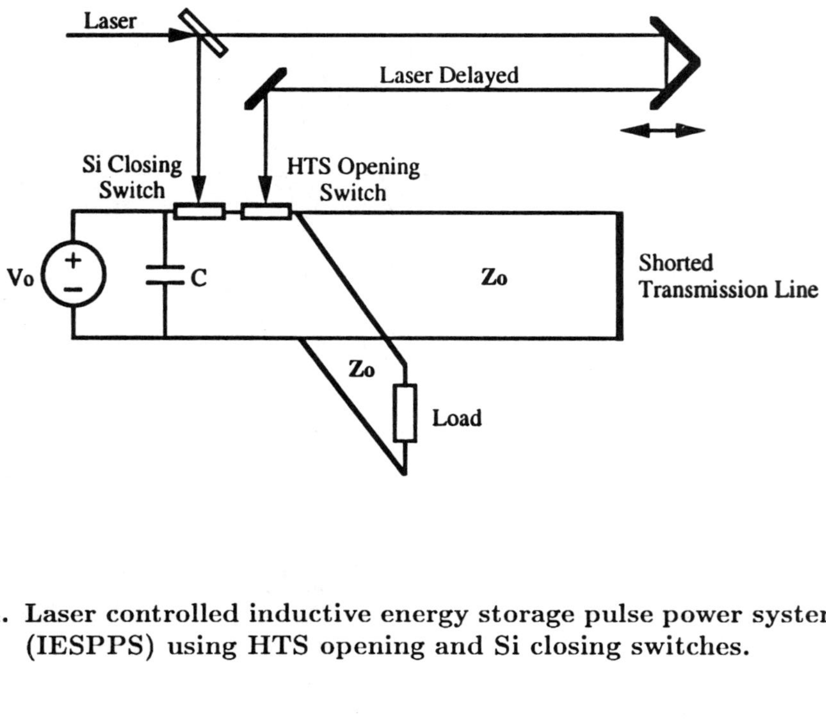

Fig. 14. Laser controlled inductive energy storage pulse power system (IESPPS) using HTS opening and Si closing switches.

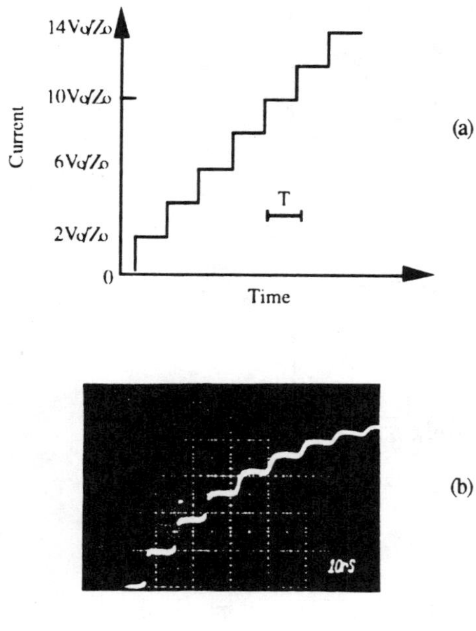

(T = 16.5 ns)

Fig. 15.
Stair-step charging current buildup: (a) theory; V_o, charging, voltage; Z_o, characteristics impedance of the transmission line; T, round-trip transit time for the current wave traveling in the transmission line. (b) experiment.

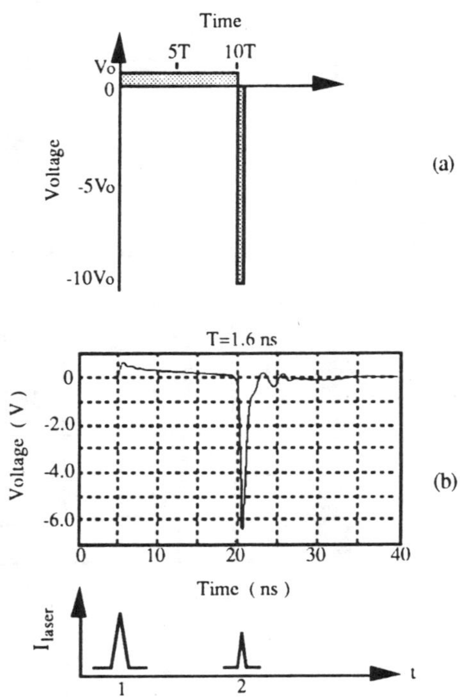

Fig. 16.
Electrical pulse with power gain obtained from the IESPPS with HTS opening switch: (a) theory; V_o, charging voltage; T, the round trip transit time; (b) upper trace, experimental result; lower trace: 1, laser pulse to close the Si-switch and 2, laser pulse to open the HTS-switch.

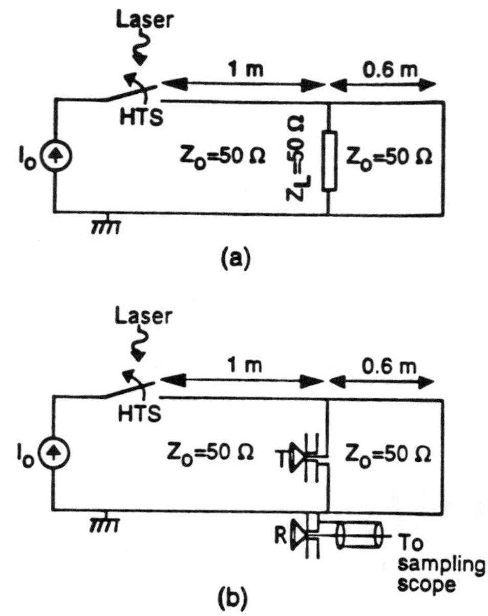

Fig. 17. Experimental setup for ultrawideband radiation.

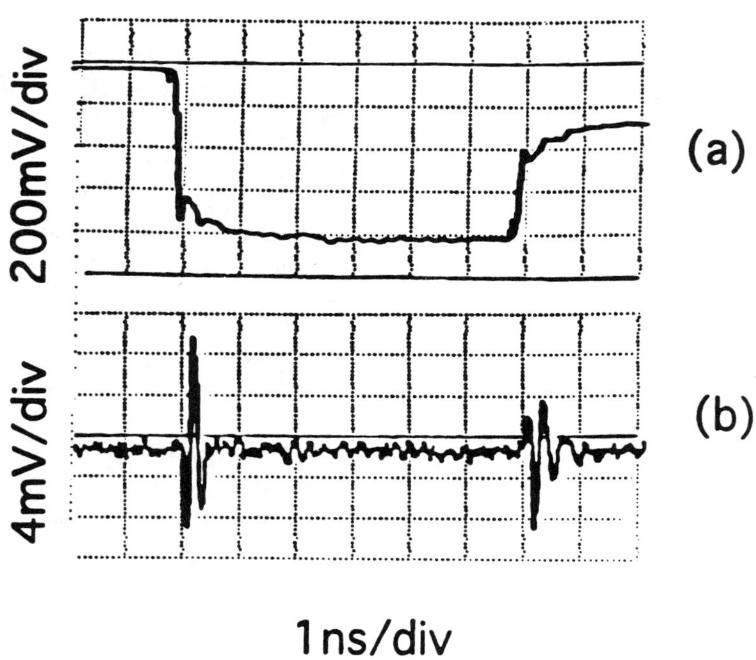

Fig. 18. (a) Pulse delivered to 50 Ω load from pulse forming network, (b) received voltage waveform.

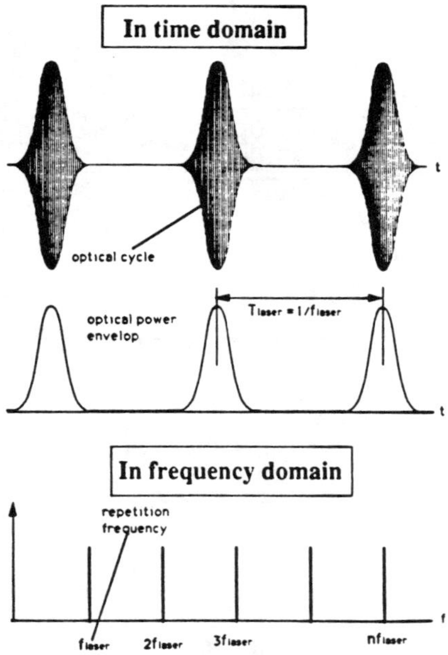

Fig. 19. Comb signals generated by the mode-locked laser.

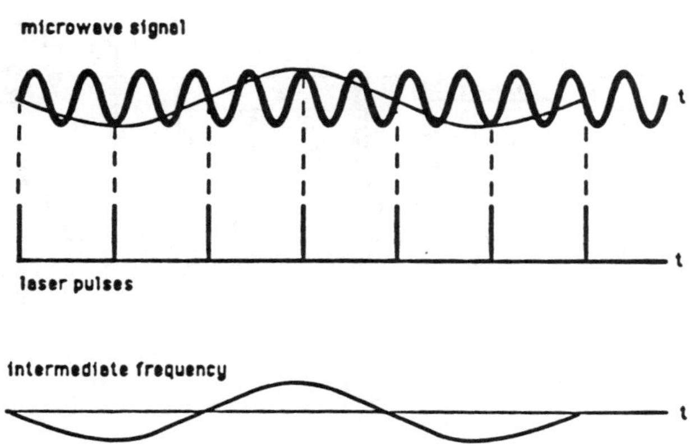

$$F_{RF} = NF_{Laser} \pm F_{IF}$$

Fig. 20. Time domain relationship between microwave signal, optical pulse train and IF waveform.

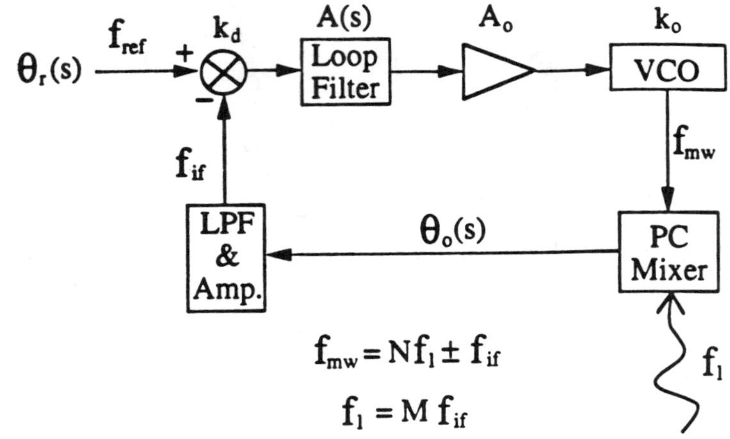

Fig. 21. Schematic diagram of the OMPLL; f_{mw}: VCO freq; f_q: laser repetition rate; $\theta_r(s)$, phase of reference f_{ref}; $\theta_o(s)$: phase of intermediate freq, f_{if}; k_d; gain of phase comparator; k_o, VCO gain; A_o: amplifier gain; $A_o(s)$, transfer function.

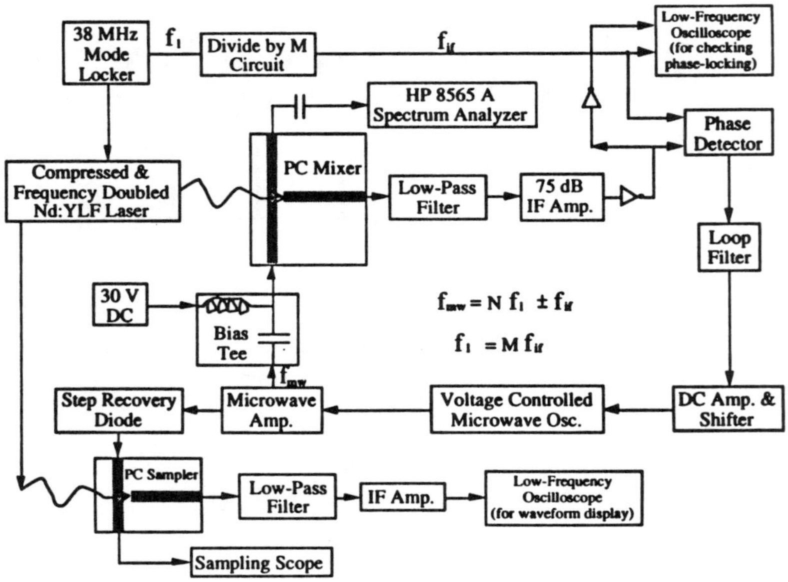

Fig. 22. Broadband and real-time waveform sampling set-up using OMPL technique.

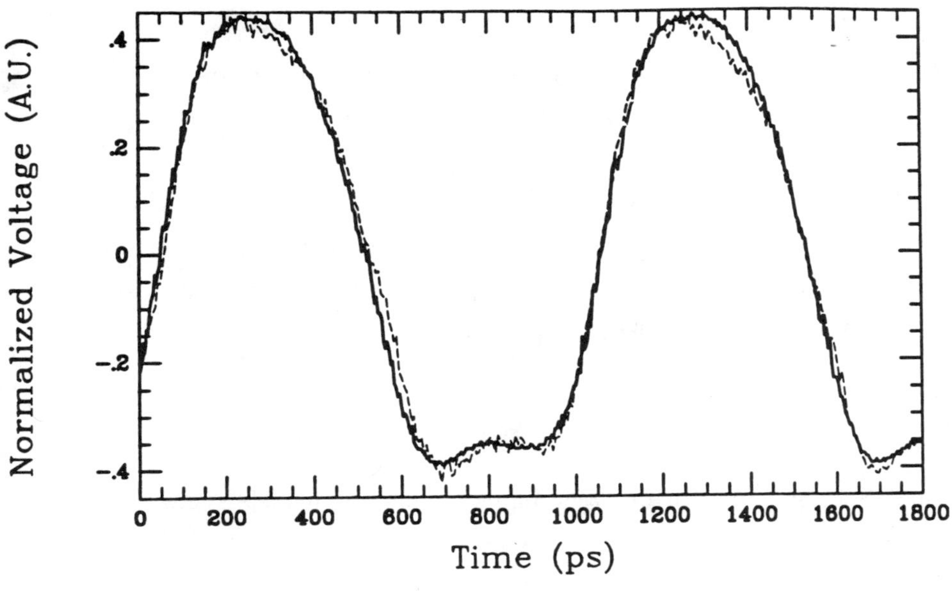

Fig. 23. Measured waveform comparison using a conventional scope (solid-line) and the OMPL technique (dashed line).

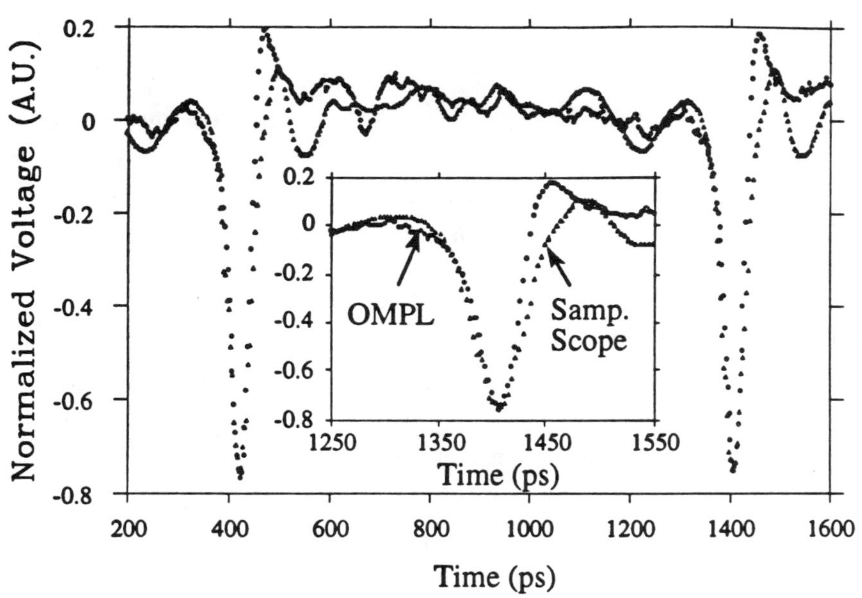

Fig. 24. Measured waveform comparison using a conventional sampling scope and the OMPL technique. Inset shows a time scale enlargement at 1400 ps.

Modeling and design of millimeter-wave components in the time domain with TLM techniques

Wolfgang J.R. Hoefer

NSERC/MPR Teltech Research Chair in RF Engineering
Dept. of Electrical and Computer Eng., University of Victoria
Victoria, B.C., Canada V8W 3P6

ABSTRACT

This paper provides an overview of techniques for modeling and simulation of millimeter-wave components with the Transmission Line Modeling (TLM) method. It is divided into three parts: Theoretical and algorithmic foundations, computer implementation and validation, and design/optimisation.

1 INTRODUCTION

In recent years, discrete time domain modeling of electromagnetic structures has reached a high level of sophistication. The increasing prominence of time domain methods is due partly to the success in solving a growing number of complex structures with this approach, and partly to the rapid expansion of computer power available to researchers and practitioners alike. These developments have resulted in considerable progress towards the realization of design tools based on time domain field analysis.

Two distinct time domain solution approaches are presently used. One is the family of Finite Difference - Time Domain (FD-TD) methods which can be derived by discretizing Maxwell's or Helmholtz's equations and thus, evolve naturally from classical electromagnetic theory. Among them, the approach formulated by Yee[1] in 1966 is by far the most frequently used by microwave engineers. It is based on a discretization of Maxwell's two curl equations which are solved stepwise in a leapfrog fashion. In contrast, Transmission Line Modeling (TLM) - invented by Johns and Beurle[2] in 1974 - employs a spatial transmission line network or mesh; its wave properties emulate those of continuous space as the mesh discretization approaches the infinitesimal limit. Time domain modeling is performed by exciting the TLM network with Dirac-like voltage impulses (time sampling) and tracking their propagation and scattering throughout the network on a computer.

FD-TD and TLM methods are similar, but the principal difference is that FD-TD schemes employ discretized differential (or integral) equations, while TLM is a network-based impulse scattering formulation. In all time domain methods, frequency characteristics are extracted from the time response by Fourier transform.

TLM algorithms have been described extensively in the literature. A detailed bibliography on TLM can be found in Ref. 3. In the following, we will focus on some recent innovations in TLM modeling, on the computer implementation and validation of TLM, and finally on microwave and millimeter-wave design, optimisation and synthesis techniques using TLM.

2 THEORETICAL AND ALGORITHMIC FOUNDATIONS

The theoretical foundations of TLM include general 2D and 3D TLM algorithms for discretized time domain representation of electromagnetic fields, techniques for modeling dispersive and nonlinear boundaries and materials in the time domain, and extraction of circuit parameters such as S-parameters, from time domain field simulations.

2.1 Generalized TLM algorithms

Traditional TLM algorithms are based on a square discretization of space for 2D, and cubic discretization for 3D field modeling. Hence, continuous space is modeled by a 2D square, or a 3D cubic lattice of transmission lines interconnected at so-called nodes. Dirac-like voltage impulses are then injected into the network. Their scattering at the nodes and spreading throughout the lattice is tracked step by step by the computer, yielding the time response of the structure. Fig. 1 shows the unit cells of the three most frequently used TLM discretization schemes.

One of the limitations of traditional mesh algorithms is their restriction to square (in the 2D case) or cubic elementary cells. The aspect ratio of cells can, of course, be changed within limits by introducing impulse-delaying techniques in the form of stubs or impulse storage, but these techniques require additional computer resources and introduce unwanted dispersion errors. This is a drawback when modeling guiding structures in which the transversal and longitudinal components of the propagation vector differ considerably in magnitude. When discretizing such structures for numerical modeling one would thus prefer a scheme which allows for different discretization steps which are commensurate with the required resolution of the fields. Consider, for example, a microstrip line with rather small transverse dimensions and fine features such as finite metallizaton thickness and thin dielectric layers. In such a case the discretization must be much finer in the transverse than in the longitudinal directions. Traditional TLM schemes with square/cubic mesh size[2,3] become very uneconomical in such cases because they impose essentially the same mesh dimensions in both transversal and longitudinal direction.

This has motivated the development of new TLM models with mesh cells of arbitrary aspect ratio[4]. In two space dimensions, isotropic propagation space can thus be modeled by a rectangular anisotropic mesh as well as by a square mesh. To illustrate this point, Fig. 2 shows equivalent square and rectangular 2D TLM shunt meshes side-by-side (aspect ratios $\alpha = 1$ and $\alpha = 1/2$, respectively). The same ideas have been applied to 3D TLM schemes and will be published shortly.

The features of a generalized TLM mesh will be described briefly using the example of the 2D rectangular mesh (Fig. 2, right-hand side). The desired properties of this rectangular mesh are the following:

Property 1: Impulse Synchronism:

The velocities of impulses on the dispersionless link lines must be such that the transit time Δt is the same for all lines. This implies that the line velocities must be proportional to the line lengths:

$$\frac{v_x}{v_z} = \frac{k_z}{k_x} = \frac{\Delta x}{\Delta z} = \alpha \tag{1}$$

where α is the aspect ratio of the rectangular cell.

Property 2: Isotropic Wave Properties:

In the infinitesimal limit ($\Delta x,\ \Delta z \ll \lambda$), the plane wave network velocity v_n must be the same in x and z-directions (as opposed to the line velocities v_x and v_z which are different in view of the synchronism requirement). This implies:

$$v_{nx} = v_{nz} \quad and \quad k_{nx} = k_{nz} \tag{2}$$

This property must hold for all aspect ratios α. (At higher frequencies where $\Delta x,\ \Delta z$ are no longer very small compared to λ, these equalities fail due to dispersion errors which occur as a result of the finite mesh size).

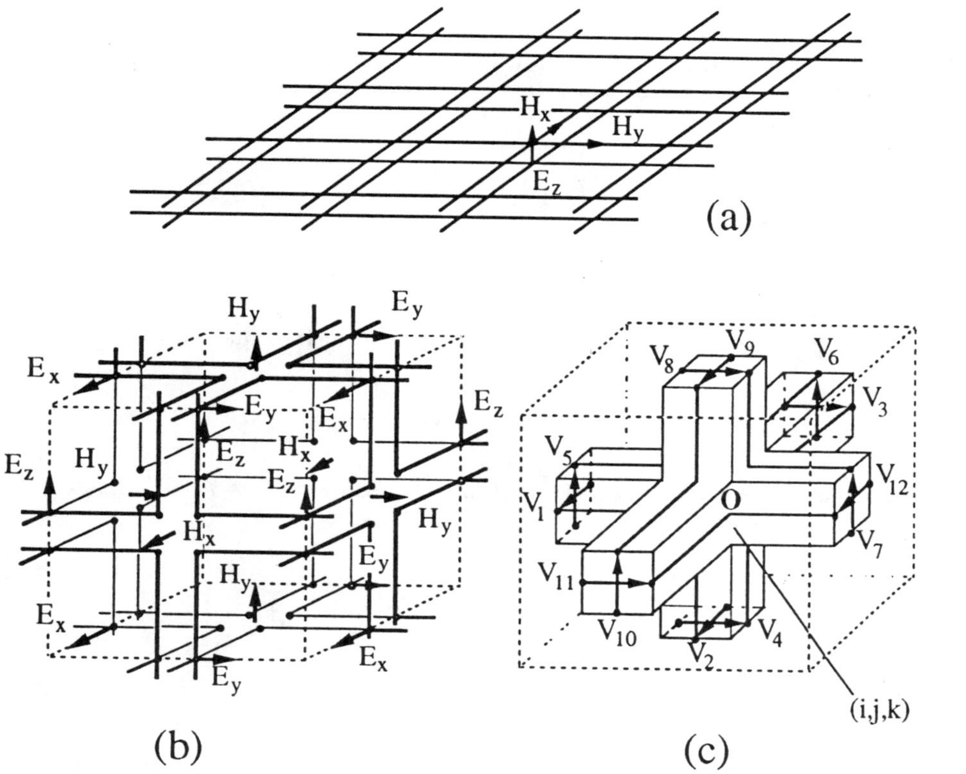

Fig. 1 Traditional 2D and 3D TLM nodes. (a) 2D shunt network with shunt nodes; (b) 3D distributed node; (c) 3D symmetrical condensed node

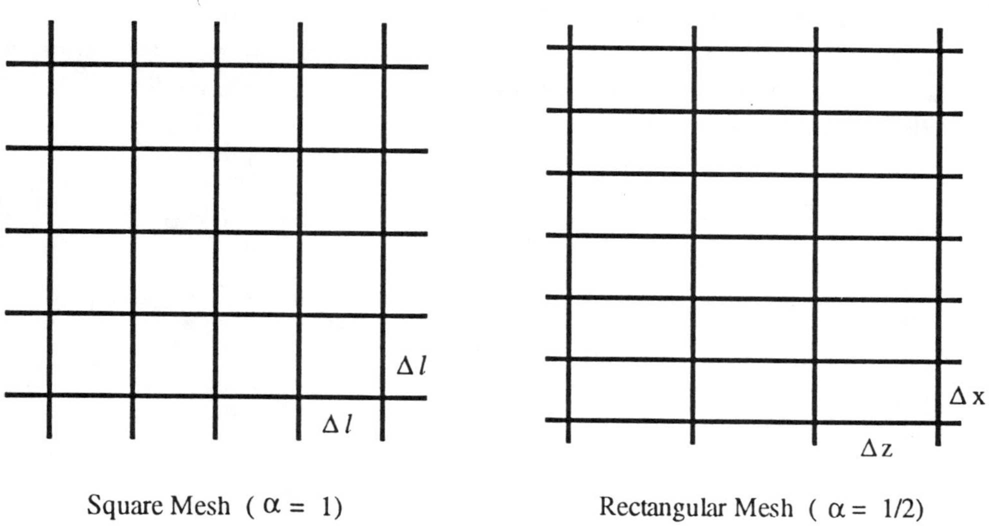

Fig. 2 Square mesh ($\alpha = 1$) and equivalent rectangular TLM mesh ($\alpha = \frac{1}{2}$). The aspect ratio is defined as $\alpha = \Delta x/\Delta z$

It can be shown that both conditions are satisfied if the inductances and capacitances per unit length, and the characteristic impedances of the mesh lines have the following ratios:

$$\frac{L_x}{L_z} = \alpha \quad ; \quad \frac{C_x}{C_z} = \frac{1}{\alpha^3} \quad ; \quad \frac{Z_{0x}}{Z_{0z}} = \frac{Y_{0z}}{Y_{0x}} = \frac{\sqrt{L_x C_z}}{\sqrt{L_z C_x}} = \alpha^2 \tag{3}$$

The general dispersion relation which governs the relationship between the propagation vectors in the network (k_{nx}, k_{nz}) and in free space (k_0), is

$$cos(k_{nx}\Delta x) + \alpha^2 cos(k_{nz}\Delta z) = (1+\alpha^2)cos(k_0 \Delta l) \tag{4}$$

where Δl is the equivalent square mesh parameter defined as

$$\Delta l = \Delta x\sqrt{\frac{2}{1+\alpha^2}} = \Delta z\sqrt{\frac{2}{1+\frac{1}{\alpha^2}}} \tag{5}$$

The dispersion for arbitrary propagation directions is illustrated in Fig. 3 (dispersion circle). The magnitude of the normalized propagation vector is plotted vs. the angle it forms with the z-axis. In the infinitesimal limit, the locus described by this vector is the unit circle. However, for a rather coarse discretization of $\Delta z/\lambda = 0.15$ the vector becomes larger in the axial directions, and more so along the longer mesh dimension Δz. For comparison, the same function has been plotted for the "*equivalent square mesh*" defined in Eq. (5) above. Note that there is no dispersion when the propagation vector forms an angle $\phi = arc\ tan\ \frac{1}{\alpha}$ with the z-axis. This can be verified by analyzing Eq.(4). For a square mesh ($\alpha = 1$) this angle becomes 45° as expected.

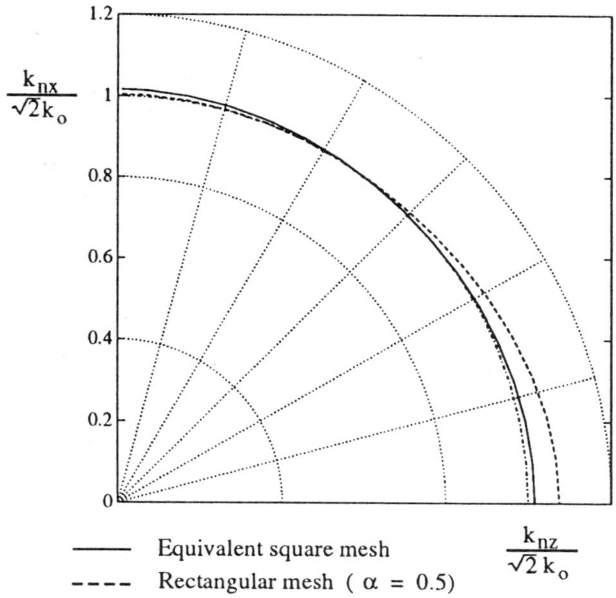

Fig. 3 Dispersion in arbitrary direction in a square and in a rectangular mesh ($\alpha = 0.5$, $\Delta z/\lambda_0 = 0.15$)

The extension of these modeling techniques to three dimensions is providing sufficient flexibility of discretization for efficient time domain modeling of most passive millimeter-wave structures and components.

2.2 High performance absorbing boundaries

In order to exploit the full inherent bandwidth of a TLM simulation, high performance absorbing boundary conditions are crucial. "High performance" means that such boundaries must be numerically stable, and capable of absorbing hybrid fields over a very wide frequency range with a very low return loss. Two types of boundary conditions have been developed and tested extensively for TLM applications: a) Multi-modal Johns matrix absorbing boundaries for separable structures such as homogeneous waveguides[5]; b) Second and third-order one-way boundary conditions based on Higdon's scheme for non-separable structures[6,7]. Examples are inhomogeneous waveguides such as microstrip, finline and coplanar lines, or antenna problems.

2.2.1 Multi–modal Johns Matrix absorbing boundaries

The method of absorbing the dominant mode in a waveguide using the numerical convolution of the incident impulses at the center of the waveguide with a modal impulse response of a semi-infinite waveguide has been described in detail in [8,9], and implemented in the 2D-TLM simulator software included in Ref. 8. This is possible because in uniform waveguides with homogeneous dielectric the transverse field distribution is independent of frequency and hence, of the time behaviour of an electromagnetic signal. This approach can be extended to higher order modes as well. In a linear environment, an overmoded signal can therefore be separated into modes which may be absorbed individually by an appropriate boundary condition. The dominant and higher order modal impulse responses (modal Johns matrices) are determined either by a recursive TLM analysis of a waveguide slice [9], or by discretizing the analytical Green's function of the structure (if it is known).

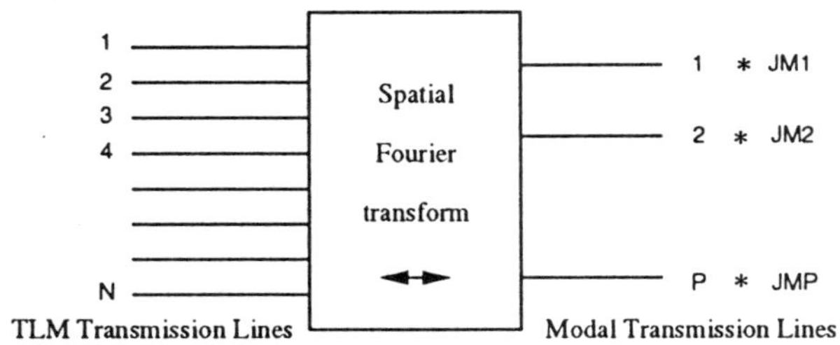

Fig. 4 Network representation of the transformations between the TLM transmission lines in the physical structure and the fictitious uncoupled modal transmission lines which are terminated by modal Johns matrices

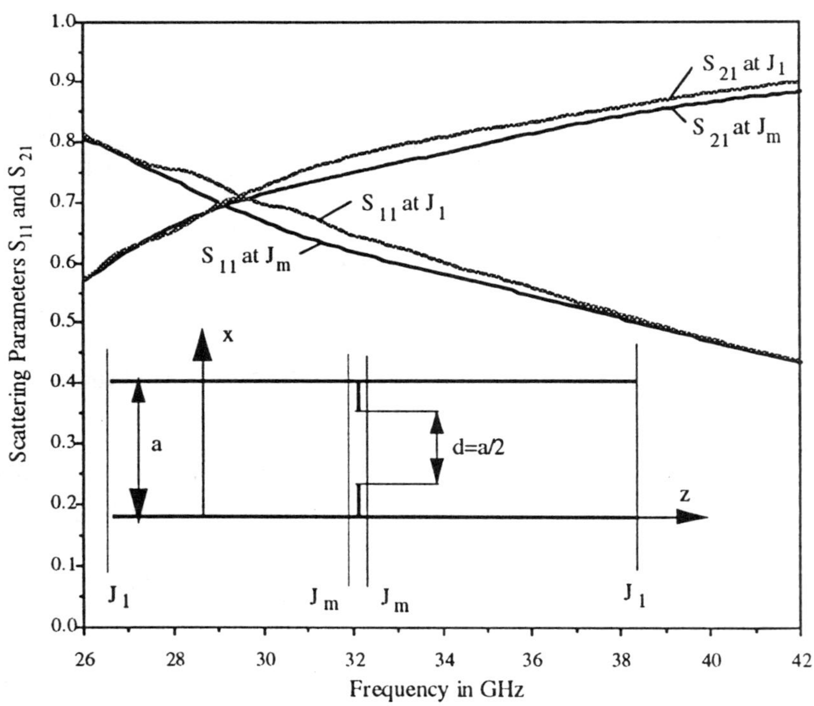

Fig. 5 Scattering parameters of an inductive iris in rectangular waveguide using single-mode Johns matrix ABC's at reference planes J_1 and multi-mode Johns matrix ABC's at J_m.

To this end the field incident upon the boundary is spatially Fourier transformed to obtain the amplitudes of the modes contained in it. Each modal impulse stream is then convolved numerically with its corresponding Johns matrix term, and the result is recombined and injected back into the TLM computational domain at the boundary [5]. The process is schematically represented in Fig. 4.

Since the modal Johns matrix is diagonal, this procedure is extremely efficient. Furthermore, such a boundary may be placed very near a discontinuity since it absorbs higher order modes even if they are below cutoff. In this way the computational domain can be dramatically reduced (typically by a factor ten) in return for inclusion of a few higher order modes in the boundary model. Fig. 5 demonstrates such an arrangement and shows S-parameters computed for an inductive iris in a rectangular waveguide[5].

2.2.2 One–way absorbing boundary conditions

Structures with inhomogeneous cross-sections such as partially dielectric-filled transmission lines cannot be terminated in this way since the modal transverse field distributions are frequency dependent. Structures of this type are best terminated with absorbing boundaries modeled by one-way equations known from Finite Difference – Time Domain models. We have adapted Higdon's scheme[11] for TLM applications. However, we apply the boundary operators to the reflected impulses rather than the total field values in front of the boundary. In the 3D Condensed Symmetrical Node scheme, it can be applied simultaneously to both polarizations of the boundary arm, resulting in the boundary condition being imposed simultaneously on the tangential electric and magnetic fields. One of the principal problems affecting one-way boundary conditions in Symmetrical Condensed Node TLM is the presence of low-frequency spurious modes which are excited at discontinuities and which are amplified by the boundary operators due to their high spatial frequency. We have found that the numerical stability of the differential boundaries could be considerably improved by a judicious determination of their characteristic parameters[6].

The performance of a third-order Higdon-type absorbing boundary in a microstrip line is demonstrated in Fig. 6. The boundary has a return loss better than 30 dB from 0 to 200 GHz, a range that exceeds by far the dominant operating range of the line. It is thus appropriate for absorbing very short transients and for nonlinar modeling where higher harmonic frequencies must be properly terminated. Such boundaries also work well in waveguides of homogeneous cross-section, and they may be used as free-space radiating boundaries in antenna modeling[7]. Fig. 7 shows the return loss of a microstrip patch antenna computed with variable-mesh SCN 3D-TLM and employing a stabilized second order Higdon absorbing boundary (see Ref. 7 for details of the computation]. The TLM result is in good agreement with results obtained by Wu et al.[12] with the spectral domain technique.

2.2.3 Refinements in boundary modeling

Since the dispersion of the three-dimensional condensed node mesh is almost negligible, it is preferable to maintain a uniform Cartesian grid throughout the computational domain and modify the properties of the mesh only in the vicinity of features that create strong local non-uniformities, or that are not conformal to the grid. Examples are sharp corners, edges, and slanted or curved boundaries.

Sharp edges and corners are the principal sources of so-called coarseness error in the TLM analysis of structures. Since this error affects all strip and slot-type lines it is particularly serious. It is due to the insufficient resolution of the edge field by the discrete TLM network resulting in a lack of "stiffness" at the edge, and results obtained are always shifted towards lower frequencies. Traditionally, the remedy for this problem was to use a finer mesh in the vicinity of the edge, but this introduces additional complications and computational requirements. Two possible solutions have been proposed recently. One amounts to adding a branch to the node situated in front of a corner to provide a direct interaction of this node with the boundary[14]. A second is to place a node directly into the edge or corner. Both approaches require the modification of the impulse scattering matrix of the corner node, which represents only a minor increase in computational resources, particularly when parallel processing is used. The effect of the first type of corner correction is demonstrated in Fig. 8 which shows typical results for the first resonant frequency of a cavity containing a sharp edge as a function of the mesh parameter Δl. The parameter p is proportional to the fraction of power carried by the additional branch of the corner node and is equal to half the characteristic admittance of the corner branch when normalized to the link line admittance . For $p = 0$ (no corner correction) the coarseness error increases almost linearly with increasing mesh size while for $p = 0.1$ the frequency remains accurate even for a very coarse mesh.

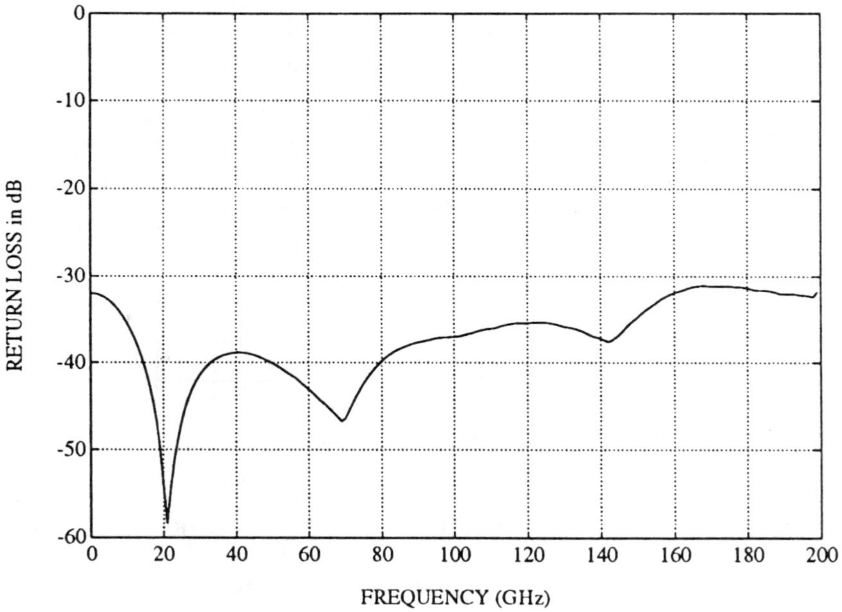

Fig. 6 Return loss of a third-order Higdon absorbing boundary condition in an open microstrip with ϵ_r = 10.2, w = 0.508 mm, h = 0.635 mm).

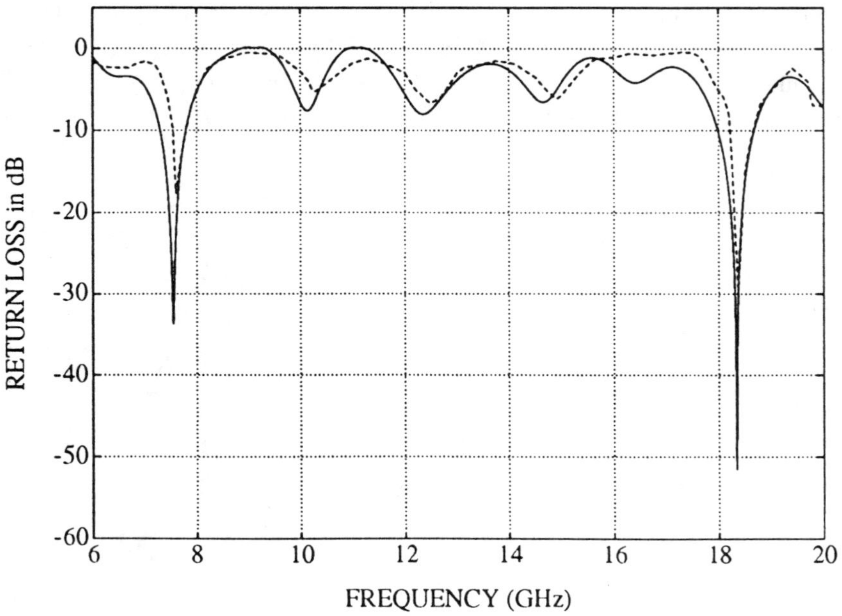

Fig. 7 Return loss of a microstrip patch antenna using second-order Higdon absorbing boundary conditions in the microstrip feed and the remaining walls[9]. TLM results agree well with spectral domain results (dotted line) by Wu et al. [12].

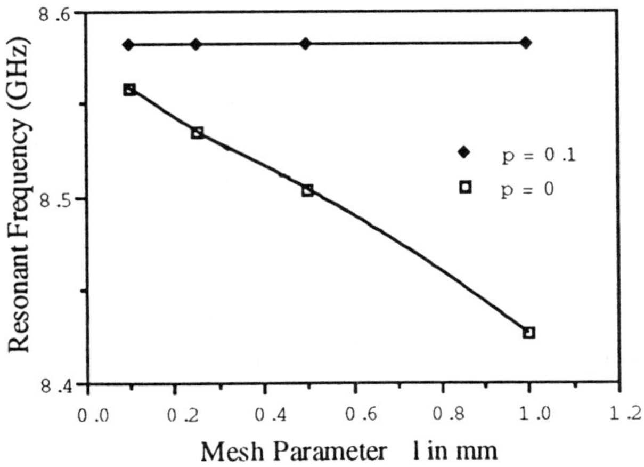

Fig. 8 Effect of corrective corner branch in a 2D-TLM simulation of a sharp edge.

Regarding the positioning of boundaries in the discrete TLM mesh, accurate modeling of millimeter-wave components, discontinuities and junctions requires a precision in the positioning of boundaries that is identical to, or better than the manufacturing tolerances. If boundaries can only be introduced either across nodes or halfway between nodes, then the mesh parameter would have to be very small indeed, leading to unacceptable computational requirements. Similar considerations apply when curved boundaries with very small radii of curvature must be modeled. It is therefore important to provide for arbitrary positioning of walls. The basis for this feature has been described already in 1973 by Johns and Slater[13] who, at the time, thought that the advantage of this procedure over stepped contour modeling was too small to warrant the additional complexity of the algorithm. However, this

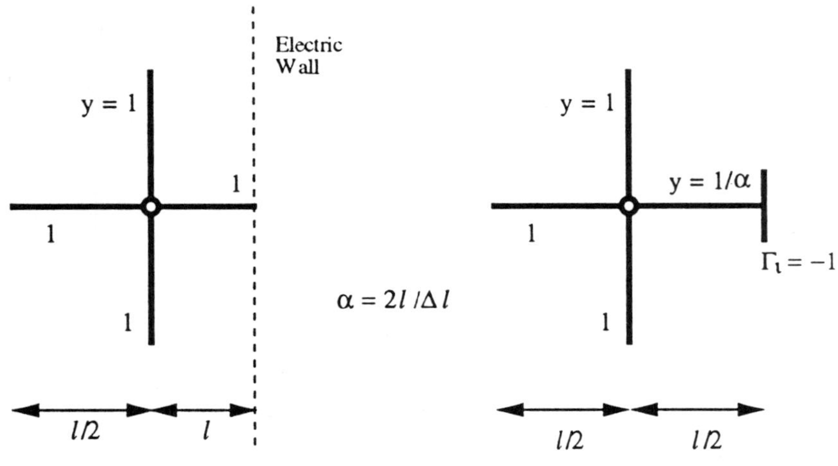

Fig. 9 Modification of boundary node for arbitrary wall positioning.

is not true when analyzing, for example, narrowband waveguide components such as filters. Fig. 9 shows the concept of arbitrary wall positioning in 2D TLM. The boundary branch which has a length different from $\Delta l/2$ is simply replaced by an equivalent branch of length $\Delta l/2$ having the same input admittance. This ensures synchronism, but requires either a different characteristic admittance for the boundary branch and hence, a modification of the impulse scattering matrix of the boundary node [8], or some additional signal processing in the form of a recursive reflection algorithm [9]. In this way, the boundary branch can be continuously tuned over a range of more than one mesh parameter length Δl without appreciable error. This important technique has definitely removed the restriction that dimensions of TLM models can only be integer multiples of the mesh parameter. Since the position parameter $\alpha = 2l/\Delta l$ can be continuously varied even during a simulation, this feature can also be used to model moving boundaries[15–17].

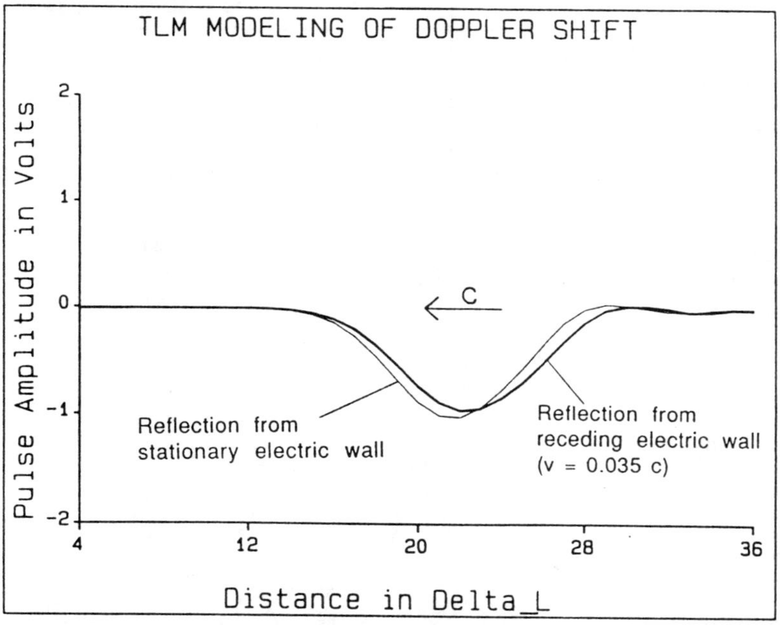

Fig. 10 Reflection of a Gaussian pulse from a moving (receding) electric wall, simulated with TLM.

Fig. 10 shows the delay and broadening of a Gaussian pulse reflected by a conducting boundary that is moving away from the source at a speed of 0.035c. The modification of the pulse shape results in the Doppler shift of its spectrum. These techniques of modifying wall positions during a TLM simulation can be used for boundary adjustments in optimization problems.

3 COMPUTER IMPLEMENTATION AND VALIDATION OF TLM

TLM being computationally intensive, it is essential to properly implement the algorithms to ensure optimal use of computational resources. Furthermore, a user-friendly interface and compatibility with other CAD tools are important. Programming is thus performed preferably in C++ on computers which incorporate special graphic processing hardware in order to free the main memory for TLM computation. Particularly impressive computational speeds can be obtained with SIMD-type massively parallel processors[18] such as the Connection Machine or the DEC 12000 (MasPar). In 2D simulations, each processor can be associated with one TLM node, and all nodes are updated simultaneously, while in 3D the computer can process one vertical or horizontal slice of the structure at a time. Typically, a parallel computer with 8K processors performs 250,000 3D scatterings per second.

Computers	CPU time in seconds	2D scatterings/second
Toshiba PC, Model T5200/100	6,500	5,000
IBM Model 90 XP486	1,250	26,000
DEC RISC, Model 5100	352	93,000
IBM RS6000, Model 350	117	280,000
HP 9000, Series 7000, Model 755	88	370,000
DECmpp 12000 with 8K processors	12	2,730,000

Table 1 Execution time and scatterings per second for a 2D-TLM simulation on various computers. The problem comprizes 128 x 64 mesh cells and 4000 iterations.

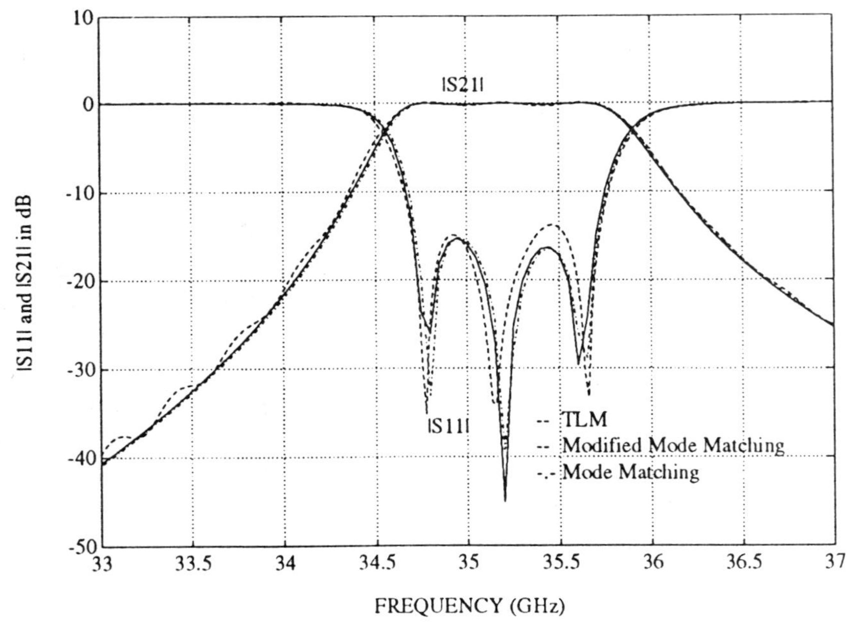

Fig. 11 S-Parameters of waveguide filter computed with TLM and mode matching.

Table 1 compares the performance of various well-known computer platforms when solving a 2D problem with 128x64 mesh cells and 4,000 iterations.

Validation of computer codes is an important issue. To demonstrate the accuracy that can be achieved with TLM, Fig. 11 shows the S-Parameters of a waveguide filter computed with TLM, together with results computed with two different mode-matching methods[19]. These results are in excellent agreement.

4 CIRCUIT DESIGN AND SYNTHESIS WITH TLM

Design and synthesis of electromagnetic structures are the ultimate engineering applications of a modeling technique. Two possible approaches to this theme have been demonstrated. The first – more classical – approach combines time domain field analysis controlled by an optimization program. By using datapipe techniques the TLM analysis can be performed on a powerful processor such as MasPar or the Connection Machine [20], or the task can be divided among several workstations[21] connected via a network. The other approach makes use of time reversal to synthesize a structure geometry directly from its desired frequency response. This process is similar to inverse time domain scattering.

4.1 Design by analysis and optimization

Fig. 12 sketches an arrangement which combines a commercially available CAD pacage, OSA90/hope^{TM}, with a TLM Electromagnetic Field Simulator[20]. The optimization is performed on a workstation which pipes updated geometrical data to a massively parallel computer. The new structure is then analyzed in the time domain with TLM, S-parameters are extracted by Fourier transform and piped back to the optimizer. This process permits design without the need for equivalent circuit representation and includes all parasitic effects. It is also much more flexible as far as geometrical complexity is concerned.

To demonstrate his process the design of a mm-waveguide bandpass filter has been performed with such an arrangement. Fig. 13 shows the geometry of the filter. Fig. 14 shows the initial response, the response after 30 random optimization steps, and the final response after 180 random optimization steps.

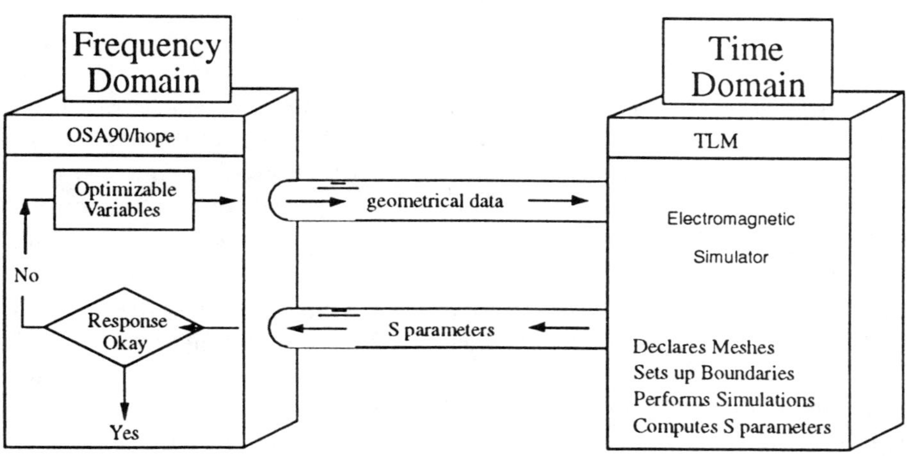

Fig. 12 Design by TLM electromagnetic field analysis controled by an optimization program: OSA90/hope^{TM} and TLM communicate through UNIX high speed datapipe

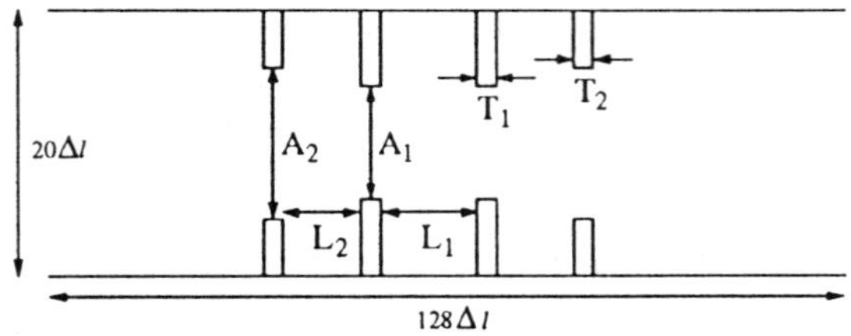

A bandpass filter in the WR28 rectangular waveguide Δl = 0.3556 mm. Using single precision floating operation, the mesh requires 50K of RAM.
The optimization goal is:

30.0 to 32.0 GHz step 0.1 GHz	S11=1	S21=0	weight=1
32.5 to 33.5 GHz step 0.1 GHz	S11=0	S21=1	weight=2
34.0 to 36.0 GHz step 0.1 GHz	S11=1	S21=0	weight=1

Fig. 13 Top view of an iris-coupled waveguide bandpass filter optimized with the arrangement in Fig. 12

Each step takes about 30 seconds on a RISC workstation, yielding a complete field analysis and the S-parameters of the filter. While there are faster and more efficient methods available for the analysis of this type of structure, the computational effort of the TLM simulator is virtually insensitive to the complexity of the structure, provided that the number of discretizing cells remains the same. The method thus combines accuray with maximum flexibility and generality as far as the geometry of the structure is concerned.

It is also worth mentioning that several measures allow us to accelerate the analysis if the bandwidth of the structure is narrow. Firstly, the excitation can be bandlimited by injecting, for example, a sine function weighted with a Gaussian envelope. This enhances convergence and reduces truncation error. Secondly, signal processing of the TLM signal, such as Prony's method, can be used to extract the S-parameters directly from the time response without the need for a Fourier transform. This leads to a reduction in computation time by typically a factor five.

4.2 Synthesis by time reversal

The reconstruction of a scatterer from the scattered field is demonstrated in Fig. 15. A diamond-shaped conducting obstacle is placed in a parallel plate waveguide. A uniform plane wave excitation with a Gaussian time profile is injected into the structure from the left side. The time response is picked up at all points of the absorbing walls at the left and the right. From this response (total field solution) is subtracted the response of the empty guide to the same excitation (homogeneous solution), yielding the scattered field (particular solution). The latter is then re-injected in reverse time sequence at these boundaries into the empty waveguide, and the resulting field distribution in the guide is is processed, for example by finding the minimum of the Pointing vector profile in the computational domain. We can thus reconstruct the exact shape of the scatterer within the resolution of the TLM grid, as shown in Fig. 15.b.

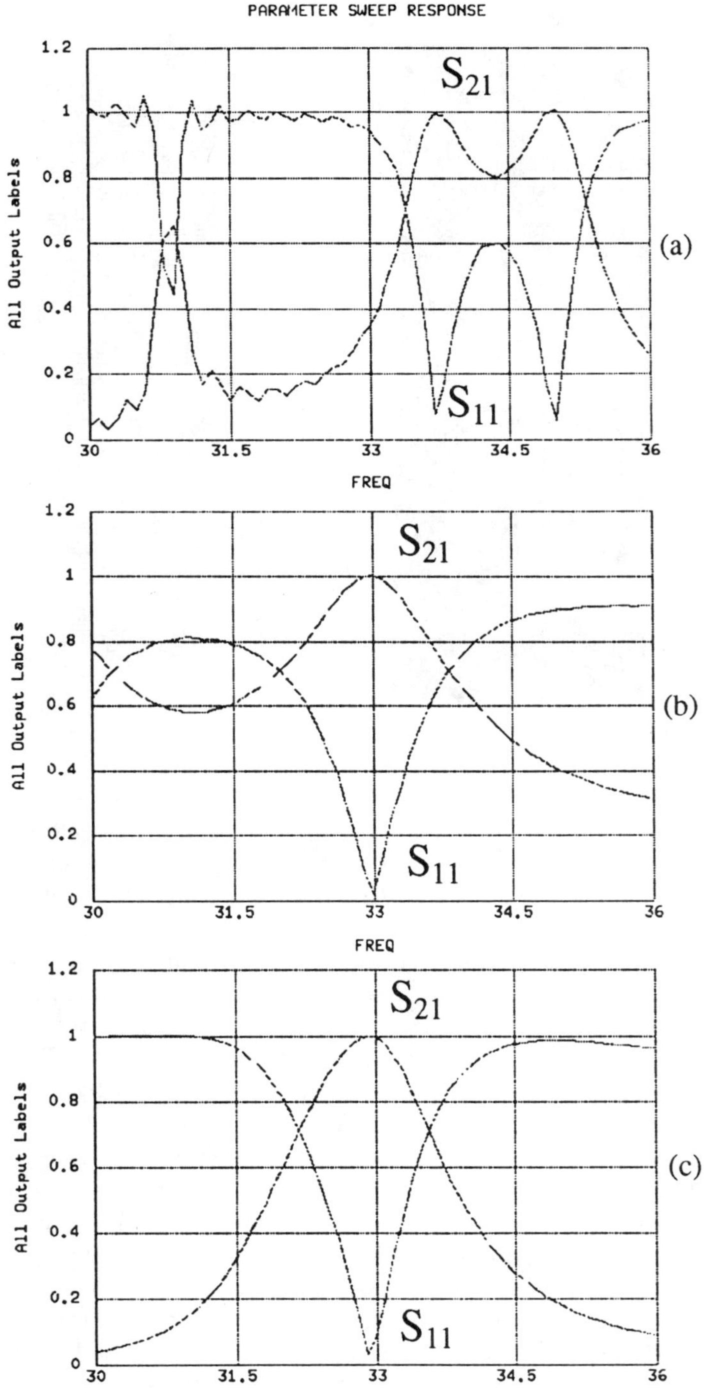

Fig. 14 Response of a mm-wave waveguide bandpass filter. a) Initial response, b) after 30 optimisation steps, c) final response after 180 steps.

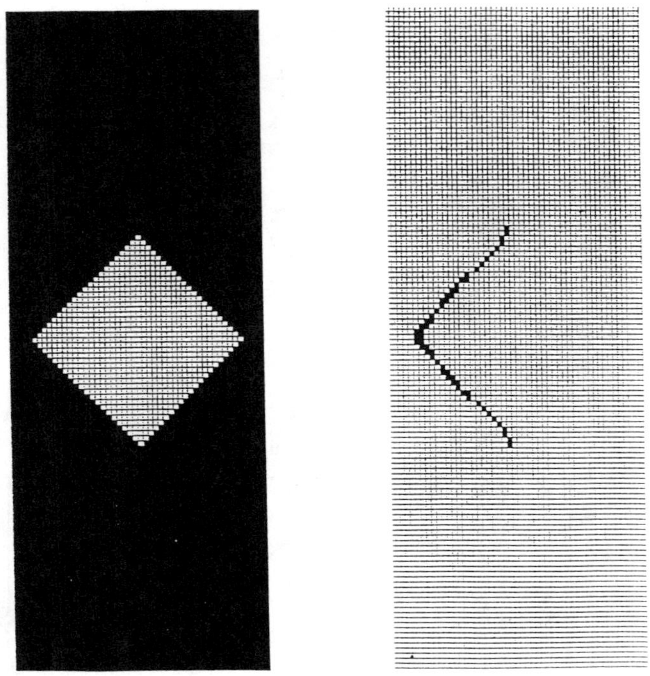

(a) (b)

Fig. 15 Example of shape reconstruction of a metallic obstacle in a parallel plate waveguide through time inversion. (a) Shape of the scatterer, (b) reconstructed shape of left side of the obstacle. The electric field is perpendicular to the paper.

The principle of synthesis of electromagnetic structures by time inversion is the following: Since the sources of a scattered field reside on the surface of the scatterer, this reverse process allows us to construct the geometry of scatterers from a desired scattered field response. This procedure is not unique, but the difficulties associated with the non-unicity of the inverse problem can be removed by selecting *a priori* an approximate topology which has the desired geometrical features of the structure we want to design. These features are then mapped into the time domain by a forward TLM analysis. The Fourier transform of this response is subsequently modified in the lower frequency range (operating range) of the structure, while the high frequency content (which defines the shape of the structure) is preserved. The updated and processed time response is then injected back in reverse time sequence into the computational domain to yield an improved topology. The analysis-synthesis cycle can be repeated until the desired response is achieved. Experience has shown that such a process converges very quickly, typically after two or three cycles only[22,23]. The flowchart of this process is given in Fig. 16. It is much faster und more efficient than optimization by repeated forward analysis. The process is best clarified by a simple example. Suppose we want to synthesize a shunt inductance in a parallel plate waveguide. The specification given to us are the S-Parameters of a shunt inductance in the frequency domain (Fig. 17.a)

There are many ways of realizing a shunt inductance in a waveguide - the solution is not unique. We must choose a basic topology. We select a centered thin transverse septum and guess its width required to yield the desired inductance, say 41 Δl (Fig. 17.b). The approximate dimensions can be found in many cases from closed-form expressions in the literature, linking the dimensions of discontinuities to their equivalent lumped element circuits.

The approximate structure is then analyzed forward-in-time, and the dominant mode content of the first response is extracted (Fig. 17.c). We see that S_{11} is already quite close to specifications, but S_{21} is not.

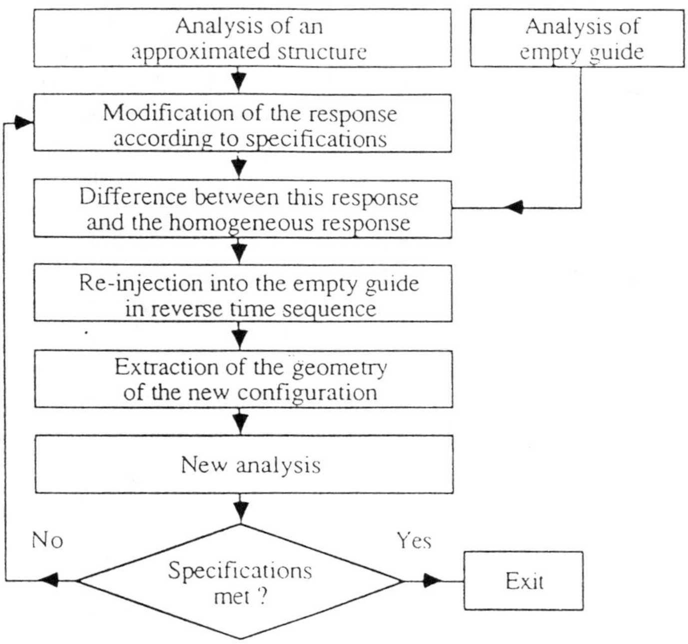

Fig. 16 Algorithm of the synthesis method using alternate forward and backward-in-time TLM processes.

We thus replace the low-frequency part of S_{21} by the desired (green) characteristic and keep the high frequency content the same. The modified total reponse is then converted back into the time domain and, reduced by the homogeneous response of the empty waveguide, re-injected into the computational domain in the inverse time sequence. The resulting synthesized scatterer geometry is an obstacle of width 31 Δl.

A new analysis of this obstacle yields the S-Parameters shown in Fig. 17.d. Obviously, the result is much better for S_{21} now. If we are still not satisfied, we may repeat the synthesis cycle again.

5 CONCLUSION AND FUTURE TRENDS

While the important issues and numerous recent developments in TLM modeling could only be scanned very briefly in this paper, it is obvious that TLM has evolved into a powerful, accurate and numerically efficient tool for general purpose electromagnetic analysis, design and synthesis. However progress is still being made on many fronts. The most dynamic areas of development are: improvements in the basic TLM algorithms, refinement of high performance boundary modelling, implementation on massively parallel computers, combination of TLM with optimizers for direct geometry/field based CAD, and synthesis of structures by time reversal. Other important developments which have not been discussed in this paper include signal processing to reduce the required number of computation steps, and modelling of nonlinear and dispersive materials as well as of semiconductors. Given the rapid growth in computer performance and the increasing sophistication of TLM modeling it seems likely that TLM based electromagnetic simulators will soon become highly versatile and powerful engineering design tools indeed.

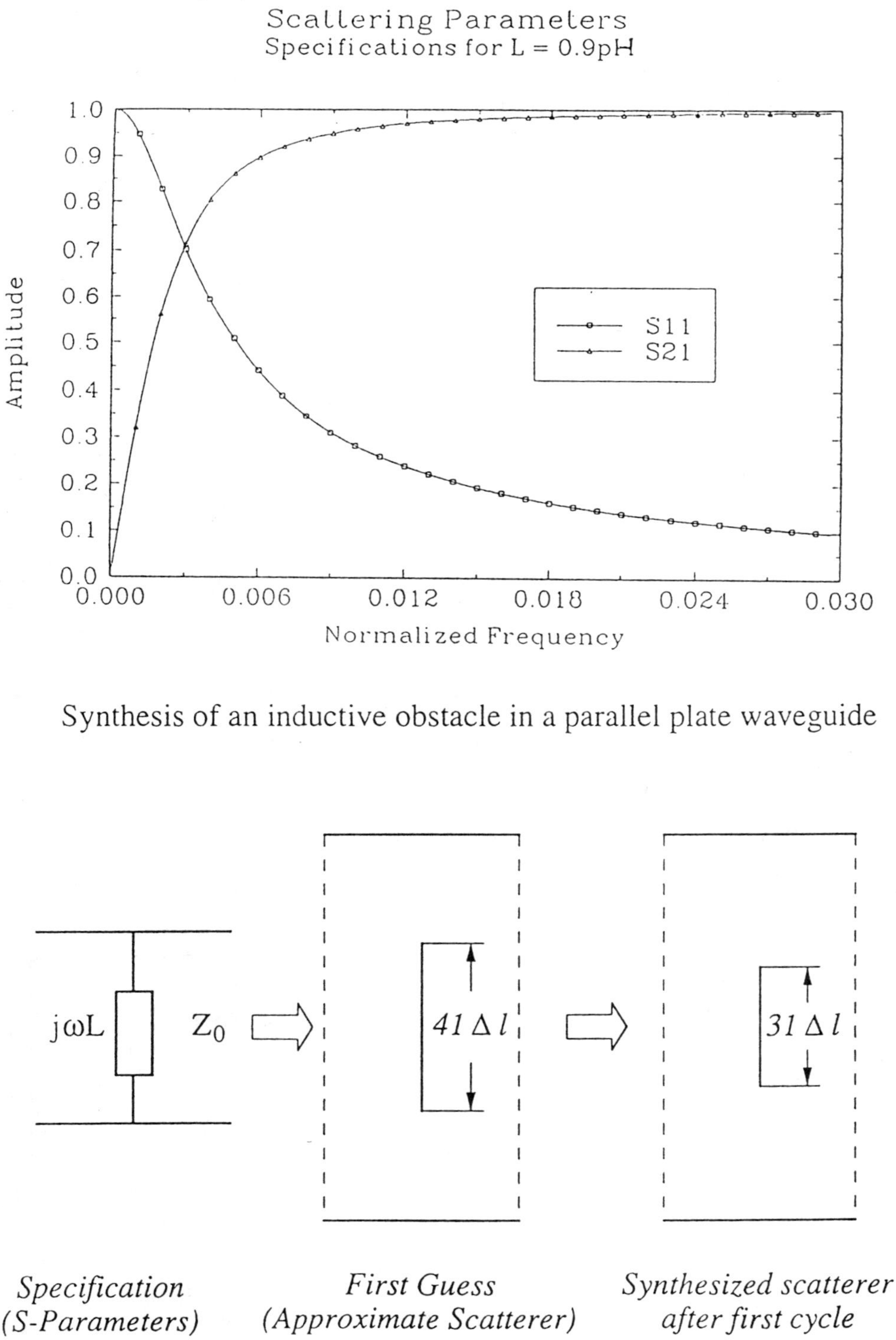

Fig. 17 (a) Desired S-parameters of a shunt inductance in a parallel plate waveguide; (b) possible topology for its practical realization.

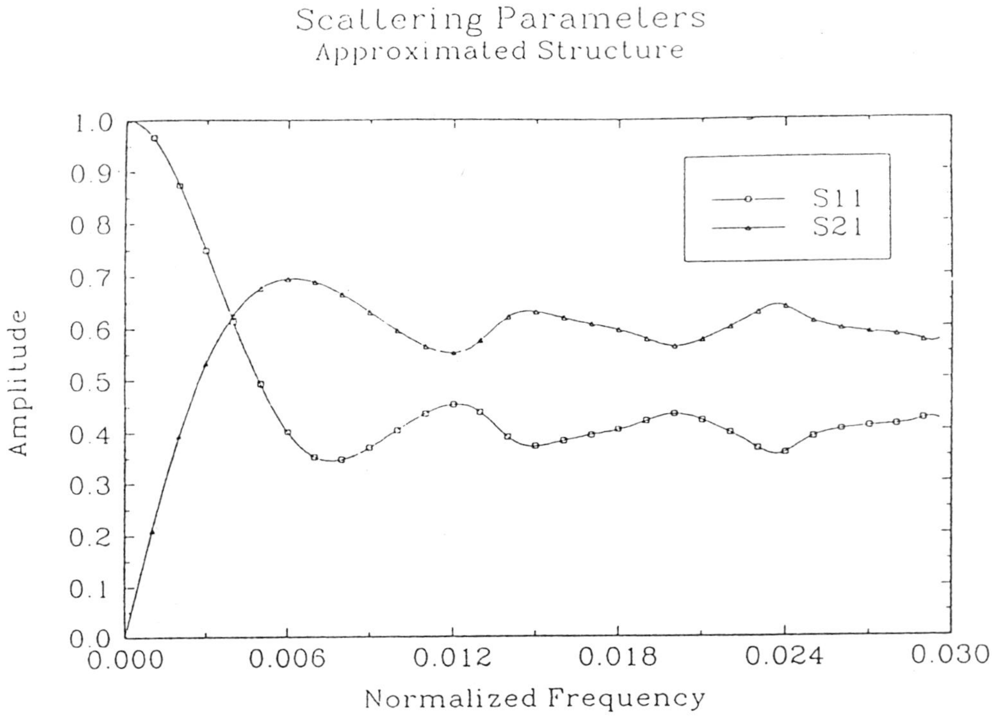

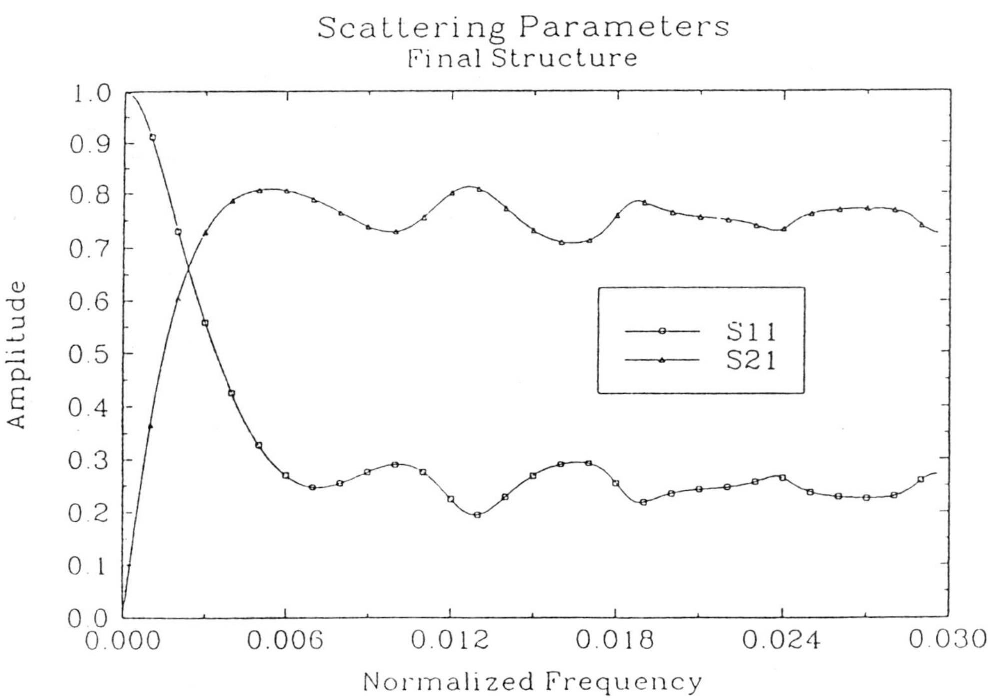

Fig. 17 continued (c) S-parameters obtained by analysis of the first guess of width 41 Δl ; (d) S-parameters obtained by analysis of the new geometry obtained by inverse time synthesis from the combination of the characteristics in Fig. 17.a and 17.c

6 ACKNOWLEDGEMENTS

Financial support by the Natural Science and Engineering Research Council of Canada, MPR Teltech, and the Science Council of British Columbia is gratefully acknowledged.

7 REFERENCES

1. K.S. Yee, "Numerical solution of initial boundary value problems involving Maxwell's equations in isotropic media", *IEEE Trans. Antennas Propagation*, vol. AP-14, no. 5, pp. 302-307, May 1966.

2. P.B. Johns, and R.L. Beurle, "Numerical solution of 2-dimensional scattering problems using a transmission-line matrix", *Proc. IEE*, vol.118, no.9, pp. 1203-1208, Sept. 1971.

3. W.J.R. Hoefer, "The Transmission Line Matrix (TLM) Method", Chapter 8 of "Numerical Techniques for Passive Microwave and Millimeter-Wave Structures," edited by T. Itoh, John Wiley & Sons, New York 1989, pp. 496-591.

4. W.J.R. Hoefer, and P. Sautier, "Characteristics of the General Rectangular 2D-TLM Network", *International Journal of Numerical Modelling*, vol 6., (in press).

5. M. Mongiardo, M. Righi, R. Sorrentino, and W.J.R. Hoefer, "Efficient TLM diakoptics for separable structures", *in 1993 IEEE Intl. Microwave Symp. Dig.*, pp. 425-428, Atlanta, Ga, June 14-18, 1993.

6. C. Eswarappa, W.J.R. Hoefer, "Wideband absorbing boundary conditions using third order boundary operators for time domain TLM analysis", *in 1993 IEEE AP/S Symposium and URSI Meeting Dig.*, pp. 10-13, Ann Arbor, MI, June 27 - July 2, 1993.

7. C. Eswarappa, W.J.R. Hoefer, "Absorbing boundaries for 3D-TLM modelling of microstrip patch antennas",*in 23rd European Microwave Conference Dig.*, pp. 291-292, Madrid, Spain, 6-9 Sept. 1993.

8. W.J.R. Hoefer, and P.P.M. So, "The Electromagnetic Wave Simulator, A Visual Electromagnetics Laboratory based on the 2D TLM Method", *John Wiley & Sons*, Chichester, U.K., 1991.

9. P.P.M. So, C. Eswarappa, and W.J.R. Hoefer, "A two-dimensional TLM microwave field simulator using new concepts and procedures", *IEEE Trans. Microwave Theory Techniques*, vol. MTT-37, no. 12, pp. 1877-1884, Dec. 1989.

10. M. Mongiardo, M. Righi, R. Sorrentino, and W.J.R. Hoefer, Rigorous and Fast Computation of Modal Johns Responses, *in Second International Workshop on Time Domain Modelling of Fields and Networks Dig.*, Berlin, October 28-29, 1993.

11. R.L. Higdon, "Numerical absorbing boundary conditions for the wave equation," *Math. Comp.*, vol. 49, no. 179, pp. 65-91, July 1987.

12. S. Wu, N.G. Alexopoulos, and O. Fordham, "Feeding structure contribution to radiation by patch antennas with rectangular boundaries", *IEEE Trans. on Antennas and Propagation*, vol. 40, no. 10, pp. 1245-1249, Oct. 1992.

13. P.B. Johns, and G.F. Slater"Transient Analysis of Waveguides with Curved Boundaries," *Electronics Letters*, vol. 9, no. 21, 18th Oct. 1973.

14. U. Mueller, P.P.M. So, and W.J.R. Hoefer, "The compensation of coarseness error in 2D TLM modeling of microwave structures", *in 1992 IEEE Intl. Microwave Symp. Dig.*, pp. 373-376, Albuquerque, NM, June 1-5, 1992.

15. F.J. German , "Infinitesimally adjustable boundaries in symmetrical condensed node TLM simulations",*in 9th Annual Review of Progress in Applied Computational Electromagnetics Dig.*, pp. 482-490, Monterey, CA., 22-26 March, 1993.

16. U. Mueller, A. Beyer, and W.J.R. Hoefer, "The implementation of smoothly moving boundaries in 2D and 3D TLM simulations", *in 1992 IEEE Intl. Microwave Symp. Dig.*, pp. 791-792, Albuquerque, NM, June 1-5, 1992.

17. U. Mueller, A. Beyer, and W.J.R. Hoefer, "Moving Boundaries in 2D and 3D TLM Simulations Realized by Recursive Formulas", *IEEE Trans. Microwave Theory Techniques*, Vol. MTT-40, No. 12, pp. 2267-2271, Dec. 1992.

18. So, P.P.M., Eswarappa, C., and Hoefer W.J.R., Transmission Line Matrix Method on Massively Parallel Processor Computers, in *9th Annual Review of Progress in Applied Computational Electromagnetics Dig.*, pp. 467-474, Monterey, Ca, March 22-26, 1993.

19. W. Menzel, R. Sorrentino, C. Eswarappa, P.P.M. So, and W.J.R. Hoefer, "Analysis of a millimeter-wave filter using transmission line matrix and mode matching methods and comparison with measurements",*in 9th Annual Review of Progress in Applied Computational Electromagnetics Dig.*, pp. 289-296, Monterey, Ca, March 22-26, 1993.

20. P.P.M. So, W.J.R. Hoefer, J.W. Bandler, R.M. Biernacki, and S.H. Chen, "Hybrid frequency/ time domain field theory based CAD of microwave circuits", *in 23rd European Microwave Conference Dig.*, pp. 218-219, Madrid, Spain, 6-9 Sept. 1993.

21. P.P.M. So, and W.J.R. Hoefer, "Distributed Computing for Transmission Line Matrix Method", *in Second International Workshop on Time Domain Modelling of Fields and Networks Dig.*, Berlin, October 28-29, 1993.

22. M. Forest and W.J.R. Hoefer, "TLM synthesis of microwave structures using time reversal",*in 1992 IEEE Intl. Microwave Symp. Dig.*, pp. 779-782, Albuquerque, NM, June 1-5, 1992.

23. W.J.R. Hoefer, and M. Forest, "TLM modelling of electromagnetic scattering in forward and inverse time sequence", *in 1992 International Symposium on Antennas and Propagation (ISAP) Dig.*, pp. 733-736, Sapporo, Japan, Sept. 22 - 25, 1992.

Future of Electromagnetics for Millimeter Wave Circuit Design

B. Houshmand and Tatsuo Itoh

Electrical Engineering Department
University of California, Los Angeles
Los Angeles, CA 90024-1549
U. S. A.

ABSTRACT

As the requirement for microwave engineering has been quickly changing, the research community must also change. Rather than observing the entire microwave research, this paper picks up a subjective view in the form of challenge to the electromagnetics for new requirement for microwaves. This challenge comes from quickly changing environment prompted by a renewed interest in "radio" engineering combined with defense cut. The classical electromagnetic theory must change to be able to provide constructive roles toward the modern radio engineering. Some of the features needed toward this goal is to capture interdisciplinary areas from the electromagnetics point of view. They include close couple to microelectronics, visualization, and system applications. Several subjective views and examples will be presented.

1. INTRODUCTION

For quite some time, the electromagnetic field theory has been considered to be a mature subject without much excitement except for those individuals, typically academics, directly involved with the subjects. In many instances, the problems typically dealt with, particularly in the academic community, have been neither theoretical nor application oriented. Many projects dealt with the analytical and/or numerical exercises. There have been very few fundamental theoretical developments throughout the history of electromagnetics. Many "theoretical" papers have dealt with pre-solved problems by other means or improved methods. This type of exercise is not constructive to the new environment in which efficient solutions to practical problems are needed.

It is somewhat fortunate that the arrival of the renaissance of radio engineering due to a significant increase in interest in personal and mobile communications and intelligent automobile and highway technologies all over the world coincides with the need for the dual use of microwave technology in the U.S. and other parts of the western world. Government and industries in the U. S. have invested a large sum of resources in microwave and millimeter-wave technologies mainly for military use. Some countries, typically Japan, have advanced this segment of technology essentially through commercial and other non-military applications. Both sides need to learn from experiences of the other in many respects for realization of dual use technology. The dual use of the technology must be appealing to a greater segments of engineering community with relative ease. This is a significant challenge to technology which must provide a low cost solution to many complicated problems . One key element is the enhanced role of electromagnetics. This is because the electromagnetics is fundamental to any design of microwave and millimeter-wave structures. However, the electromagnetics community must provide a swift solution to practical problems, which may be far from the idealized "theoretical" problems. In order to accomplish this assignment, the role of electromagnetics must be enhanced by making its process more cost effective, more responsive to real world problems by importing the recent advances in computing technology, more interaction with related subjects such as microelectronics and nonlinear phenomena.

As the new era starts, more innovative technology, more cost effective applications of existing technology and exploitation of higher frequencies become important. Significant opportunities exist for those engaged in electromagnetics if they recognize the changing needs and find new approaches. Some representative examples toward implementing this change will be described in this paper.

2. INTERACTIONS WITH ACTIVE DEVICES

In most microwave circuit design, the solid state devices are treated as lumped elements represented by S parameters. It is on the other hand well recognized that the passive elements are considered as distributed elements and characterized as such. Although they may be represented by lumped element models, this is done after electromagnetic analysis.

The interaction of the electromagnetic wave with active devices needs to be considered carefully and must be understood well in the new microwave circuits yet to be proliferated. As the cost effectiveness is emphasized in the coming dual use technology, it is desirable to design the entire circuit with a single design process. In addition, in the case of the MMIC the chip size must be as small as possible for cost reduction. Therefore, the electromagnetic interaction on the chip is very important. For such a purpose, the enhanced electromagnetics should play an important role. The wave-device interaction must be taken into account. An example is the active antenna. The antenna acts not only as a radiating element but also provides a circuit function such as diplexer or resonator. This latter function becomes an integral part of the active circuit such as the oscillator.[1,2] For the design, it is necessary to characterize the entire electromagnetic behavior of such a circuit including the dynamic behavior of the active device.

It is well known that the antenna is one of the bottleneck issues for personal communication. The concept of an active antenna can add some "intelligent" function right at the antenna. Several examples are presented in the paper. Some efforts are described in the next section. In the active antenna, the antenna element is not only a radiating element but also a part of active circuit. Often the antenna acts as a resonant feedback element for an oscillator or a diplexer.

3. NEW DIRECTIONS OF NUMERICAL ELECTROMAGNETICS

As the operating frequency is increased and the packing density is increased in the integrated circuits, interactions among the circuit elements and between the chip and the environment needs to be the integral part of the analysis and design.

At the moment, most of the CAD software treats the passive and active components separately and environmental effects such as packaging issues are not well integrated. This lack of coordination will be fatal for next generation of the microwave and millimeter wave integrated circuits or chips which may be multifunctional. The cost reduction of such complex structures is enormous challenge and is particularly important for commercial applications. The design cycle must be reduced and the chance of the first path design success must be enhanced. Hence, all of these effects mentioned above need to be addressed and attacked simultaneously in an integrated and coherent fashion.

There are a number of numerical techniques for passive components including the integral equation type and differential equation type both in the frequency domain and time domain. Most integral equation types including the spectral domain method require highly analytical preprocessing and reduced computational effort.[3] On the other hand, the time domain methods such as FDTD and TLM require minimal amount of analytical preprocessing. Instead a large memory and a long computation time are required.[4,5] As the computational power is increasing, the latter problems are being overcome, particularly by parallel processing machines. Some efforts are being implemented to make the time domain method more computationally efficient as we will see later.

One of the appealing features of the time domain methods as applied to a practical problem is that they provide good physical insight on the wave phenomena in a complicated circuit structure by way of easy visualization which is evolving in time. However, their greatest appeal, in the opinion of the author, is its potential to model the entire circuit including the active devices and packaging effect, although not enough has been done in these areas. It is noted that many active device characteristics are formulated in the time domain. Therefore, their characteristics can be integrated into the methods such as FDTD. A recent effort along this line has provided a remarkably interesting results for a number of circuits including active devices such as the active antennas. The circuit is modeled in its entirety by the FDTD with the device under the large signal condition. The electromagnetic field pattern in the circuit can be visualized in this manner. It is clear that the results are very useful for providing enhanced information for the circuit designer by visualizing the wave phenomena in a practical circuit.

Another useful feature of the FDTD and TLM is that they are very well suited for parallel computation machine, because the space calculations are quite localized. Nevertheless, there still remain a significant effort to make these methods computationally efficient. For instance, in many circuits, the major computational volume is spent for the passive structure which are not nonlinear. Therefore, such a subarea may be replaced by the terminal characteristics of its interface with active devices. This concept can be accomplished by the use of Diakoptics which is essentially the time domain numerical Green's function.[6]

4. PACKAGING AND MULTIMODING EFFECTS

Recently, the effect of packaging has drawn attention in the circuit design. In many cases, the packaging effect implies the deviation of the designed circuit characteristics when the circuit is placed in a package. However, the packaging effect needs to be interpreted in a wider sense. Multilayer configuration and flip chip mounting are just two examples related to the packaging technique and need to be characterized as such.

In addition, in a highly complicated circuit configurations particularly those involved with multilayers, possible existence of undesired "modes" needs to be recognized. One of the highly publicized phenomena is the lateral leakage in a planar transmission lines such as the conductor backed coplanar waveguide or slot line.[7] This leakage is essentially due to the parallel plate TEM modes and is now well understood. Although this leakage is detrimental in many cases, a possibility of making use of such a leakage phenomenon and constructing a possible new component has been reported.[8] It is possible to create a directional coupler which takes advantage of the leakage mechanism. In the next section an overview of some advances in the computational capabilities for analyzing the complex microwave structures is presented.

5. COMPUTATIONAL ISSUES

It is well known that the Finite-Difference Time-Domain (FDTD) is a powerful method for analyzing the electromagnetic wave behavior in a complicated geometry. This method, however, is a memory intensive and time consuming operation. Recently, we have utilized several techniques to alleviate these deficiencies. Specifically, we have implemented the FDTD Diakoptics method to use numerical Green's function to replace large computational volume with its impulse response. Hence, the memory requirement is drastically reduced. For reducing the computational time, we have implemented a method based on the system identification (SI) technique. A reduction of computation time of a factor of ten can readily be attained. In addition, we have analyzed large volumes containing active and nonlinear regions (or devices, plasmas, etc.) by means of FDTD environment.

In this section, an overview of the FDTD Diakoptics, application of system identification to the FDTD algorithm, and application of FDTD to nonlinear and active regions are presented. Several examples which illustrates these methods are included.

5.1. FDTD Diakoptics

Analysis of a computationally large circuit can be accomplished by dividing the circuit structure into several small modules. Each module can be analyzed separately, and the mutual interaction of the modules can be included by proper treatment of the circuit boundaries. Based on this methodology, the time-domain Diakoptics method has been developed in TLM[1,2], and in FDTD[3,4]. The time domain Diakoptics can include all the mutual interactions among modules through a convolution interface without invoking any approximations.

Time-Domain Diakoptics originates from the linear circuit theory. Once input and output ports are identified, the system output Y(n) of a passive structure can be determined from the convolution of the system impulse response h(n) and the input X(n). This indicates that the complete two-port linear passive structure can be replaced by its impulse response h(n). Similarly, multi-port linear passive region in the field calculation can be replaced by an impulse response matrix [g], see Fig. 1. This operation is similar to the transverse resonance method. The multi-port convolution is defined

$$Y_m(k)=\sum_{n=1}^{N}\sum_{k'=0}^{K} g(m,n,k-k')X_n(k') \qquad (1)$$

where g(m,n,k') is the impulse response (or the time-domain Green's function) at port "m" at time t=k' due to the unit excitation at port "n" at t=0.

The computation of the numerical Green's function is performed by applying an impulsive source at the input port of the passive structure. If the impulse response over a limited frequency range is required, the frequency band-limited response can be computed by applying a deconvolution process between the structure output and the input signal which spans the frequency range of interest[14].

The numerical behavior of the FDTD Diakoptics is demonstrated by applying it to a one-dimensional circuit . Fig. 2 shows an infinitely long parallel-plate transmission line loaded with lumped elements and material discontinuity. Only the TEM mode of operation is considered due to the dimensions of the problem. As such, the results produced by the time-domain Diakoptics can be directly compared with those produce by conventional time-domain circuit simulators, like Microwave SPICE (MWSPICE). Fig. 3 shows the agreement between the MWSPICE and the FDTD Diakoptics method.

5.2. System Identification

The Finite-Difference Time-Domain algorithm is an effective computational method for full vector analysis of microwave structures[15]. The theoretical formulation directly follows the Maxwell's equations, and algorithm implementation is simple and flexible for general structures of interest. The computation requirements, however, is excessive due to the spatial-temporal discretization. Recently, digital signal processing methods have been used to reduce the computational requirements of the time-domain methods[16,19]. For example, The Prony's method is used to estimate the time signal in terms of the previously computed values[16], also a covariance based system identification (SI) algorithm has been used to reduce the computation cost of the Transmission Line and FDTD Methods by employing a stochastic ARMA model[17,18]. In this paper a Least-Squares based system identification projection algorithm for a deterministic Auto-Regressive Moving Average model is applied to the FDTD algorithm[20]. The application of this algorithm to the partially filled rectangular cavity has demonstrated excellent numerical results. Savings in the computation requirements are achieved by replacing the computationally intensive FDTD algorithm by the ARMA model for output signal computation, after the system parameters converge to their final values. In addition, the frequency response is evaluated directly from the computed system parameters, thus eliminating the need for Fourier Transformation.

The computed time signal at an appropriate location in the computational volume and the corresponding input signal can be interpreted as the input and output signals of a discrete linear system. This linear system description is

$$y(n) = -\sum_{k=1}^{K} a_k \, y(n-k) + \sum_{m=0}^{M} b_m \, x(n-m) \qquad (2)$$

The output signal is completely known when the model parameters (a_k, b_m) are computed. The parameter space is taken to be large enough to allow the convergence of the model output to the FDTD simulated field values. Equation (2) can be written in a compact form

$$y(n) = \Phi^T(n-1)\,\Theta_0 \tag{3}$$

where T stands for Transpose, and F is a vector containing the present and past values of the input and output which can be considered as data. The vector Θ_0 contains the system parameters and uniquely defines the properties of the linear system such as the resonance frequencies. Equation (3) represents the output of a linear system as the inner product of the F and the parameter vector. Using the available data vector F, the output signal can be estimated in terms of the estimated system parameters

$$\hat{y}(n) = \Phi(n-1)^T\,\hat{\Theta}(n-1) \tag{4}$$

The difference in Equations (3) and (4) is minimized with respect to the system parameters to arrive at a parameter update law

$$\hat{\Theta}(n) = \hat{\Theta}(n-1) + \frac{P(n-1)\,\Phi(n-1)}{\Phi(n-1)^T\,P(n-1)\,\Phi(n-1)}\,[e(n)] \tag{5}$$

$$P(n) = P(n-1) - \frac{P(n-1)\,\Phi(n-1)\,\Phi(n-1)^T\,P(n-1)}{\Phi(n-1)^T\,P(n-1)\,\Phi(n-1)},\ P(0) = I \tag{6}$$

where P(n) provides an orthogonal projection search in the parameter space which results in rapid parameter convergence[20], Θ(n) is the computed parameter vector, and e(n) is the discrepancy between the estimated output and the FDTD computed field value. Computation of Equations (5) and (6) requires only vector addition and multiplication, and results in minimal additional cost to the FDTD computation. We note that the system parameters converge to their final values when the output error is sufficiently small.

The numerical behavior of this method is demonstrated by applying it to the cavity problem. An ARMA model with system parameters K=40, M=40, see Equation (2), is used to obtain the resonance frequencies of a rectangular cavity. The cavity is excited at the center plane by imposing a TE_{10} mode distribution with impulsive temporal dependence. Fig. 4 shows the parameter convergence for a number of system coefficients. The initial condition for the parameters is set to the origin of the parameter space, and the parameter values are updated at each sampled interval.

The resonance frequencies of the cavity can be derived directly from the poles of the ARMA model. They can also be recovered from the spectrum of the output signal. This spectrum is computed by the Fourier Transform of the output signal, or directly by evaluation of the Z-Transform of Equation (2) on the unit circle which is defined in terms of the system parameters. Fig. 5 shows the spectrum of the output signal using the system parameters and the Fourier Transformation of the output signal. The recovered resonance frequencies of the first three odd modes are illustrated. The location of the observation point coincides with the null position of the even modes as the result these modes can not be recovered form this time signal. In this example, the ARMA based spectrum is computed using 140 output samples. Similar spectrum is obtained by applying the Fourier Transformation to 500 output samples. Fig. 5 also shows the Fourier Transform of the 140 output samples which are augmented with zero padding to provide sufficient spectral resolution for locating the spectrum peaks. This spectrum, while qualitatively locates the resonance frequencies, is distorted and might not provide sufficient resolution where the resonance frequency separation is small. We note that the first two resonance frequencies are predicted accurately for this example, while higher order modes are underestimated due to frequency dispersion of the FDTD spatial grid. This method is used to obtain the resonant frequencies of a partially filled rectangular cavity. Fig. 6 shows the shift in the resonant frequency as the permittivity is gradually changed from $\varepsilon=1$ to $\varepsilon=2$.

5.3. Modeling of Nonlinear and Active regions with the FDTD method

The FDTD method can be extended to include nonlinear and active regions embedded in distributed circuits. Recently, the two dimensional FDTD was extended to include active, passive and possibly nonlinear lumped circuit elements[21]. In that work the incorporation of the lumped elements into the FDTD algorithm is described, and transmission lines with various lumped element loads were simulated. In the above paper only one active load was modeled, which was linear, and the modeling follow the same procedure as the passive lumped elements. The inclusion of active nonlinear regions, however, requires additional procedures in order to preserve the numerical stability of the FDTD algorithm.

The TLM method[22] was used to model active nonlinear subregions of distributed circuits, and it was noted that regions of negative conductivity may cause spurious oscillations at the TLM mesh cut-off frequencies. To circumvent this problem, the TLM mesh cut-off is chosen to be well above the active device cut-off frequency. Similar condition holds for the FDTD algorithm. In addition, we have found that in many cases the FDTD algorithm will become unstable unless some care is taken in incorporating the active device model. This is due to the fact that realistic active devices can produce extremely large local currents. These local currents, in turn, produce large fields that are fed back into the device model which can create unstable behavior.

Here we describe the steps we have implemented to produce a stable algorithm, and we use this algorithm to simulate an active antenna. This method is used to simulate a three-dimensional microwave circuit containing an active and nonlinear device. Fig. 7 shows a two element active antenna which is examined. Each patch is excited by a separate Gunn Diode and therefore the circuit really consists of two oscillators. However, the two oscillators are strongly coupled through a length of transmission line and there are several possible modes of operation for the entire coupled circuit. An extensive model[23] to analyze this effect, which determines the most stable mode of operation, has been developed. We will demonstrate here that the FDTD is able to correctly predict this model as well, with the additional advantage that the complete electromagnetic behavior of the circuit is obtained.

As mentioned above, special care must be taken to simulate nonlinear and active elements in the FDTD or TLM. First, it is necessary that the cut-off frequency of the active device be well below the mesh cut-off frequency to prevent the unstable oscillations at these frequencies. We therefore model each Gunn diode by the equivalent circuit shown in Figure 8. The active current is given by the polynomial

$$F(V_s) = -G_1\, V_s + G_3\, V_s^3 \tag{7}$$

The coefficients were determined experimentally from measurements at 10.48 GHz (the patch resonance frequency) to be G1 = 0.0252 ohm^{-1} and G3 = 0.0265 ohm^{-1} V^{-2}, and the capacitance was determined to be C= 0.2 pF. The series resistance was estimated to be R=1.0 ohm .Note the instead of using complicated model for Gunn diode which would incorporate the correct dispersive behavior, we are using a simplified model which is approximately correct over a narrow frequency range and ensures that the active device cut-off frequency is bellow the mesh cut-off frequency. This simplification is justified by the highly resonant nature of the circuit, which limits the possible frequencies of interest.

To incorporate the package diodes into the FDTD mesh, we use an equivalent active region which extends over three vertical cells between the microstrip and the ground plane (Fig. 7), and occupies only one cell in the horizontal or x-y plane. Note that we cannot assume each mesh cell in the active region is dependent only on the local field at the cell, as was done in [21,22,24]. This is due to the fact that each cell will then act as a separate diode, with aggregate effect of tree diodes in series. Instead, we model the entire active region as a single diode. The total voltage across this diode is given by

$$V_z(t) = \frac{V^{n+1} + V^n}{2} = \frac{\Delta z}{2} \sum_{k=1}^{3} [E_z^{n+1}(i_s, j_s, k) + E_z^n(i_s, j_s, k)] \tag{8}$$

Here n represents the time step increment, and (i_s, j_s) are indices in the x, y plane for the two active regions (s= 1, 2). This time average voltage is then fed into our active device model (Fig. 8) which then calculates the total current by

$$A_0 I^{n+1} = A_1 I^n - A_2 F(V_s^n) - A_3 V^{n-1} - A_4 V^n + A_5 V^{n+1} \tag{9}$$

with

$$A_0 = 2RC + \Delta t + R\,\Delta t\,\dot{F}(V_s^n),\ A_1 = 2RC - \Delta t + R\,\Delta t\,\dot{F}(V_s^n),\ A_2 = 2\,\Delta t,$$

$$A_3 = C,\ A_4 = \Delta t\,\dot{F}(V_s^n),\ A_5 = C + \Delta t\,\dot{F}(V_s^n)$$

where (.) denotes the derivative of the dependent current source with respect to the voltage. A forward differencing scheme with time averaging has been used in order to produce stable oscillations. This process is described in more detail in[25]. The current is then fed back as a source into the FDTD cells in the active regions. For each active region (s= 1, 2),

$$\frac{\varepsilon}{\Delta t} E_z^{n+1}(i_s, j_s, k) = \frac{\varepsilon}{\Delta t} E_z^{n}(i_s, j_s, k) + L[H_x, H_y] - \frac{I^{n+1} + I^{n}}{2\Delta x\, \Delta y} \tag{10}$$

The term $L[H_x, H_y]$ is

$$\frac{H_x^{n+\frac{1}{2}}(i_s, j_s-1, k) - H_x^{n+\frac{1}{2}}(i_s, j_s, k)}{\Delta y} + \frac{H_y^{n+\frac{1}{2}}(i_s, j_s, k) - H_y^{n+\frac{1}{2}}(i_s-1, j_s, k)}{\Delta x} \tag{11}$$

Equations (8) and (9) are then used in (10) in order to obtain a stable FDTD algorithm in the active region[25]. This algorithm is stable for circuits embedded with nonlinear active regions which we have considered.

By using the modified FDTD algorithm described above, we have simulated the two element active antenna shown in Fig. 7. A small amount of numerical noise is introduced into the FDTD mesh, and oscillations build up until a steady state frequency of 12.4 GHz is achieved (Fig. 9). The measured frequency is 11.8 GHz and the frequency predicted by the modal analysis[26] is 12.2 GHz. The 5 percent discrepancy in the predicted frequency can be attributed to modeling errors in the geometry description and measurement of the Gunn diode parameters. The stable mode is an odd mode which can be seen clearly in Fig. 10, where we show the steady state voltage across each diode as a function of time. Fig. 11 shows the distribution of the z component of the electric field at the dielectric-air interface.

6. CONCLUSION

Some examples were presented from the point of view of new directions needed for electromagnetic engineering which can attack practical microwave problems required for microwave technology under a changing but exciting environment.

7. ACKNOWLEDGMENT

The contents reported in this paper have been supported by a number of research contracts including those from ARO under DAAH04-93-G-0068, ONR under N00014-91-J-1651 and Joint Services Electronics Problem by AFOSR under F49620-92-C-0055.

8. REFERENCES

1. J. Birkeland and T. Itoh, "Planar FET oscillators using periodic microstrip patch antennas," IEEE Trans. Microwave Theory and Techniques, Vol. 37, No. 8, pp. 1232-1236, Aug. 1989.

2. J. Lin, T. Itoh and S. Nogi, "Mode switch in a two-element active array," 1993 IEEE AP-S Symposium Digest, pp.664-667, June 28-July 2, 1993, Ann Arbor, MI.

3. T. Itoh and R. Mittra, "Spectral domain approach for calculating the dispersion characteristics of microstrip lines," IEEE Trans. Microwave Theory and Techniques, Vol. MTT-21, No. 7, pp. 496-499, July 1973.

4. K. S. Yee, "Numerical solution of initial boundary value problems involving Maxwell's equations in isotropic media," IEEE Trans. Antennas & Prop., vol. AP-14, No.,-307, May 1966.

5. W. J. R. Hoefer, "The transmission line matrix method - Theory and applications," IEEE Trans. Microwave Theory and Tech., Vol.MTT-33, No.10, pp.882-893, Oct. 1985.

6. T. W. Huang, B. Houshmand and T. Itoh, "The FDTD Diakoptics method," IEEE MTT-S International Microwave Symp. Dig., pp.1435-1438, June 1993, Atlanta, GA.

7. H. Shigesawa, M. Tsuji and A. A. Oliner, "Conductor-backed slot line and coplanar waveguide: dangers and full-wave analysis," IEEE MTT-S Int. Microwave Symp. Dig., pp.199-202, May 1988, New York, NY.

8. Y. Liu, T. Hirota and T. Itoh, "Coupling phenomena in conductor-backed slotline structures," IEEE MTT-S Int. Microwave Symp. Dig., pp.1203-1206, Atlanta, GA.

9. P. B. Johns, and K. Akhtarzad, "The use of time domain diakoptics in time discrete models of fields," Int. J. Num. Methods Eng., vol. 17, pp. 1-14, 1981.

10. W.J.R. Hoefer, "The discrete time-domain Green's function or Johns matrix - a new powerful concept in transmission line modeling (TLM), " Int. J. Num. Modeling, vol. 2, pp. 215-225, 1989.

11. T.W. Huang, B. Houshmand, and T. Itoh, "The FDTD Diakoptics Method," IEEE MTT-S Int. Microwave Symp. Digest, 1993, pp. 1435-1438.

12. T.W. Huang, B. Houshmand, and T. Itoh, "The Sequential FDTD Diakoptics method for modular computation" Asia-Pacific Microwave Conference Proceeding, Oct. 1993.

13. C.W. Huo, and T. Itoh, "A new time-domain transverse resonance method in solving guided wave problems," Int. J. Num. Modeling, vol. 3, pp. 229-234, 1990.

14. T.W. Huang, B. Houshmand, and T. Itoh, "The Implementation of Time-Domain Diakoptics in the FDTD Method," Submitted to the IEEE Trans. Microwave Theory Tech.

15. X. Zhang and K.K. Mei, "Time-Domain finite-difference approach to the calculation of frequency dependent characteristics of microstrip discontinuities", IEEE Trans. Microwave Theory Tech., vol. MTT-36, no. 12, pp. 1775-1787, Dec. 1988.

16. J.L. Dubard, D. Pompei, J. Le Roux and A. Papiernik, "Characterization of Microstrip antennas using the TLM simulation associated with a Prony-Pisarenko Method," Int. J. of Numerical Modeling: Electronic Networks, Devices and Fields, Vol. 3, 269-285, 1990.

17. W. Kuempel and I. Wolff, "Digital signal processing of time domain field simulation results using the system identification method", IEEE MTT-S Digest,

June 1992.

18. W. Kuempel and I. Wolff, "System Identification Method for Transient Analysis of (M)MIC-Components Using Time Iterative Methods," 22ND European Microwave conference, Finland, 24-27 August 1992.

19. B. Houshmand, and T. Itoh, "Parameter estimation for characterization of microwave structures," presented at the National Radio Science Meeting, Jan. 5-8, 1993.

20. P. Strobach, Linear Prediction Theory: A Mathematical Basis for Adaptive Systems, Springer Verlag, Berlin, 1990.

21. W. Sui, D.A. Christensen, and C.H. Durney, "Extending the two dimensional FDTD method to hybrid electromagnetic systems with active and passive lumped elements," IEEE Trans. Microwave Theory Tech., vol. MTT-40, no. 4, pp. 724-730, April 1992.

22. P. Russer, P.M. So, and W.J.R. Hoefer, "Modeling of nonlinear active regions in TLM," IEEE Microwave and Guided Wave Letters, vol. 1, no. 1, pp. 10-13, Jan. 1991.

26. C.C. Yang and D.S. Pan, " Theoretical investigations of a proposed series integration of resonant Tunneling Diodes for millimeter-wave power generation," IEEE Trans. Microwave Theory Tech., vol. MTT-40, no. 3, pp.434-441, March 1992.

24. B. Toland, B. Houshmand, and T. Itoh," Modeling of nonlinear active regions with FDTD Method," to be published in the IEEE Microwave and Guided Wave Letters.

25. B. Toland, J. Lin, B. Houshmand, and T. Itoh," FDTD Analysis of an Active Antenna," to be published in the IEEE Microwave and Guided Wave Letters.

26. S. Nogi, J. Lin, and T. Itoh, "Mode analysis and stabilization of a spatial power-combining array with strongly coupled oscillators," to be published in the special issue on Quasi-optical Techniques of the IEEE Trans. Microwave Theory Tech., Oct. 1993.

Figure 1.	A multi-port linear passive region is replaced by its impulse response matrix.
Figure 2.	A multi-layered circuit is analyzed by the FDTD diakoptics. The impulse response of segmentations are combined sequentially.
Figure 3.	Comparison of the transient current flow through the circuit resistor. The FDTD diakoptics compare well with the MWSPICE simulation.
Figure 4.	Parameter convergence of the ARMA model with 80 coefficients. The evolution of the first four output coefficients is demonstrated.
Figure 5.	FDTD generated spectrum. 140 samples (at a rate of 1 sample per 5 FDTD output values) are used to obtain the spectrum by the SI method. Similar spectrum is obtained by Fourier Transformation of 500 samples. The Fourier Transform of 140 samples with zero padding is distorted.
Figure 6.	Resonance frequency computation of a partially filled rectangular cavity using the ARMA model. The first resonance frequency shifts at the permittivity is changed gradually from 1 to 2.
Figure 7.	The layout of the two element patch array.
Figure 8.	Circuit model of Gunn Diode.
Figure 9.	The time development of the total voltage across one oscillator.
Figure 10.	The steady state time variations of total voltage across each oscillator.
Figure 11.	The steady state electric field distribution (z component) at one instant in time.

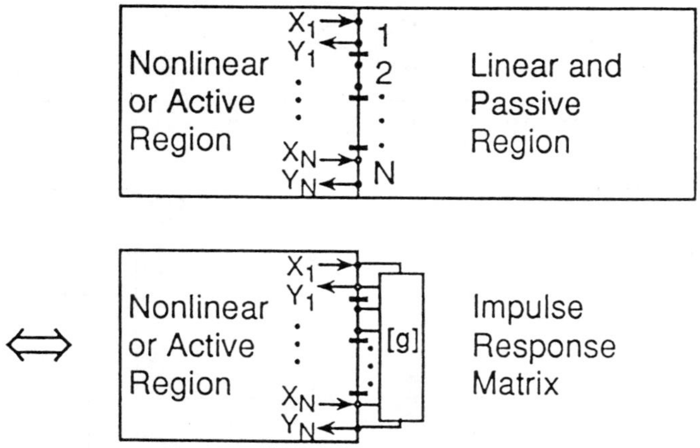
Nonlinear or Active Region
X_1
Y_1
X_N
Y_N
1
2
N
Linear and Passive Region
Nonlinear or Active Region
X_1
Y_1
X_N
Y_N
[g]
Impulse Response Matrix

Figure 1

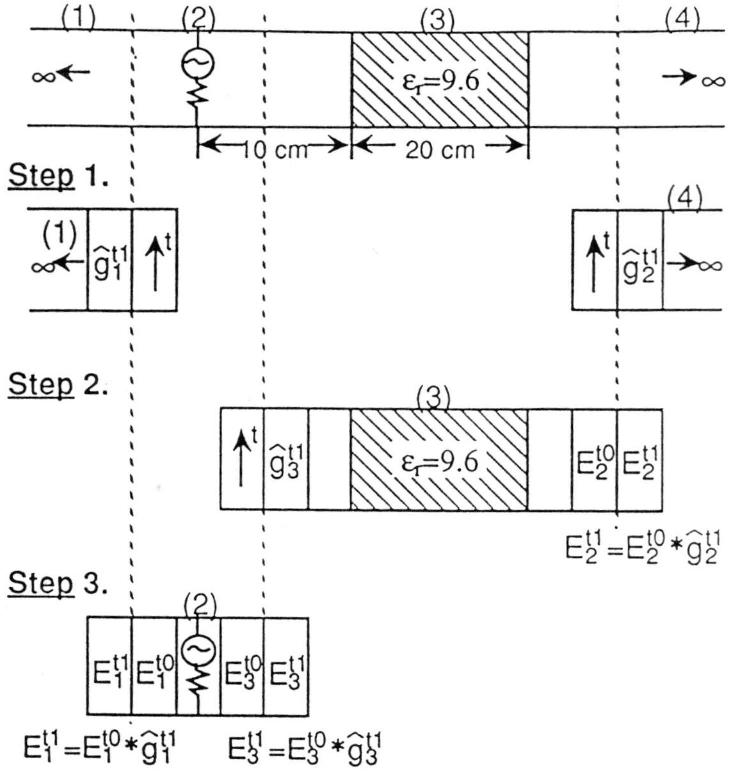
(1)
(2)
(3)
(4)
εr=9.6
10 cm
20 cm
Step 1.
Step 2.
Step 3.
E2t1=E2t0*ĝ2t1
E1t1=E1t0*ĝ1t1
E3t1=E3t0*ĝ3t1

Figure 2

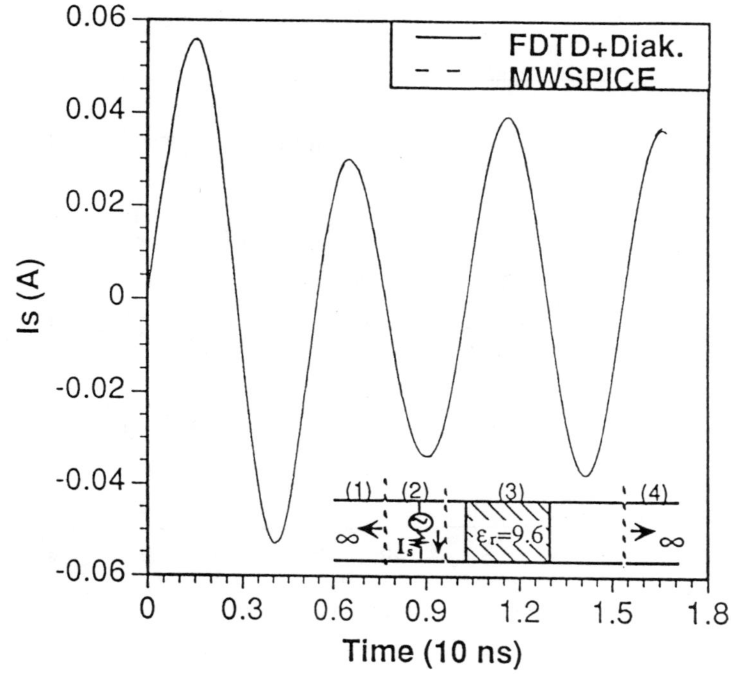
0.06
0.04
0.02
0
-0.02
-0.04
-0.06
Is (A)
0
0.3
0.6
0.9
1.2
1.5
1.8
Time (10 ns)
FDTD+Diak.
MWSPICE
(1)
(2)
(3)
(4)
εr=9.6

Figure 3

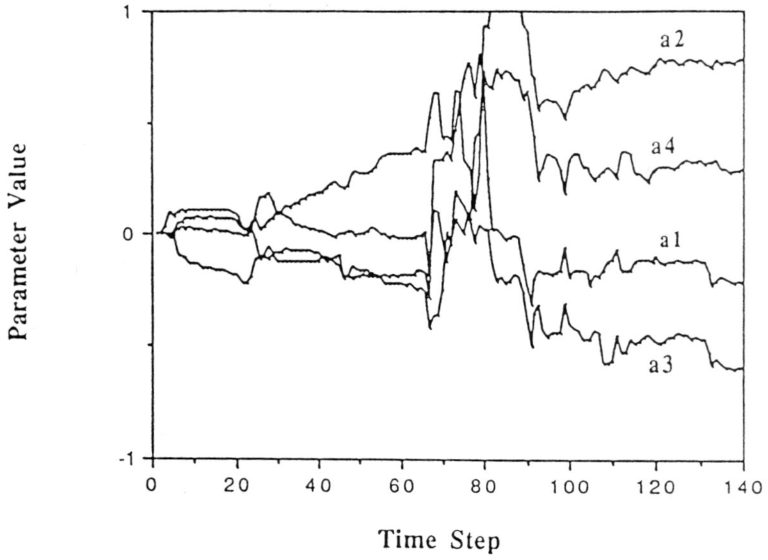
a2
a4
a1
a3
1
0
-1
Parameter Value
0
20
40
60
80
100
120
140
Time Step

Figure 4

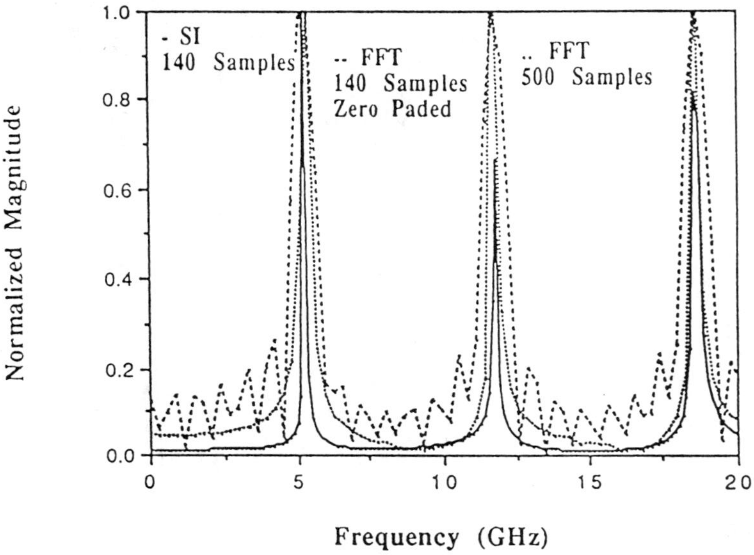

1.0
0.8
0.6
0.4
0.2
0.0
0
5
10
15
20
- SI
140 Samples
-- FFT
140 Samples
Zero Paded
.. FFT
500 Samples
Normalized Magnitude
Frequency (GHz)

Figure 5

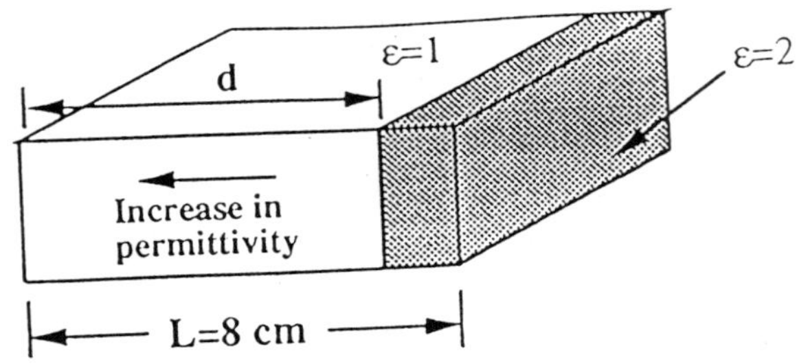
ε=1
ε=2
d
Increase in permittivity
L=8 cm

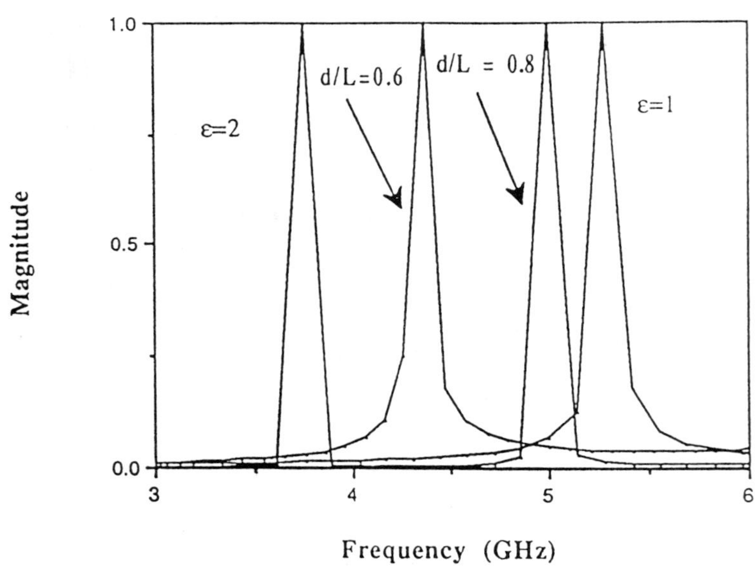
1.0
0.5
0.0
Magnitude
d/L=0.6
d/L = 0.8
ε=2
ε=1
3
4
5
6
Frequency (GHz)

Figure 6

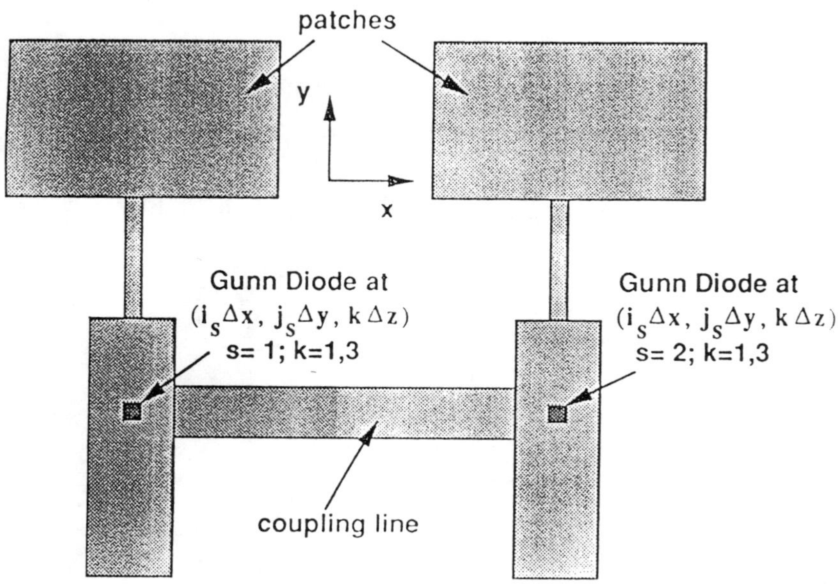

Metallization on dielectric subtstrate ($\varepsilon = 2.33$). Gunn Osc. are between strip and groundplane ($H = 3\Delta z$).

Figure 7

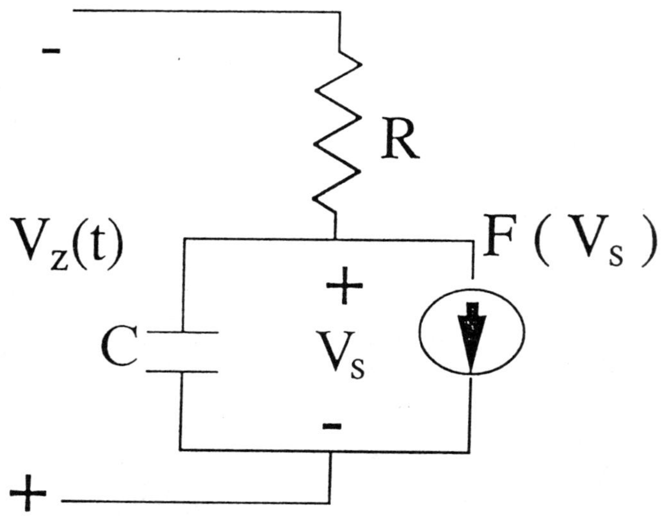

Figure 8

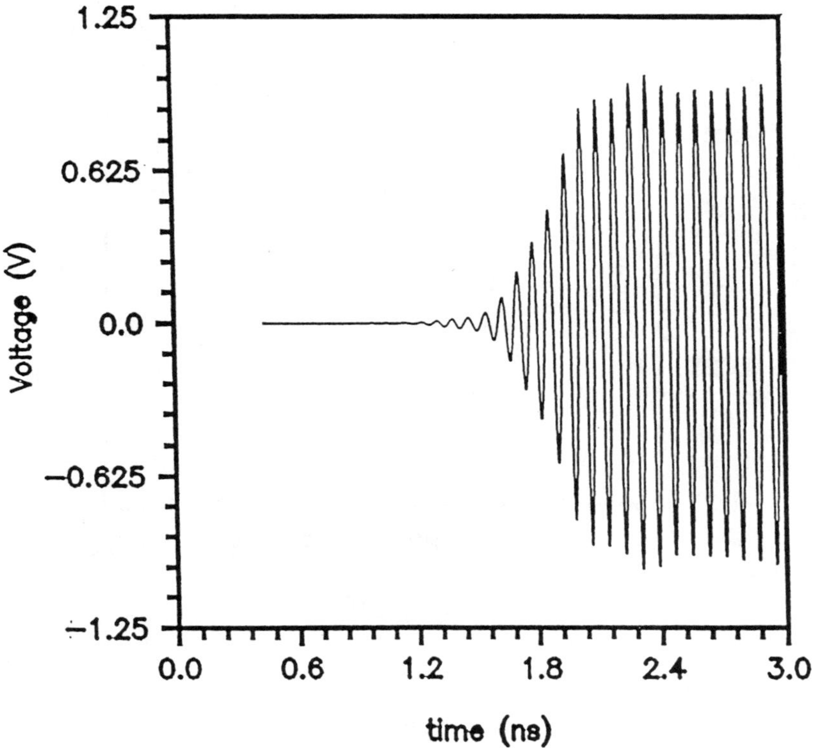

Figure 9

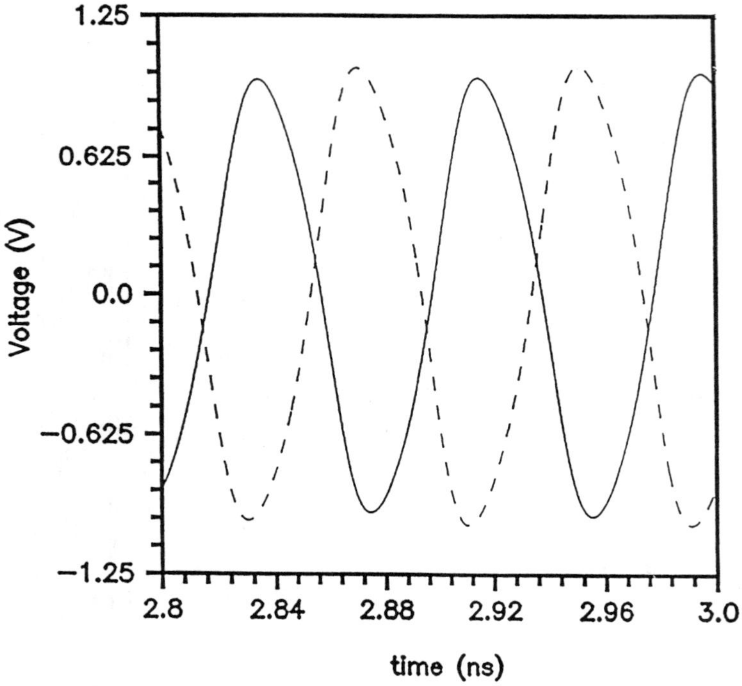

Figure 10

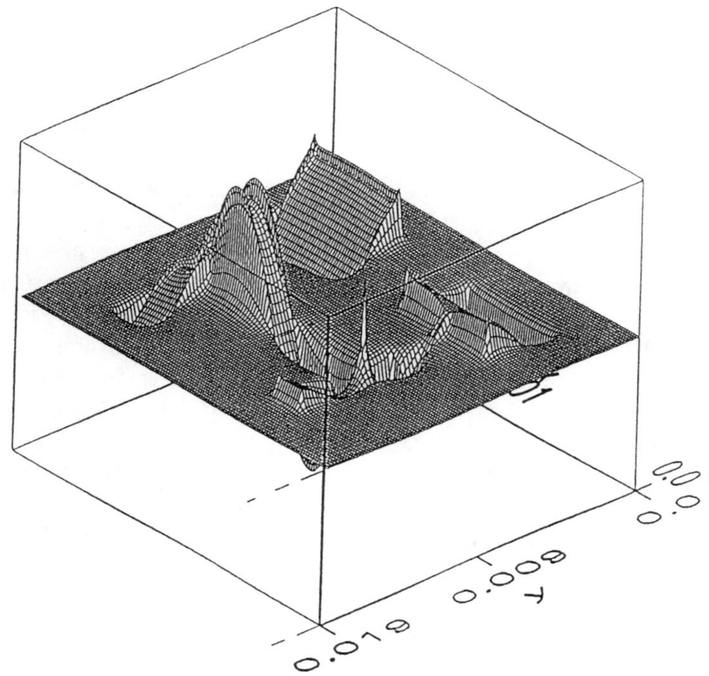

Figure 11

Millimeter wave MMIC technology for smart weapons

Charles R. Seashore

Alliant Techsystems Inc., Defense Systems Group
Hopkins, Minnesota 55343

ABSTRACT

Millimeter wave MMIC component technology has made dramatic progress over the last ten years largely due to funding stimulation received under the ARPA Tri-Service MIMIC program. In several smart weapon systems, MMIC components are now specified as the baseline approach for millimeter wave radar transceiver hardware. Availability of this new frontier in microelectronics has also enabled realization of sensor fusion for multispectral capability to defeat many forms of known countermeasures. The current frequency range for these MMIC-based components is approximately 30 to 100 GHz. In several cases, it has been demonstrated that the MMIC component performance has exceeded that available from hybrid microstrip circuits using selected discrete devices. However, challenges still remain in chip producibility enhancement and cost reduction since many of the essential device structure candidates are themselves emerging technologies with a limited wafer fabrication history and accumulated test databases. It is concluded that smart weapons of the future will rely heavily on advanced microelectronics to satisfy performance requirements as well as meeting stringent packaging and power source constraints.

1. INTRODUCTION

Smart weapons generally refer to systems that are required to acquire, track, target, and kill a class of military threats. The target sensing and signal processing required to provide these functions are contained within the smart munition itself making operation fully autonomous. Many such weapons are delivered to the general area of potential targets within a single "bus vehicle" which can be an aircraft, artillery shell, or guided missile. Once in the targeting area, the weapons are ejected from the bus and are deployed to detect and subsequently destroy targets of opportunity in a fire-and-forget fashion. Autonomous smart weapons are projected to have great tactical value because they allow the targeting of valuable enemy assets well behind the forward troop line while preserving the security and flexibility of the respective launching platform. The "smart" aspect of such a weapon resides in its inherent capability to detect targets autonomously, discriminate targets from clutter, and prioritize the targets according to type, status, and identity. To be useful in a military scenario, such weapons must be low priced, small sized, immune to mutual interference, and capable of overcoming a variety of countermeasures.

The current primary sensors for smart weapons include millimeter wave radar, radiometry, and passive infrared. Radar provides all-weather, longer detection range capability. Radiometry provides a precise target aimpoint based on reflection of the cold sky temperature. Infrared, operating either as a hot spot detector or in an imaging mode, provides enhanced target detection and the means to overcome both active and passive millimeter wave countermeasures. All the sensors described must rely heavily on advanced microelectronics to meet packaging and cost objectives. For the case of millimeter wave radar or radiometer sensors, careful application of GaAs MMIC technology has provided a means to meet those system requirements.

Significant progress has been made over the past ten years in millimeter wave MMIC components and subsystems which are now being inserted in radar sensors for smart weapons. These developments have primarily focused on the unique front-end needs of expendable weapon systems with their inherent critical interaction between the sensor hardware and the lethal mechanism. To meet this challenge, the millimeter wave radar transceiver has been carefully partitioned to define MMIC development candidates which will provide producibility enhancements and innovative packaging. Depending upon the operating frequency, these circuits have been structured into single-function and multifunction chip sets. It should be noted that on the Defense Department Critical Technologies listing during recent years, semiconductor materials and microelectronic circuits have consistently been ranked #1 on the priority list. This is fully supportive of the ongoing millimeter wave MMIC technology base development initiative.

The millimeter wave frequency range for smart weapon millimeter wave sensors is shown in Figure 1. Traditional window frequencies of 35 GHz and 94 GHz are still used to define MMIC developments although system operation up to 42 GHz and 95 GHz, respectively, have been investigated to sharpen antenna resolution within weapon diameter constraints. The current smart weapon system candidates are summarized in Figure 2. They fall into two categories; sensor-fuzed munitions and terminally-guided munitions. The former is characterized by a strapdown sensor package with fixed beam, staring antennas, and a very critical interface with the lethal mechanism. A millimeter wave radar of FMCW architecture is used to generate ranging and target detection data. The latter is slightly more sophisticated in performance and mechanization. Since the munition is terminally guided, there is need for guidance and control functions, as well as target tracking commands. These are derived from a conical scan or monopulse antenna mounted on a two-axis gimbal platform. In either case, targeting requirements include stationary, slow moving, relocatable, and camouflaged engagements. In almost all cases, these smart weapon systems utilize millimeter wave radar and/or radiometry as the primary sensor. The SADARM submunition is shown in Figure 3. It represents the classic sensor-fuzed munition with sensor, lethal mechanism, and safe and arm, all self-contained in a 155mm gun-fired projectile. A millimeter wave radar and a quasi-imaging passive infrared array are used to achieve sensor fusion for improved performance in a tactical countermeasure environment. The radar operates in Ka-band and utilizes innovative packaging and interconnect technologies to meet the stringent submunition volume constraints. This radar utilizes MMIC technology in its transceiver. Figure 4 shows a Ka-band seeker head for a defense mission terminally guided munition. The radar is based on waveguide component technology to demonstrate effectiveness of a conformal antenna array with an ogive geometry for producing forward looking beams with target detection and tracking capabilities. This seeker head illustrates the keen interest in conformal arrays for projectile and missile applications. MMIC development has now advanced to the point where many of the array antenna T/R and control functions can be directly implemented using chip technology to realize an active array.

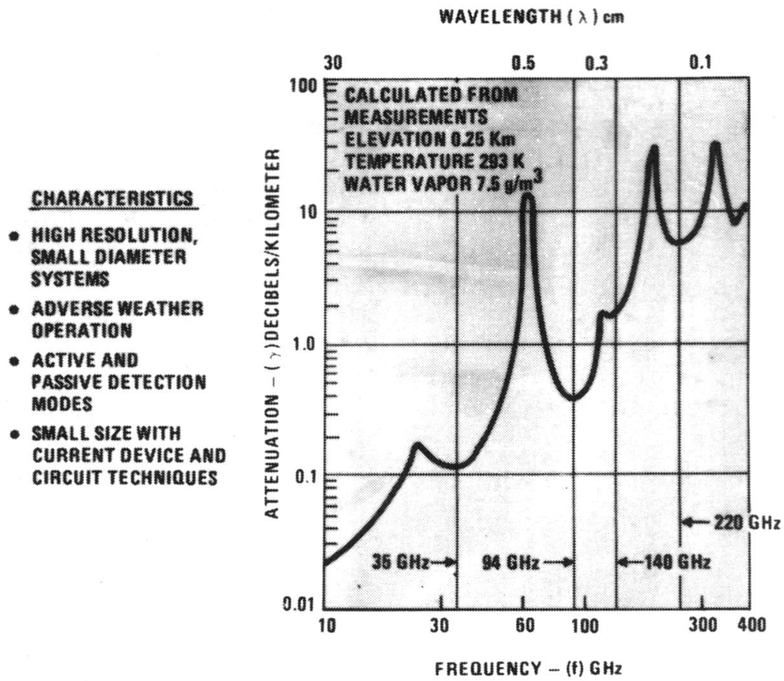

Figure 1. Millimeter Wave Frequency Range for Smart Weapons is 30 to 100 GHz

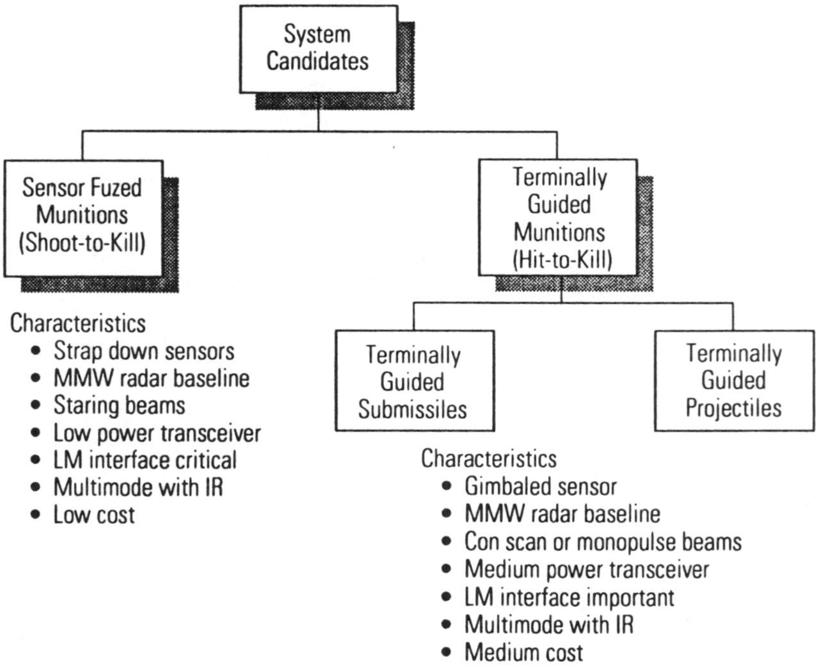

Figure 2. Smart Weapon System Candidates

In the two examples described above, both planar and conformal antenna geometrics are illustrated to demonstrate their importance to the smart weapon family. The concept of millimeter wave, small diameter, tactical antennas for smart weapons is tied closely to key MMIC developments to achieve a highly compact, low cost radar frontend which can meet both strapdown and scanning requirements. In this sense, MMIC technology is enabling since it finally permits active arrays to be directly applied to a variety of smart weapon requirements.

An assessment of MMIC content in current and projected smart weapons is presented in Figure 5. For radar sensors operating in Ka, Q, and V-bands, the transceiver is fully implemented with a MMIC chip set. Typically, this is either MESFET or HEMT based depending on cost/performance requirements. For W-band operation, the transceiver can be fully MMIC, however, for transmit power requirements in excess of about 150 milliwatts, a hybrid approach based on diode oscillators, amplifiers, and frequency multipliers becomes necessary. In D-band applications, a hybrid approach using discrete components is necessary. Recent reports of InP circuit performance at frequencies above 100 GHz makes it feasible to assume a full MMIC transceiver implementation is forthcoming.

In summary, MMIC transceivers for smart weapon millimeter wave radars have progressed from the laboratory bench to system insertion in recent years. It is anticipated that the following system benefits will be derived from this very important technology development:

- Cost and producibility leveraging
- Demonstrated performance margin
- Radar functional enhancements
- Planar and conformal packaging flexibility
- Bus power reduction from three-terminal device utilization.

Figure 3. Army SADARM Submunition Uses Dual-Mode Sensors

Figure 4. BMD Missile Seeker Uses Conformal Active Array

2. TECHNOLOGY STATUS

Millimeter wave MMIC technology has made tremendous progress over the past ten years in the entire frequency range 30 to 100 GHz. CAD techniques, device and circuit modeling tools, as well as automated wafer testing have now evolved to a point where a first-pass chip design success is frequently achieved. Figure 6 illustrates the dramatic shift in millimeter wave hardware during recent years, from a series of interconnected black boxes to planar and conformal surfaces with MMIC based transceivers for smart weapons applications. The black box approach with waveguide interconnects tended to be costly and have suspect reliability due to all the mechanical interfaces. A positive feature of this approach was the ability to test each subsystem component as an individual item in the trouble-shooting process. Such is not the case with the highly integrated MMIC structures. Here, the antenna and transceiver become a singular subassembly, with overall performance defined by measuring a predefined Figure of Merit.

A historical look at millimeter wave component development is shown in Figure 7. The current technological focus is on analog MMIC development with appropriate system insertions. As we look forward into the coming years, attention will be directed to mixed analog MMIC and digital microelectronics for the radar transceiver and initial signal processing. Photonics integrated circuits with fiber optic links will be utilized to supply control signals to the transceiver module and for antenna beam steering. The synergism between MMIC PHEMT device structures and those structures required for photonic transmitter and receiver functions will be thoroughly exploited. The ongoing multiyear ARPA MIMIC program plan is presented in Figure 8. This program represents one of the most successful and broad-based technology developments in the past decade. The technical advances and sweeping economic advantages for so many military systems makes MMIC designs an essential foundation of current and future platforms. This program has been very productive in developing and demonstrating millimeter wave GaAs MMIC technology with both single and multifunction chip architectures. Device structures investigated have included MESFET, HEMT, PHEMT, and HBT. One of the objectives for the MIMIC program has been to significantly reduce chip costs through materials, design, fab, and test cost control. Figure 9 shows the positive trend in this important parameter as the MIMIC program has progressed through its phases. From a starting cost of $20 per square millimeter, the MIMIC program has demonstrated a reduction in cost to approximately $1 per square millimeter as referenced to October 1994. Although Figure 9 primarily reflects microwave MMICs, the cost reduction trend experienced with relevant millimeter wave MMICs has been more gradual, primarily due to the emerging fabrication technology experienced with PHEMT structures. The critical factors for a successful MMIC insertion in a radar transceiver are identified in the F^3 diagram of Figure 10. Acceptability primarily relates to meeting the overall system specifications or requirements document. This document contains all electrical, mechanical, physical, and environmental guidelines for the radar hardware. The environmental factors remain a major challenge for millimeter wave MMICs, particularly electrical performance over temperature. Availability relates to foundry capability in responding to a MMIC volume need within budgeted schedule and dollars. Foundry services for millimeter wave MMICs are much more limited than the more common microwave products. Since only a single foundry source is too high risk for a

Band Designation	Frequency Range	Application	MMIC Status
Ka	26.5 – 40 GHz	Term. Guided & Sensor Fuzed Munitions	Full Transceiver
Q	33 – 50	Sensor Fuzed Munitions	Full Transceiver
U	40 – 60	Open	Components
V	50 – 75	Sensor Fuzed Munitions	Full Transceiver
W	75 – 110	Term. Guided Munitions	Full Transceiver
F	90 – 140	Open	Components
D	110 – 160	Very Small Diameter Munitions	Components

Figure 5. MMIC Assessment for Current Smart Weapons

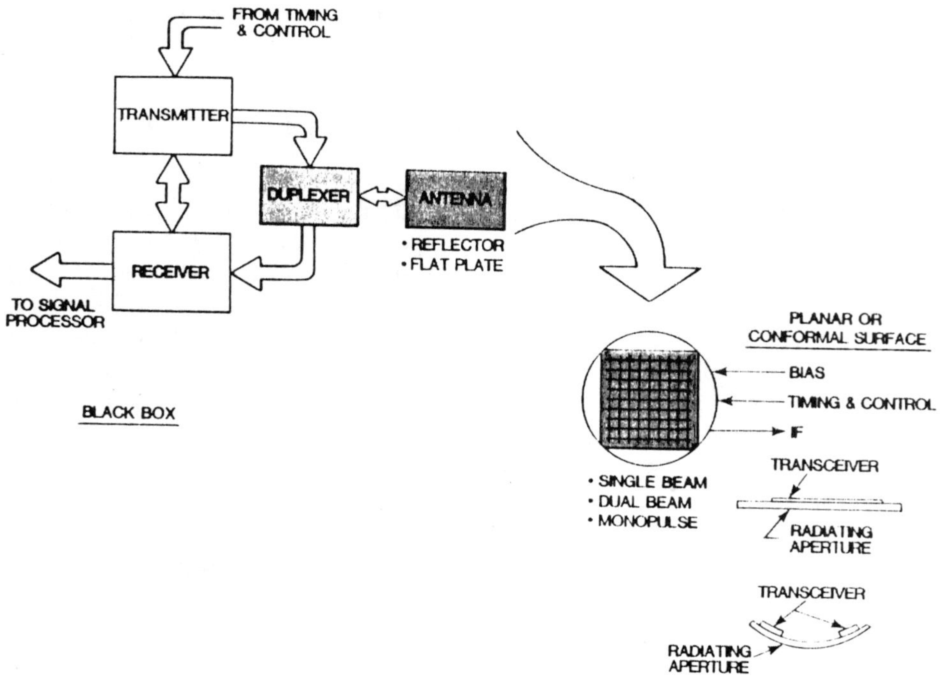

Figure 6. Evolution of Millimeter Wave Radar Hardware

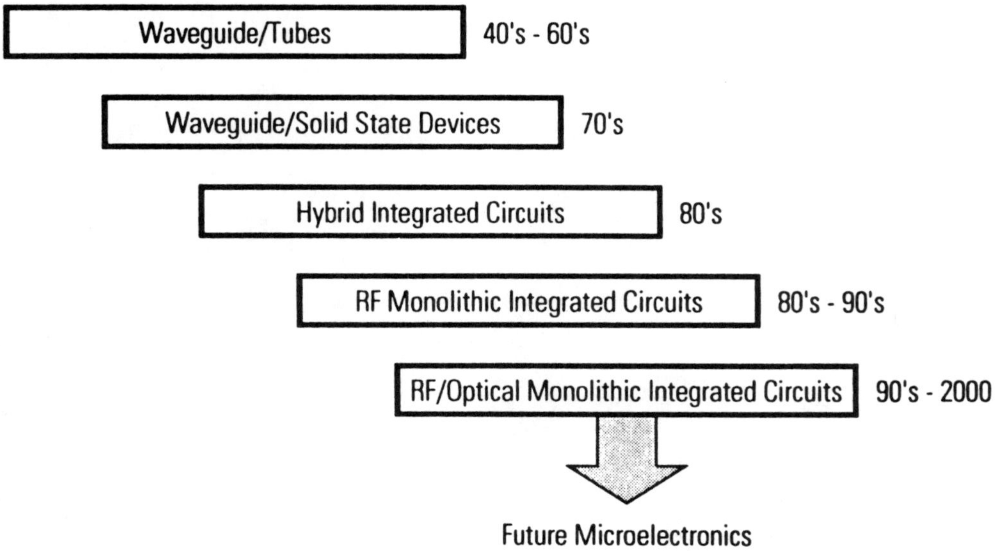

Figure 7. Historical Look at Millimeter Wave Component Development

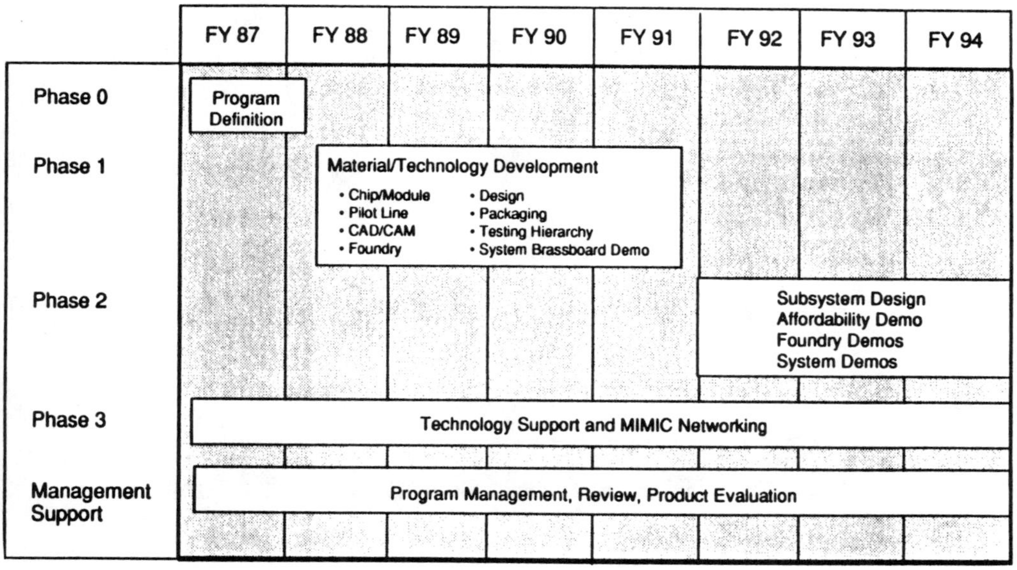

Figure 8. ARPA MIMIC Program Plan

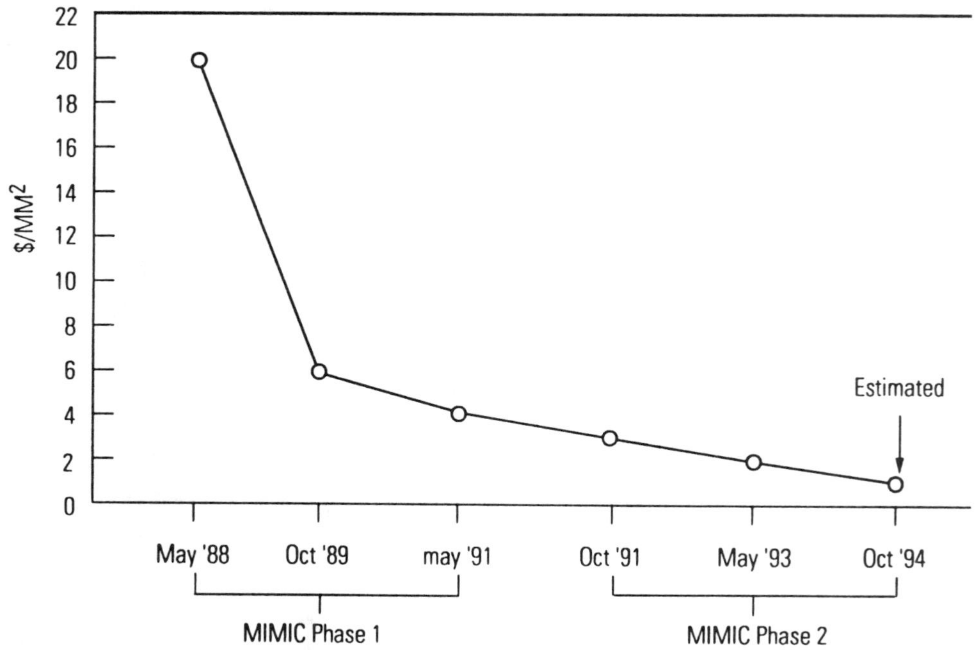

Figure 9. Chip Cost Reduction During MIMIC Program

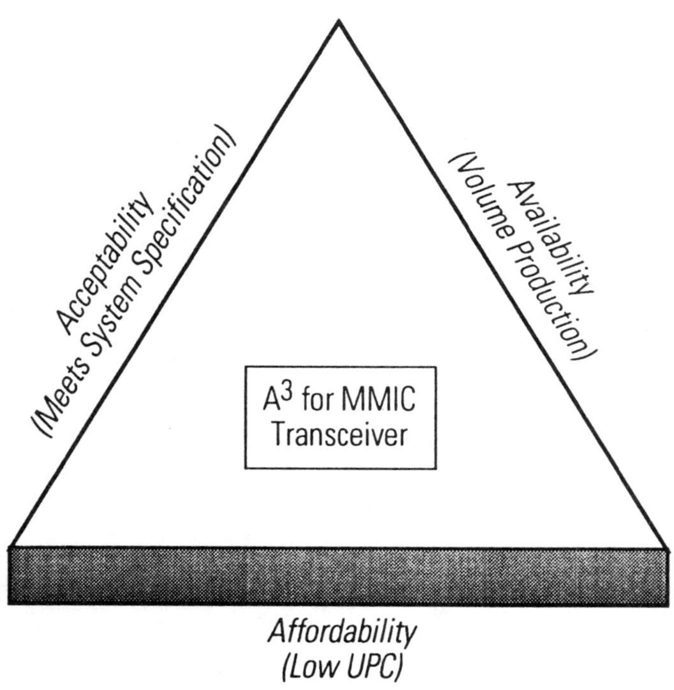

Figure 10. A^3 for Smart Weapon Radar Transceivers

weapons production program, a secondary source must be qualified and brought on-line as soon as possible to mitigate this risk of a wafer fab line shutdown. The third factor, affordability, is perhaps the most critical in smart weapons applications, where system costs must be tightly controlled and reflect learning curve experience. These costs relate to the chips themselves and the finished transceiver module. If the cost goals cannot be met, the MMIC insertion is in jeopardy regardless of any other positive features. Figures 11 and 12 show some early MIMIC program Ka-band chips based on the 0.25 micron MESFET. This development was very important since it forced a MESFET modeling understanding to be developed in Ka-band and demonstrated the feasibility of multifunctions being integrated on a single chip. The transmitter chip contains a fundamental VCO directly coupled to a multiple-stage power amplifier. Tuning is accomplished by applying a control voltage to the VCO gate. The receiver chip contains a low noise amplifier directly coupled to a mixer (active or passive), to isolate the critical receiver noise threshold circuits. Since the amplifier has gain and the mixer conversion loss or gain can be controlled, the overall system noise figure is determined by the amplifier. This has produced as much as a 3 db improvement in transceiver noise figure, compared to the more standard mixer input receiver design.

Figure 13 and 14 illustrate the important HEMT structures and their impact on receiver performance. HEMT emerged during MMIC Phase 1 as a technology of much promise and has matured during the Phase 2 program to become a viable MMIC insertion candidate. The pseudomorphic HEMT has now been applied to a number of circuit functions in the 30 to 100 GHz range. Impact on receiver noise figures has been dramatic, as depicted in Figure 14, with PHEMT structures of gate length 0.15 microns. It has recently been demonstrated that a single PHEMT wafer fab process can also support both low noise and high power. This removes the need to separately process receiver and transmitter MMICs and subsequently reduces chip cost. All functions for a Ka-band transceiver have now been fabricated using the PHEMT device process. Indeed for some applications, PHEMT has displaced MESFET as the device baseline, since it provides more design margin and higher wafer yields for individual circuits. A Ka-band HEMT multifunction receiver chip is shown in Figure 15. Electrical testing of this MMIC has consistently produced lower noise figures than its MESFET equivalent, although PHEMT chip size is slightly larger.

Millimeter wave MMIC technology has rapidly matured and appears to be focusing on the versatile PHEMT device structure and the HBT, particularly where noise performance is critical. It is anticipated that emerging InP PHEMT circuits will extend even further the achievable millimeter performance as observed from the data in Figure 14. Technology challenges remain in the following areas relative to smart weapon MMIC insertions:

- Foundry capability and availability
- Device technology maturity and reliability
- Chip design margin
- Transition from three-inch to four-inch wafer diameters for obtaining more chips per wafer at a lower cost.

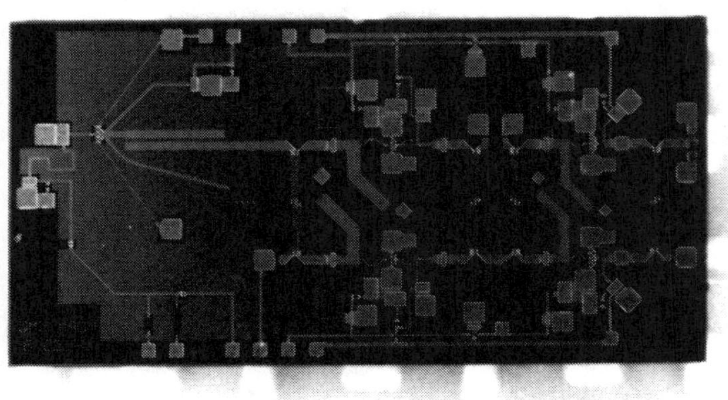

Figure 11. Ka-Band MESFET Multifunction Transmitter Chip

Figure 12. Ka-Band MESFET Multifunction Receiver Chip

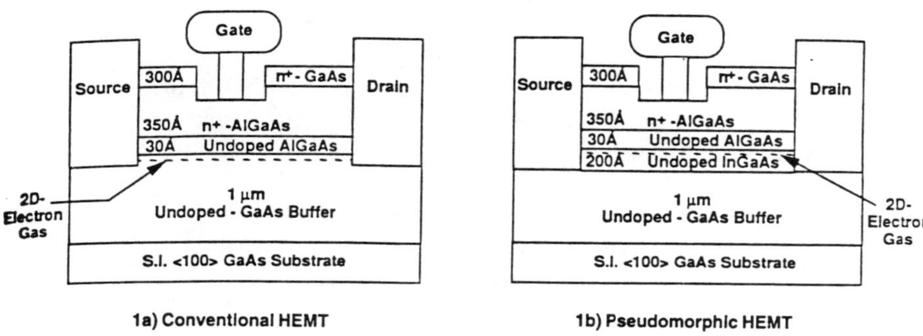

Figure 13. HEMT Device Channel Structures

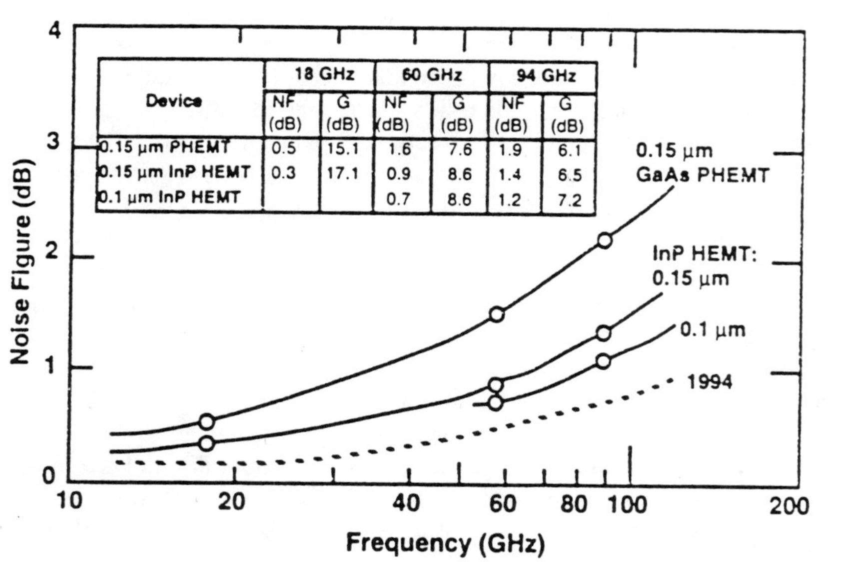

Device	18 GHz NF (dB)	18 GHz G (dB)	60 GHz NF (dB)	60 GHz G (dB)	94 GHz NF (dB)	94 GHz G (dB)
0.15 µm PHEMT	0.5	15.1	1.6	7.6	1.9	6.1
0.15 µm InP HEMT	0.3	17.1	0.9	8.6	1.4	6.5
0.1 µm InP HEMT			0.7	8.6	1.2	7.2

Figure 14. Measured Noise Performance of HEMT Devices at Room Temperature

3. PRODUCIBILITY CHALLENGES

For many radar MMIC transceiver functions at millimeter wave frequencies, producibility is only now being seriously addressed. This is particularly true of the PHEMT and HBT designs which are being demonstrated under the MIMIC program scope of work. Producibility of a MMIC transceiver is tied very closely to the level of insertion. General insertion levels are summarized as follows:

- MMIC Replacement
  - No major architecture changes in transceiver
  - Selective replacement of hybrid circuits, for example a balanced mixer.
- MMIC Supercomponents
  - Modified transceiver architecture, for example adding a LNA to the receiver
  - Partition transceiver into multifunction blocks for MMIC design
  - Possibly some remaining hybrid functions, for example a specialized transmitter.
- Total MMIC Insertion
  - All transceiver functions based on MMIC
  - Single chip architecture for complete radar frontend.

Current approaches for smart weapon radar MMIC insertion are at the second level, namely supercomponents. These generally take the form of partitioned transmitter and receiver circuit functions, with the resulting chips being interconnected with microstrip transmission lines and wire bonds. Figure 16 shows a view of a three-inch diameter wafer containing Ka-band MMIC components for a Ka-band transceiver. All the transceiver MMIC designs are contained on the main reticile, which is distributed around the wafer surface area from the center to outer edges. The wafer center is usually described as a sweet spot since circuit processing yields are very high. This diminishes significantly as one moves to the wafer edges with only partial reticle coverage.

Once the wafer fabrication steps have been completed, both DC and RF wafer probing techniques are used to determine the utility of the individual MMIC functions. Wafer probe stations are now available for selected RF parameters up to 110 GHz. These measurement results are made available for the radar systems engineer such that chip selection can be made before wafer dicing and sorting occurs. Wafer level testing at millimeter wave frequencies has moved from being an art to a reasonably well established engineering technique, particularly for both low noise and power amplifiers. Great care must be taken at these frequencies to avoid inducing amplifier oscillations and other spurious pickup into the measurement setup. From a producibility standpoint, a very experienced engineer is needed to facilitate the wafer RF testing. This translates into time and adds cost to the MMIC chip before it arrives at the user for transceiver insertion. Another form of MMIC chip performance measurement is frequently carried out in a test fixture. Figure 17 illustrates a Ka-band fixture design with the test chip mounted on a carrier and wire bonds made to the carrier such that a non-destructive evaluation may be carried out. It is not unusual to observe improved chip electrical performance in the fixture compared to that obtained from wafer probing simply due to an improved groundplane and impedance matching to all RF ports. A major task which

Figure 15. Ka-Band HEMT Multifunction Receiver Chip

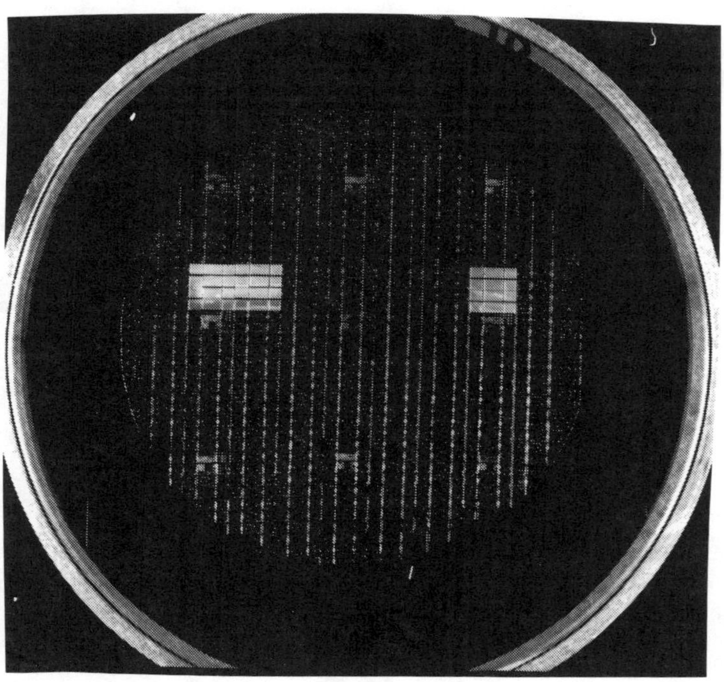

Figure 16. 3-Inch GaAs Wafer Containing Ka-Band MMIC Components

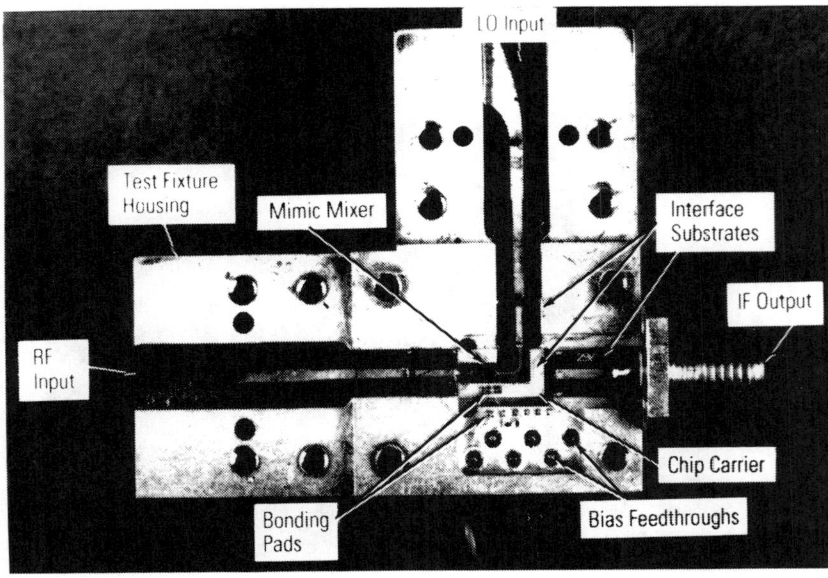

Figure 17. Ka-Band MMIC Mixer Mounted in Test Fixture

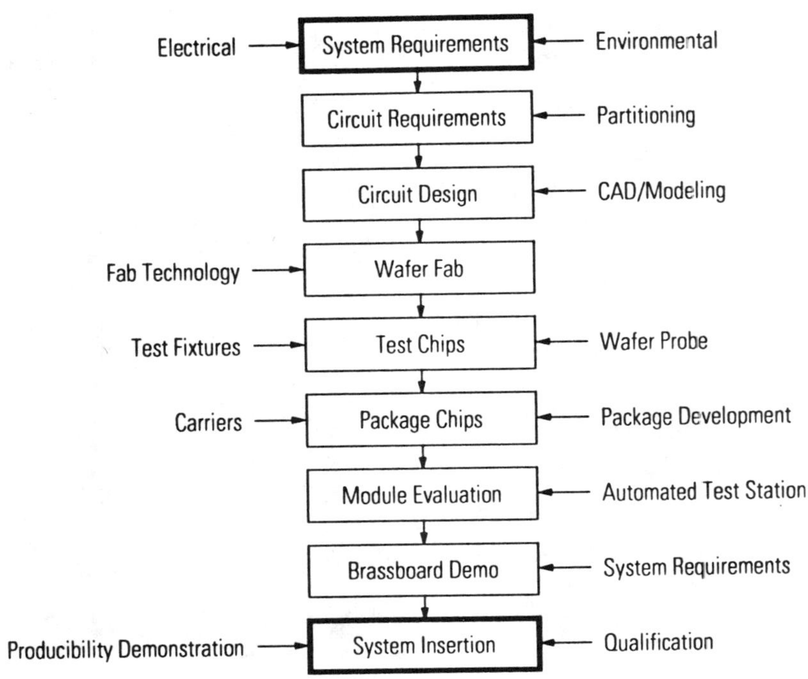

Figure 18. MMIC Technology—Requirements to Insertion

is ongoing is the correlation of fixture and wafer test data to accurately define MMIC performance.

The MMIC path from system requirements to system insertion is illustrated in Figure 18. Producibility assessments can be made starting at wafer fab, since it is assumed that a stable, reproducible design layout has been demonstrated. Once the tested MMIC chips are available, they are embedded in the transceiver module. The weapon system physical constraints dictate the module size, geometry, and bus power interface. Figure 19 shows a Ka-band transceiver module containing a two-chip architecture supporting all the transmit and receive functions. This radar is based on a single antenna FMCW design for strapdown radar sensor applications. From Figure 19, it is noted that the discrete component IF amplifier dwarfs the two MMIC transceiver chips. These chips are essentially bare die, since any package parasitics at these frequencies can significantly degrade transceiver overall electrical performance. A minimal number of wire bonds are required to interface with the chip set. A hermetic seal cover plate is laser welded in place on the module to complete the assembly process. This single module is then interconnected to a printed circuit antenna in a planar assembly to form a complete radar frontend, as shown in Figure 20. A single microstrip transmission line is used for the antenna/transceiver interface, while a flexible tape brings bus power in and IF signal out of the transceiver module to maintain a planar profile for the complete subsystem. In this case, the antenna beam pattern is formed by a comb array of radiators specifically tailored for the desired beamwidth, gain, and sidelobe structure.

This millimeter wave radar subsystem based on a full MMIC chipset lends itself to automated assembly and test technique in a production environment. Vision aided pick and place machine with automatic bonding machines, as well as optimized parts handling, have been investigated to minimize the touch labor in the module and antenna assembly process. A fully automated test station under software control provides radar subsystem testing and evaluation. The included list of references includes several good papers addressing producibility and cost issues for those interested in probing deeper into this topic. Actual MMIC insertion into a smart weapon system can occur anywhere in the four main phases of product evaluation to include:

- Advanced development
- Full scale engineering development
- Low rate initial production
- Full scale production.

Currently, more emphasis is being placed on designing in MMIC technology either in AD or FSED phases such that technology and engineering issues can be directly resolved and a producibliity philosophy built in to the subsystem hardware. For those systems already in production, ECP and P^3I paths are used to upgrade existing hardware with MMIC parts for overall system reliability and cost reduction.

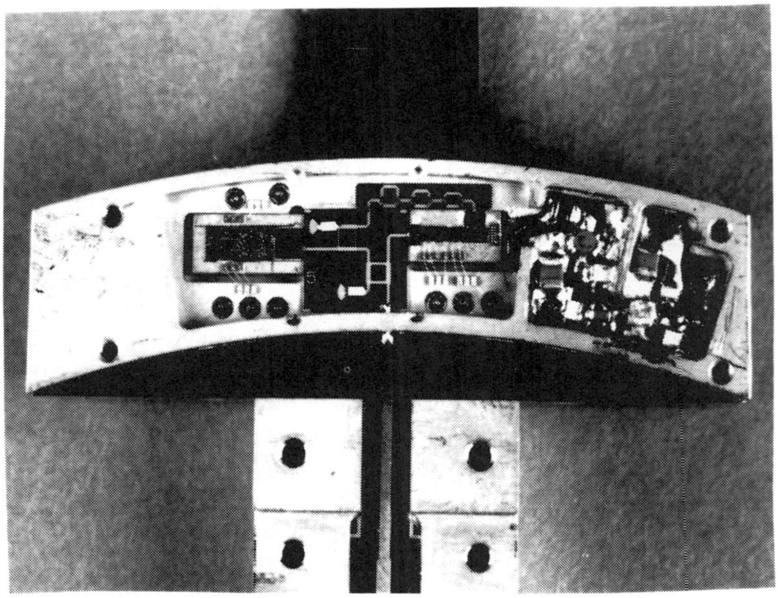

Figure 19. Ka-Band MMIC Transceiver With Two-Chip Architecture

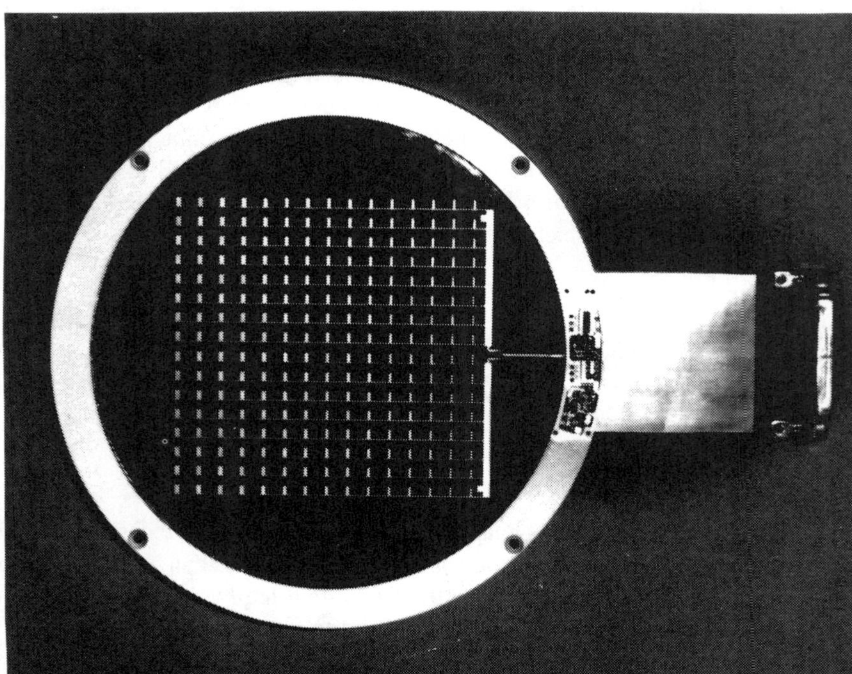

Figure 20. Ka-Band MMIC Transceiver Connected to Printed Circuit Antenna

4. FUTURE RADAR TRANSCEIVERS

As discussed earlier, several smart weapon systems are currently employing MMIC technology in their radar transceivers as a baseline configuration. This has generally been a successful endeavor, although as was observed on the VHSIC program for silicon microelectronics, there is a considerable difference between system operation in a laboratory ambient environment as contrasted with a full tactical scenario, particularly with a gun fired submunition. At this point in the MMIC technological roadmap, it is instructive to extrapolate ahead into future millimeter wave and supporting microelectronics. Figure 21 illustrates an important technological step forward for smart weapon radar. It is the realization of a small diameter, tactical, electronic scan antenna. Shown in Figure 21 is a MMIC T/R module with a single stick antenna subarray made up of a line of patch radiators. The marriage of MMIC transceivers to printed circuit antennas is indeed appropriate and a powerful tool to minimize radar frontend costs. Benefits from such an electronically-controlled antenna for terminally guided submunitions include automatic beam scan to carry out acquisition and tracking operations. This eliminates the need for gimbaled platforms and positioner subassemblies, which both have questionable reliability, high cost, and take up valuable packaging volume. The array of Figure 22 is but one element of a more complete structure, either functioning as a conventional phased array or as an active aperture phased array. The phase shifters are realized with MMIC designs including buffer amplifiers to wash out any significant insertion loss as the shifter responds to its digital control logic which can be coupled in via a photonic control link.

A future smart weapon radar functional block diagram is given in Figure 22. The millimeter wave radiating aperture and associated transceiver with signal conditioning will become a single subassembly with all RF functions implemented with MMIC technology and GaAs digital circuits used for the filtering, routing, and A/D functions. As indicated, the signal conditioner and central processor are mechanized with MCM-D modules for maximum packing density. These high levels of microelectronic insertion will bring with them testing challenges and the need for innovative heat dissipation approaches. In this configuration, the antenna and transceiver or T/R module array will be tested as a complete subassembly on an antenna range to verify beam movement, tracking accuracy, and other important RF parameters such as effective radiated power and system receive noise floor. MMIC is the enabling technology for such a radar mechanization, since its costs are prohibitive with any standard hybrid or waveguide techniques for the expendable system application. Many elements of this advanced weapon radar are either in-hand or currently being developed. It will become a demonstration brassboard in a few years, at least up through Ka-band.

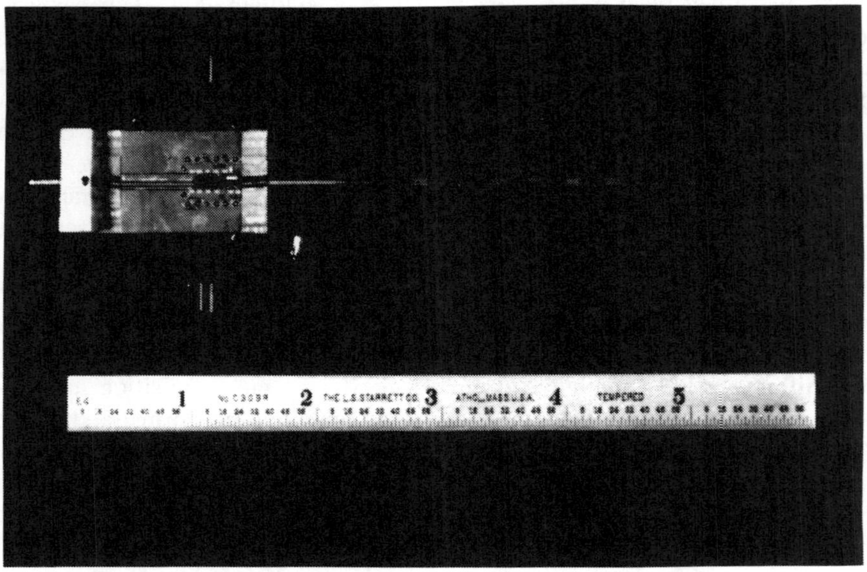

Figure 21. Ka-Band Subarray With MMIC TR Module

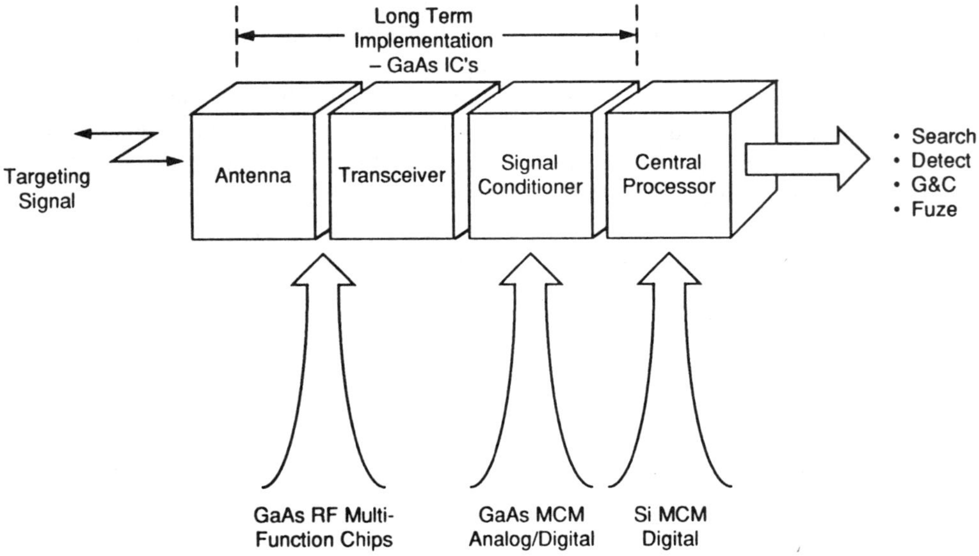

Figure 22. Microelectronics for Smart Weapon Radar Sensors

5. CONCLUSIONS

Millimeter wave MMIC technology has emerged from the laboratory benches to a place of insertion in smart weapon systems. Technical progress has been rapid with the ARPA MIMIC program serving as a catalyst for wide ranging advancements in device technology, CAD tool development, wafer-level measurements, wafer fabrication, and brassboard insertion demonstrations. The following points are appropriate to conclude this discussion of MMIC capabilities for smart weapon sensors.

- Despite DoD procurement reposturing, smart weapons remain a very important insertion vehicle for emerging MMIC technology
- Millimeter wave radar sensors are baselined for virtually all smart weapon systems
- These sensors are highly dependent on MMIC technology for achieving performance margin and production cost reduction
- MMIC second source identification and qualification are required for production system needs
- Level of MMIC insertion—component, subsystem, or system is critical to system utilization
- Cost is the most important parameter for MMIC insertion in expendable smart weapons
- PHEMT and HBT are emerging as the dominant device building blocks for millimeter wave MMICs
- The MIMIC program has provided a valuable inventory of relevant chip design layouts and associated test databases
- MMIC is an enabling technology for weapon systems moving from smart to brilliant performance capabilities.

Despite the affirmative status of MMIC technology summarized above, there are still a number of areas recommended for future work. These include:

- Develop higher levels of circuit integration and packaging for multifunction modules
- Expand CAD, CAT, and CAM tool developments to cover the frequency range up to 100 GHz
- Expand wafer-level RF testing to include all transceiver circuit functions up to 100 GHz
- Validate MMIC cost models and continue cost reduction through design compaction, improved wafer yields, and optimized testing strategies

- Early transition from three-inch to four-inch diameter wafers to make more MMIC components available per wafer fab

- Continue efforts to optimize transceiver/antenna collocation architectures with resulting higher levels of circuit integration

- Investigate and demonstrate new millimeter wave MMIC components such as InP HEMT amplifiers and HBT fundamental oscillators.

6. ACKNOWLEDGEMENTS

The author has been actively engaged in millimeter wave sensor development for smart weapons over the past 25 years. This has included MMIC design, fabrication, testing, and brassboard insertion. The work was initiated at Honeywell and most recently has been continued within the Defense Systems Group at Alliant Techsystems. It is a pleasure to acknowledge the many contributions of my colleagues in these two organizations who throughout the years have brought millimeter wave microelectronics from a science to a viable engineering technology for smart weapon radar applications.

7. REFERENCES

1. N.C. Currie, et al, "MMW System Tradeoffs," IEEE AES Magazine, October 1988, pp. 21–26.

2. Y.J. Lubkin, "Millimeter Waves," Defense Systems and Electronics, January 1987, pp. 11–15.

3. Editor, "Pentagon Identifies 21 Key Technologies for Maintaining U.S. Military Lead," Aviation Week and Space Technology, 20 May 1991, p. 57.

4. S.M. Hardy, "Missile Seekers Get Better," Journal of Electronic Defense, March 1993, pp. 36–42.

5. T. Corrado, "Smart Munitions—Are They Affordable?," Microwave System News, December 1988, pp. 32–36.

6. K.J. Sleger, "Government Systems Incorporate GaAs Monolithics," Microwave System News, September 1983, pp. 82–97.

7. D. Hughes, "Raytheon Capitalizes on Microelectronics in BAT Seeker, Other Smart Munitions," Aviation Week and Space Technology, 2 March 1992, pp. 49–51.

8. B.W. Henderson, "New MMIC Chips Could Make Millimeter Wave Radar Affordable," Aviation Week and Space Technology, 26 March 1989, pp. 78–79.

9. D.G. Fisher, "GaAs IC Applications in Electronic Warfare, Radar, and Communications Systems," Microwave Journal, May 1988, pp. 275–292.

10. E.H. Gregory, "Applicability, Availability, and Affordability of GaAs MMICs in Military Systems," Microwave Journal, March 1987, pp. 119–126.

11. C.R. Seashore, "Integrated Circuit Radar and Radiometric Sensors," SPIE Millimeter Wave Technology III, Vol. 544, 1985, pp. 157–159.

12. T.L. Duffield, "Monolithic MM-Wave ICs for Smart Weapons," Microwave Journal, April 1988, pp. 91–111.

13. F. Lalezari, et al, "Millimeter Wave Tactical/Small Aperture Antennas," Microwave System News, December 1988, pp. 41–50.

14. M.N. Cohen, "Target Recognition for Autonomous Smart Munitions," Journal of Electronic Defense, July 1990, pp. 38–42.

15. P.R. Jaszka, "Guidance Systems Advance in Air-Launched Missiles," Defense Electronics, May 1989, pp. 46–54.

16. K.J. Sleger, et al, "Trends in Solid-State Microwave and Millimeter Wave Technology," IEEE MTT-S Newsletter, Fall 1990, pp. 11–15.

17. W.H. Perkins and T.A. Midford, "MMIC Technology: Better Performance at Affordable Cost," Microwave Journal, April 1988, pp. 135–143.

18. Editor, "Pentagon's MIMIC Program Shows Potential Electronic Warfare Benefits," Aviation Week and Space Technology, 22 October 1990, pp. 50–53.

19. J.W. Rawles, "MIMIC Program Shifts into High Gear," Defense Electronics, May 1991, pp. 49–56.

20. M.R. Stiglitz and C. Blanchard, "Phase 2 of the DoD MIMIC Program: A Status Report," Microwave Journal, January 1992, pp. 90–100.

21. C. Adams, "MIMIC: Upping Yields, Cutting Costs, and Integrating Functions," Military and Aerospace Electronics, May 1992, pp. 17–19.

22. J.A. McDonald, "MMICs Step Out of the Lab and Into the Field," Military and Aerospace Electronics, 15 March 1993, pp. 21–23.

23. M.R. Stiglitz, "Unraveling the M?MIC Knot," Journal of Electronic Defense, May 1994, pp. 43–47.

24. C.D. Caposell, "A Navy View of MIMIC," Journal of Electronic Defense," July 1992, pp. 56–62.

25. D.W. Hughes, "GaAs MMICs: Some Lessons From Silicon," Journal of Electronic Defense, March 1991, pp. 51–54.

26. A. Kurdoghlian, C.S. Wu, et al, "The Demonstration of Ka-Band Multi-Functional MMIC Circuits Fabricated on the Same PHEMT Wafer With Superior Performance," 1993 IEEE Microwave and Millimeter Wave Monolithic Circuits Symposium, pp. 97–98.

27. T.I. Hsu, G. Swift, et al, "A 44 GHz HEMT Amplifier," International Journal of Infrared and Millimeter Waves, Vol. 7, No. 7, 1986, pp. 999–1004.

28. J.A. Higgins, "GaAs Heterojunction Bipolar Transistors: A Second Generation Microwave Power Amplifier Transistor," Microwave Journal, May 1991, pp. 176–193.

29. J.A. McDonald, "DARPA Funds Two HBT Fab Lines," Military and Aerospace Electronics, 15 October 1992, pp. 1 and 15.

30. S. Chen, et al, "U-Band MMIC HBT DRO," IEEE Microwave and Guided Wave Letters, Vol. 4, No. 2, February 1994, pp. 50–52.

31. E.M. Godshalk and J. Pence, "Low-Cost Wafer Probe Scales 110 GHz Summit," Microwaves and RF, March 1993, pp. 162–167.

32. A.J. Slobodnik, et al, "A Monolithic GaAs 36 GHz Four-Bit Phase Shifter," Microwave Journal, pp. 106–111.

33. K.J. Sleger, "GaAs Monolithics for Affordable Military Systems," Journal of Electronic Defense, August 1985, pp. 27–46.

34. D.N. McQuiddy, "High Volume GaAs MMICs for Military Systems," Applied Microwave, November/December 1989, pp. 35–43.

35. R.D. Skinner, "Gallium Arsenide Chip Costs," Applied Microwave, Winter 1993, pp. 19–30.

36. J. Browne, "MM Waves Aid Commercial Applications," Microwaves and RF, July 1992, pp. 113–116.

37. C. Huang, "MMICs Move into New Marketplaces," Microwaves and RF, September 1992, pp. 134–138.

38. R. Schneiderman, "GaAs Sales Gain as Markets Mature," Microwaves and RF, October 1992, pp. 31–37.

39. V.G. Gelnovatch, "More Quality at Less Cost," Microwave Journal, April 1989, pp. 26–28.

40. R.T. Kemerley and D.F. Fayette, "Affordable MMICs for Air Force Systems," Microwave Journal, May 1991, pp. 172–175.

41. J. Rhea, "MIMIC Insertion Begins with HARM," Armed Forces Journal International, June 1990, pp. 58–59.

42. Editor, "TI Wins Microwave Packaging Job," Microwaves and RF, October 1993, p. 49.

43. S.K. Wang, et al, "Producibility of GaAs Monolithic Microwave Integrated Circuits," Microwave Journal, June 1986, pp. 121–133.

44. A.K. Sharma, "Considerations in Producibility Engineering of MMICs," IEEE Trans on MTT, Vol. 38, No. 9, September 1990, pp. 1242–1248.

45. L.R. Whicker, "Two Years of Progress on the Manufacturing Technology for the Solid-State Microwave Systems Program," Microwave Journal, May 1988, pp. 295–306.

46. A.M. Pavio, "The High Volume Production of Advanced Radar Modules," Microwave Journal, January 1989, pp. 139–148.

47. H.I. Ellowitz, "World Class Manufacturing in the U.S. Microwave Industry," Microwave Journal, pp. 26–35.

48. R. Tang and R. Brown, "Cost Reduction Techniques for Phased Arrays," Microwave Journal , January 1987, pp. 139–146.

49. L. Raffaelli and E. Stewart, "A Standard Monolithic Transmitter for 38 GHz PCN Applications," Microwave Journal, October 1992, pp. 24–30.

50. L. Raffaelli, E. Stewart, et al, "77 GHz Transmitter for Automotive Radar," Applied Microwave, Winter 1993, pp. 80–86.

51. F.P. Pantuso, "What's New in Microwaves?," Journal of Electronic Defense, June 1992, pp. 42–61.

52. G.H. Heilmeier, "Microelectronics: End of the Beginning or Beginning of the End?," IEEE AESS Newsletter, March 1985, pp. 1–3.

53. K.J. Sleger, et al, "InP Monolithic Integrated Circuits for Millimeter Wave Applications," Microwave Journal, May 1984, pp. 175–189.

54. L.D. Nguyen, et al, "50nm InP High Electron Mobility Transistors," Microwave Journal, June 1993, pp. 96–101.

55. J. Berenz, "InP MMICs: Coming of Age," Microwave Journal, August 1993, pp. 113–114.

56. R. Rosenzweig, "Crossroads for the U.S. GaAs IC Industry," Applied Microwave, Winter 91/92, pp. 15–20.

57. Z. Lemnios, et al, "RF Missile Seeker Components Realized by ASMMIC Techniques," Microwave Journal, June 1988, pp. 167–176.

58. P. Herczfeld, "Monolithic Microwave—Photonic Integrated Circuits: A Possible Follow-up to MIMIC," Microwave Journal, January 1992, pp. 64–78.

59. T.F. Brukiewa, "Active Arrays: The Key to Future Radar Systems Development," Journal of Electronic Defense, September 1992, pp. 91–115.

60. J.J. Donovan and A. Firstenberg, "GaAs Integrated Circuits for Next Generation Government Electronic Systems," IEEE AESS Newsletter, November 1984, pp. 6–9.

Millimeter Radar: Current Assessment, Future Directions

Edward K. Reedy
Wayne L. Cassaday

Georgia Tech Research Institute
Atlanta, Georgia 30332, USA

ABSTRACT

After a short introduction to millimeter wavelength (MMW) radar, this paper presents a current survey and assessment of MMW technology, especially as it relates to radar applications; discusses some important recent research in MMW propagation modeling, clutter characterization, and multipath prediction; presents some examples of recent and on-going MMW radar hardware development programs in the military, phenomenology/instrumentation, and commercial areas; and concludes with a brief discussion of possible future directions for millimeter wavelength radar.

1. BACKGROUND AND INTRODUCTION

Research in and development of millimeter wavelength radar had its genesis in the component development occurring in the latter stages of World War II, especially the development of the MMW magnetron in 1946. Following the war, several experimental MMW radars were developed and evaluated providing a rich data base on which the development of much more advanced radars in the late 1950's and 60's was based. These radars were still relatively rudimentary and were evaluated for such applications as short-range search and target acquisition, weather monitoring, low-angle tracking, and ground surveillance.[1] These same applications of MMW radar, along with many others, are still under active investigation today.

Over the years following the war, support for radar research in the millimeter-wave region, especially at frequencies above 35 GHz, tended to be somewhat cyclic in nature. However, during the mid to late 1980's interest in the so-called "smart," "brilliant," and autonomous weapons systems led to a resurgence of interest in the MMW spectral region and funding, especially by the military, of research in this frequency region.

Only a cursory examination of the well-known curves for atmospheric absorption or atmospheric attenuation versus frequency, such as that shown in Fig. 1, is necessary to identify the preferred radar-operating frequencies in the millimeter band. The millimeter region is nominally considered to be that frequency band between 30 and 300 GHz. Within this region, four so-called "propagation window" frequencies exist where the atmospheric absorption is at a minimum. These occur at approximately 35, 95,* 140, and 220 GHz and define the nominal radar operating regions in the millimeter band. Most recent emphasis has been in the 35 GHz and 95 GHz bands.

* Shown in Figure 1 as 94 GHz

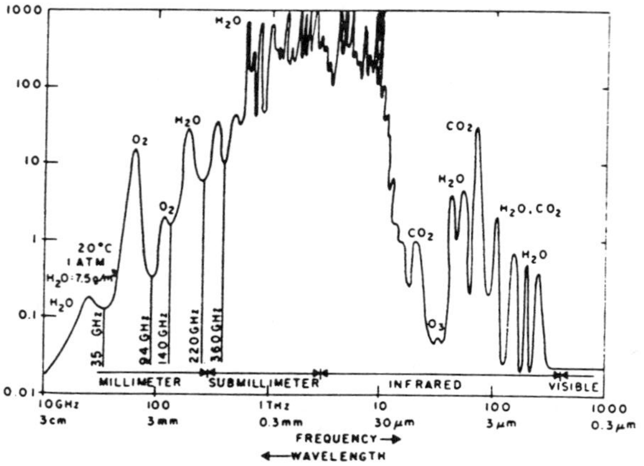

Fig. 1. Clear weather atmospheric attenuation of electromagnetic radiation as a function of frequency. (From Preissner, 1978, [2])

Applications requiring covertness or relatively short-range operation have occasionally resulted in radar or communications systems operating on or near one of the oxygen or water vapor absorption bands where molecule and electromagnetic interaction resonances produce regions of abnormally large absorption. Two of the more widely used absorption bands are near 22.5 and 60 GHz.

The MMW radar systems tradeoff considerations given in Table 1 dictate that the primary potential applications of millimeter wave radar exist in satisfying sensor requirements that involve physically small, lightweight, high resolution, relatively short-range radar possessing good Doppler characteristics. Millimeter radar is limited under adverse weather conditions when compared with microwave systems, but does exhibit higher resolution characteristics. Again, when compared to microwave radar, millimeter systems are typically much smaller physically. Although not having the extreme resolution of higher frequency sensors operating in the IR and visible regions, millimeter radar has superior propagation characteristics through smoke, haze, dust, and fog. Possessing characteristics of both the microwave and electro-optical (EO) regions, millimeter wave radar attempts to blend the advantages of both its microwave and EO counterparts in such a manner to reduce or minimize their limitations or disadvantages.

Table 1 Millimeter-Wave Radar System Tradeoff Considerations

Advantages	Limitations
Physically small equipment	Component cost high
Low atmospheric loss[a]	Component reliability/availability low
High resolution	Short range (10 - 20 km)
Angular	Weather propagation[b]
Doppler	
Imaging quality	
Classification	
Small beamwidths	
High accuracy	

Table 1 Millimeter-Wave Radar System Tradeoff Considerations (Continued)

Advantages	Limitations
Reduced ECM vulnerability	
Low multipath/clutter	
High antenna gain	
Large bandwidth	
High range resolution	
Spread spectrum ECCM	
Doppler processing	
ECM	

[a] Compared to IR and visual wavelengths.
[b] Compared to microwave frequencies.

The importance of the first two limitations should gradually be reduced as activity in the MMW area continues to increase and leads to production and large-scale manufacture and standardization of MMW components, along with the introduction of MMW circuit integration and increased use of microstrip and dielectric waveguide components.

Even with the advantages given in Table 1 notwithstanding, a MMW radar should not be considered a panacea for all radar surveillance and tracking problems. For terrestrial applications, MMW radar is basically a short-range, high-resolution sensor which may be severely degraded in some adverse weather environments. However, for many applications, it represents an excellent compromise between the performance characteristics of microwave and lower frequency radar on the one hand, and optical and infrared sensors operating at higher frequencies, on the other. Table 2, which is adapted from Bhartia and Bahl,[3] compares the relative performance of MMW radar with its microwave and optical counterparts for several important radar operating characteristics. MMW radar performance usually falls between the extremes of the microwave and optical sensors.

Table 2 Radar System Performance Comparison (Adapted from Bhartia and Bahl [3] © 1984 John Wiley and Sons)

Radar Characteristics	Microwave	Millimeter Waves	Optical
Tracking Accuracy	Fair	Fair	Good
Classification or Identification	Poor	Fair	Good
Covertness	Poor	Fair	Good
Volume Search	Good	Fair	Poor
Adverse Weather Performance	Good	Fair	Poor
Performance in Smoke, Dust, etc.	Good	Good	Poor

The remainder of this paper presents a current survey and assessment of MMW technology, especially as it relates to radar applications; discusses some recent research in MMW propagation modeling, clutter characterization, and multipath prediction; presents some examples of recent and on-going MMW radar hardware development programs in military, phenomenology/instrumentation, and commercial areas; and concludes with a brief discussion of possible future directions for millimeter wavelength radar.

2. CURRENT TECHNOLOGY ASSESSMENT

The continued development and utilization a of MMW systems is dependent upon the availability of reliable and cost effective components. Increased availability of MMW components has already spawned interest in new uses of MMW systems such as measuring fluids in zero gravity environments[4] and monitoring production in manufacturing such as measuring the fill level in blast furnaces.[5] Hughes Aircraft Company engineers have shown how off-the-shelf MMW components can be used in a cost effective power and attenuation standard to calibrate electronic warfare, radar and communication systems.[6] There continues to be large research efforts in the U.S. and abroad leading to the development of MMW components to meet the needs of current and future systems. In the U.S., research continues in active components such as high power transmitters, semiconductor receivers and transceivers using high mobility electron transistors (HEMT), microwave and millimeter wave monolithic integrated circuits (MIMIC), and multichip modules (MCM) which allow several discrete components to operate in a single high density package. Additional research into MMW passive components such as antennas and waveguide components supports new applications and more efficient use of the energy available to MMW systems.

2.1 Transmitters

Several advances in the MMW transmitter arena are under development including efficient solid state modules, helix traveling wave tubes (TWT), and fast wave devices such as the free-electron lasers (FEL), electron cyclotron masers (gyrotron), and cyclotron auto-resonant masers (CARM). The advent of these transmitters will allow new uses for MMW systems that were previously limited by space, efficiency or transmit power. More powerful and efficient solid state transmitter modules will permit the development of MMW phased array antennas for use in communications and tactical radar scenarios. Helix TWTs provide compact and efficient MMW power for use where moderate power, <1kW, is needed in a confined space. The extremely high power available from the fast wave devices (some providing more than 1 gigawatt peak and megawatt average power levels) will assist in overcoming the atmospheric absorption normally associated with MMW frequencies. A 95 GHz radar has been proposed for high resolution imaging of satellites based a gyrotron or other fast wave device.[7]

2.1.1 Fast Wave Devices

In fast wave devices, the frequency of operation is not determined by the size of the interaction region of the device. Fast wave devices use a beam of relativistic electrons injected into a strong magnetic field which creates the coherent oscillation of the electrons at the desired radiating frequency. Consequently, the same fast wave device can be used to generate different frequencies between the microwave and IR regions by changing only the magnetic field strength. High peak and average powers have been generated using fast wave devices at MMW frequencies.

The three primary fast wave devices under development are the gyrotron, the cyclotron auto-resonant maser (a type of gyrotron) and the free electron laser. These three will be discussed briefly in this paper.

The gyrotron or electron cyclotron resonant maser radiates energy from relativistic electrons "gyrating" about a magnetic field. The electrons radiate coherently because of bunching caused by the dependence on the gyration frequency determined by the strength of the magnetic field. Gyrotrons are predicted to have efficiencies as high as 90%[8,9] but currently tests on MMW gyrotrons show efficiencies of 20-40%. Operation of gyrotrons in the fundamental harmonic mode at MMW frequencies often requires a superconducting magnet. However, conventional magnets can be used in gyrotrons operating at higher harmonic modes.

The cyclotron auto-resonant maser (CARM) is a Doppler up-shifted form of the gyrotron. The CARM can operate in larger cavities than gyrotrons for the same frequency and because of the Doppler up-shift, theoretically, high power can be achieved in a highly efficient, >40%, manner with moderate magnetic field strengths.[10] A 35 GHz CARM was demonstrated by MIT to have 17 MW of output power at efficiencies of 6%.[11] However, most CARM devices have not produced energy according to their theoretical design. Consequently, these devices are not approaching either the necessary power levels (given their size and input power requirements) or reliability necessary to be useful for radar systems.

The operating principle of the FEL involves modulating a high power electron beam within an optical cavity by passing it through a high field strength magnetic undulator as shown in Fig. 2.[8] The output frequency of the FEL is determined by the period of the "wiggler." Because the period of the wiggler can be electronically controlled, real-time tunable bandwidths of 10% (24% total bandwidth) have been achieved with an FEL at 35 GHz frequencies.[12] FELs have also exhibited very high efficiencies, >60%, with respect to other types of MMW transmitters. FELs have been demonstrated with pulsewidths of up to 50 ns and PRFs of 1 kHz. While this pulsewidth may not be useful for many systems, it has been proposed that multiple pulses could be generated back-to-back to synthesize a single coherent long pulse of 1 microsecond, at reduced PRFs.

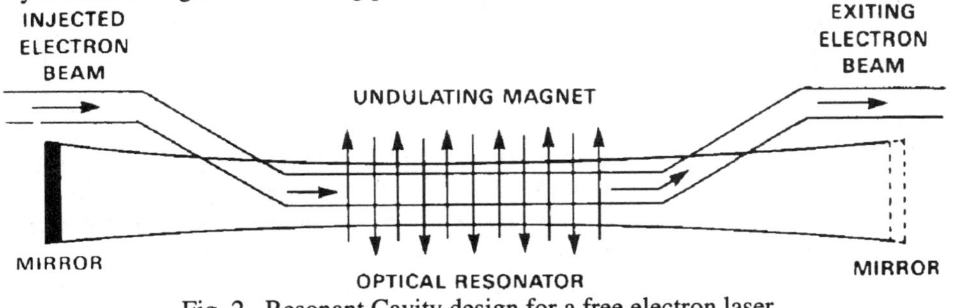

Fig. 2. Resonant Cavity design for a free electron laser.
(From Granatstein and Alexeff [8] 1987 Artech House)

2.1.2 MMW Helix TWT

Helix TWTs have been in use at microwave frequencies for more than a decade because of their inherent advantages of efficient, moderate power output over a wide bandwidth in packages smaller than standard coupled cavity approaches. Helix TWTs also provide better heat dissipation capacity and are easier to operate because the beam voltages are significantly reduced over other TWTs. However, the manufacturing tolerances needed for these devices has slowed the use of helix TWTs at MMW frequencies. Many U.S. and foreign manufacturers began development of MMW helix TWTs in the 1980s and current product lines of several tube vendors include Ka-band helix TWTs. However, research and development of W-band helix TWTs is on-going.[13,14] A 95 GHz helix TWT was recently demonstrated to have 10 W of CW power over a 3 GHz bandwidth. The amplifier provided 40 dB of gain with only a 13 kV electron beam voltage.

2.2 Integrated Circuits

The single largest impact to MMW radar which, according to C. R. Seashore, will "change system engineering thinking relating to active and passive radar architectures" is advances of integrated circuits for radar components.[15] This revolution in the radar development industry is a direct outgrowth of the ARPA supported MIMIC program which began in the mid-1980s and entered Phase II in the 1990s.[16] Integrating high electron mobility transistor (HEMT), MESFET and hybrid components onto a single substrate and co-locating subsystems such as transmitters, receivers and antennas on a single chip will allow compact, low cost and energy prudent systems useful in expendable radars. These technologies will also reduce system losses associated with waveguides and large insertion loss components to improve performance in future MMW radar systems.

The development of HEMT components utilizing gallium arsenide based materials (GaAs and InGaAs) now provides improvements of 4 to 8 dB in receiver noise figures over previously available MMW components.[17] Utilizing HEMT technology, low noise receivers (2.75 dB) with 5% bandwidth are being employed in new phased array applications[18] and a single monolithic chip FM-CW Ka-band transceiver has been demonstrated.[19] A power amplifier module used in the LONGBOW transceiver contains 16 monolithic microwave integrated circuits utilizing power HEMT (PHEMT) technology to deliver 16 watts of output power at Ka-band.[20] The modules were designed, developed and delivered in less than 12 months because of the continued research and development under Phase II of the U.S. Army's MIMIC program.

2.3 Antennas

Many applications for MMW radars place severe aperture and weight restrictions on the design engineers. Current MMW radar antenna technology relies heavily on standard dish or waveguide fed lens. Advances in microstrip,[21] planar,[21, 22] waveguide, and hybrid antennas have been able to reduce the weight, space requirements and cost of the antennas for many applications. A highly efficient quasi-integrated horn antenna has been developed for operation at W-band. The antenna shown in Fig. 3 provides 20-dB of gain (an improvement of 10 dB over a standard integrated horn) by matching a flared machined section to an integrated horn antenna.[23] The pattern for this quasi-integrated horn is shown in Fig. 4. In other research, corrugated dielectric waveguide has been used to develop an integrated antenna which can be scanned.[24] Advances continue in the designs of monopulse antennas as well. A 1.7 inch diameter five-port hybrid-integrated monopulse antenna has been demonstrated at 94 GHz to have 30.6 dB of one way gain, first sidelobe levels down 39 dB and 63% efficiency.[25]

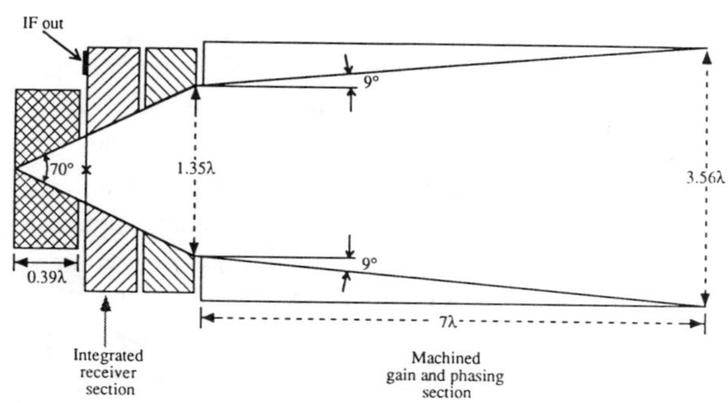

Fig. 3. Design of a quasi-integrated horn antenna.

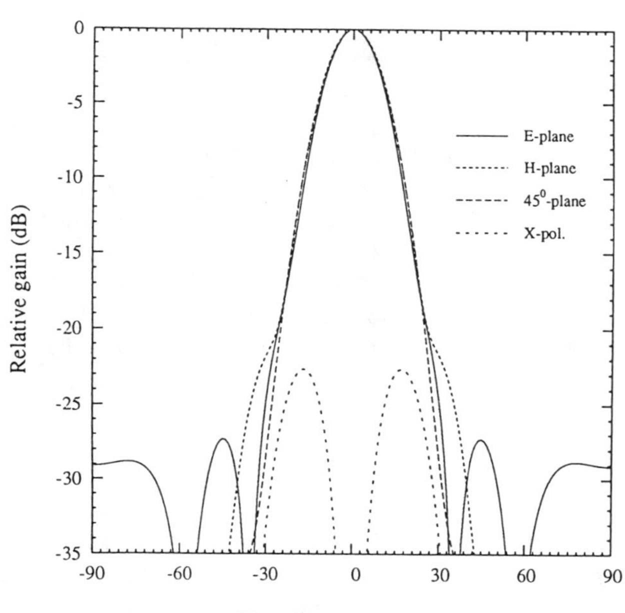

Fig. 4. Antenna patterns of a quasi-integrated horn antenna.

2.4 Passive Components

Beam waveguides (BWG) and other quasi-optical components have been considered for use in radar systems for more than 30 years. Many practical and theoretical quasi-optical components have been developed which could replace existing radar devices,[26, 27] and some quasi-optical components have been used in satellite communication applications since the early 1970's. In 1991, the installation of a quasi-optical high-power, dual-polarized angle tracking radar system completed the replacement of an existing 35 GHz microwave system at the Kiernan Reentry Measurements System (KREMS) in the Kwajalein Atoll.[28] This quasi-optical system provided more than 10 dB better sensitivity over the original microwave system while handling 60 kW peak (5 kW average) power. The BWG system replaced many major components in the radar including a circular polarizer, circulator, and polarization filters. The original microwave system and the quasi-optical components that replaced it are shown in Fig. 5 and 6. A quasi-optical power combiner was also used to coherently merge the output power of two 50 kW Ka-band TWTs. Similar upgrades are planned for the KREMS 95 GHz radar. (It should be noted that blockage by the BWG system of a Cassegrain reflector makes utilization of quasi-optical techniques most effective on systems with an aperture of 400 wavelengths or greater.[28])

3. THEORY/PHENOMENOLOGY

3.1 MMW RCS Measurements and Modeling

Because it is important to system designers to understand both the radar and the radar environment, much effort has been invested into both measuring radar targets (buildings, land, sea, etc.) and attempting to predict or model the returns from radar targets. Radar frequencies, waveforms, and even detection/classification algorithms are often chosen based on measurements and predictions of radar signatures – signatures that include amplitude, phase, and even polarization effects. The modeling of radar targets utilizing physical optics and physical theory of diffraction predictions is fairly well understood at frequencies between 500 MHz and 15 GHz. The extension of these predictions to MMW frequencies has not been extremely successful. Consequently, significant resources have been expended trying to understand the MMW wave environment from empirical measurements and computer models which may be based on either theoretical or empirical information about the targets.

3.1.1 Advances in Modeling for MMW Frequencies

A major consideration in modeling at MMW frequencies is the fact that most targets or major scatterers of interest are significantly larger than a wavelength. This leads to the rise of more complicated models because scatterers are not "discrete," as they may have been at lower frequencies, and second order effects such as surface roughness, edge effects, and surface curvature come into play. The most accurate way to include these "new" scatterers in signature predictions is to increase the fidelity of the target model by providing more scatterers and more detail on the target structure. This has the negative impacts of requiring significantly more effort in the construction of the computer models (requiring a much more detailed understanding of the target characteristics) and the

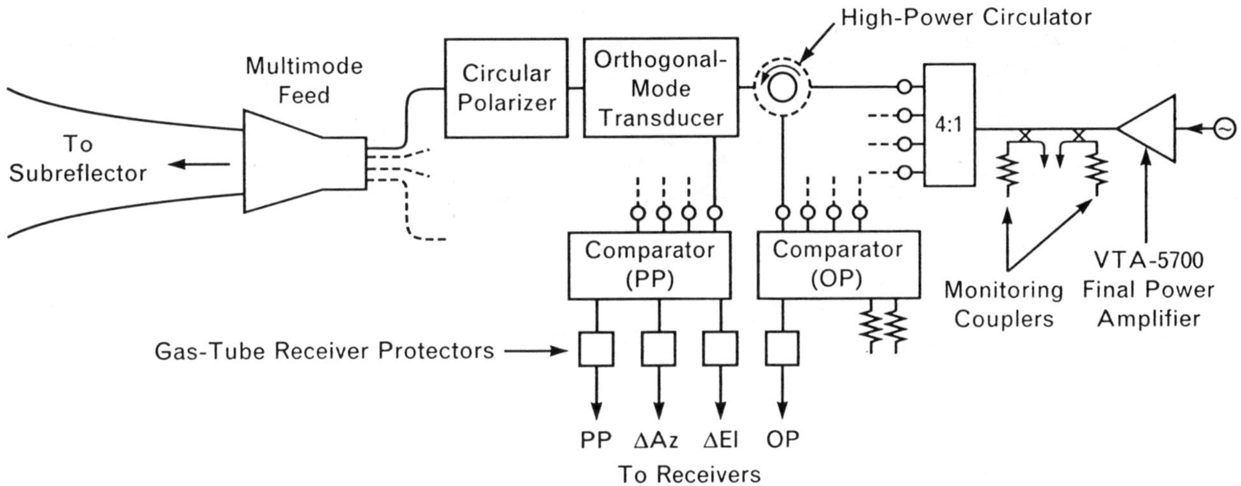

Fig. 5. Microwave system used in the Kwajalein Missile Range Ka-band radar.

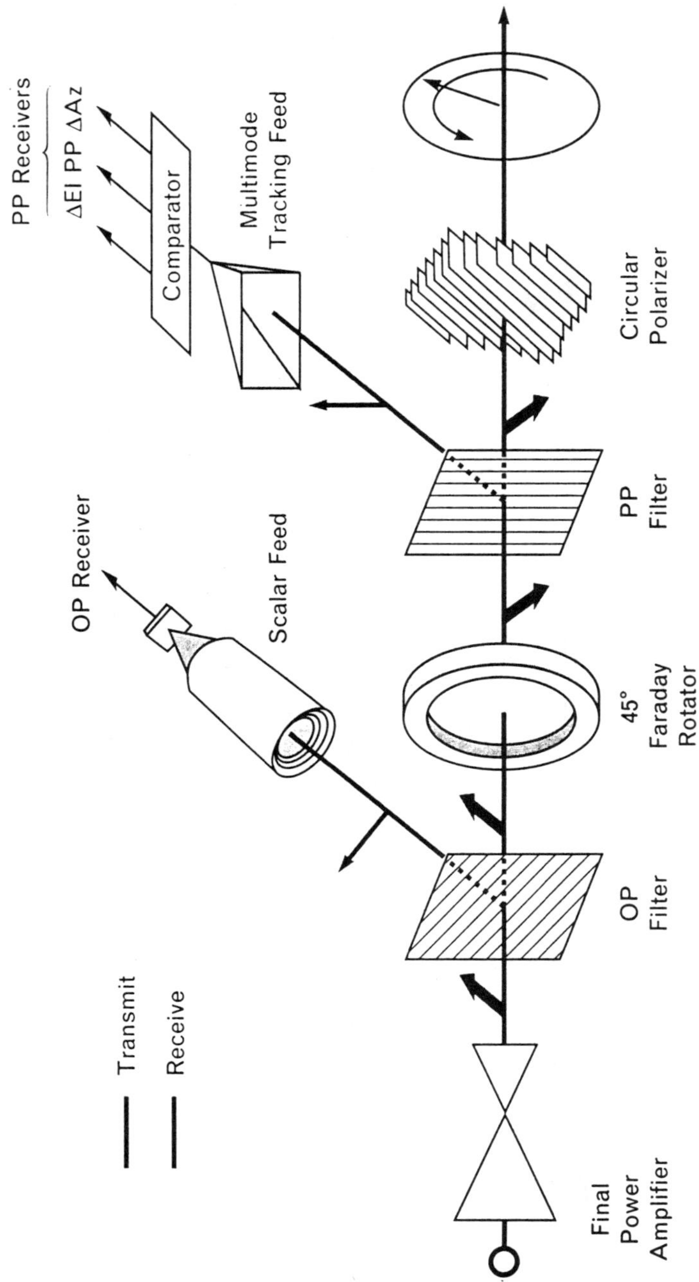

Fig. 6. Quasi-optical system implemented to replace the microwave system shown in Fig. 5.

increase in scatterers used in the prediction adds to the computation time required to calculate the signature. Many theoretical studies have been performed and computer codes written to determine the electromagnetic scattering from surface roughness, electrically large cavities, edges, and other second order scatterers. These also add to the complexity of the signature prediction codes. However, an alternative approach has been suggested to reduce the requirement for significantly increasing the number of scatterers required to define a target.[29] This approach is based on a statistical means of defining the number of "local" RCS cells which are needed for a given radar antenna beamwidth and assigning a mean and standard deviation to each cell. This reduces the need to calculate the RCS for each cell at all aspect angles and minimizes the number of computations needed. The cells are then added coherently to determine the RCS and reasonably good results are shown for the limited number of targets discussed in the paper. But, research and analysis into second and even third order scattering effects will (and must) continue in order to truly understand and predict target signatures at the higher MMW frequencies.

3.1.2 MMW Radar Measurements of Targets and Clutter

RCS measurements of targets continues to be supported for use in both military and remote sensing applications. The MIT Lincoln Laboratory Advanced Detection Technology Sensor (ADTS) 35 GHz synthetic aperture radar (SAR) has been utilized to collect data on targets and clutter in order to support target recognition algorithms. The dual polarized system supports polarimetric algorithm development[30] and clutter data is currently being used to support target detection simulations in the Patriot PAC3 seeker. (The ADTS radar is more fully described in Section 4.2.2.) A 95 GHz airborne radar measurement system (95-ARMS) is being flown to measure ground targets and large areas of clutter at various test sites including Greyling, MI and Yuma, NM. The dual-polarized, stepped frequency 95-ARMS scans ± 30° and provides 4 inches of down range resolution.[31] Currie, Hayes and Trebits have recently released a text that is a well written summary of MMW clutter, clutter effects on MMW system designs and a summary of clutter measurement results that includes both terrain and weather related clutter statistics.[32] Ulaby and Dobson also include some Ka-band and limited W-band terrain scattering statistics in their 1989 text.[33]

Two versions of a MMW seeker are in competition for the hardware upgrade to the current Patriot missile. Both systems are being designed to track tactical ballistic missiles. This development program has spawned modeling and measurement efforts aimed at understanding MMW TBM signatures.[34] The Smart Weapons Management Office at the U. S. Army Missile Command has continued to conduct both Ka-band and W-band measurements of ground vehicles which support the development of MMW smart weapons such as the MMW Hellfire and upgrades to the Brilliant Antitank (BAT) munitions. MMW signatures are being required on more targets and many RCS ranges, both indoor and outdoor, are adding Ka-band and W-band measurement capabilities in order to compete in this increasing market.

3.2 Multipath Predictions

One of the main purposes for using small beamwidths associated with MMW radars in low-angle surface-to-surface scenarios is to reduce the amount of surface illumination to minimize both ground clutter and multipath returns. Clutter returns affect the ability to detect targets by decreasing the target to clutter ratio. Multipath can reduce the ability to accurately track the location of an object because of interfering radar returns created by the various radar paths to the target. It is not always possible to eliminate the unwanted effects of multipath and some techniques have been developed to overcome this problem. Therefore, it is important to understand and be able to accurately predict the effects of multipath on tracking performance in a radar. A realistic model for predicting multipath effects at MMW frequencies has been devised and tested using a fully 95 GHz coherent monopulse radar over varying terrains.[35, 36] The program, called TKMOD5 and the modified TKMOD5B, was used to determine that individual multipath reduction techniques such as frequency agility, offset-angle tracking, etc., may not be sufficient to resolve elevation angle errors caused by multipath. However, the predictions indicate that combinations of these techniques will improve MMW tracking over all but the smoothest of terrains.

4. CURRENT APPLICATIONS OF MMW RADAR

This section will discuss several examples of MMW radar which are intended to be typical of current applications of MMW technology. Three general categories of applications will be presented: military; phenomenology/instrumentation systems; and commercial or non-military applications.

4.1 Military

For much of its history, MMW radar advances and systems have been driven by military or defense related applications. One of the earliest MMW radars developed to evaluate the military usefulness of millimeter waves was the AN/MPS-29 radar shown in Fig. 7.[1] This radar was intended as a relatively short-range battlefield surveillance and target acquisition system featuring a very rapid azimuth scan (up to 70 scans/second) coupled with high range and angle resolution. These characteristics allowed a degree of radar imaging or mapping of the terrain illuminated by the radar and provided a limited ability to navigate vehicles using the radar-generated map. Some current military applications of MMW radar are briefly described in the following discussion. Unfortunately, security restrictions prevent the detailed discussion of specific operating characteristics and parameters for most of these systems. However, these systems are indicative of current MMW systems development.

4.1.1 LONGBOW Targeting and Missile System

One of the U.S. Army's current major weapons system development programs is the LONGBOW Targeting and Missile System,[17, 37, 38] shown in Fig. 8. Engineering and Manufacturing Development (EMD) of the LONGBOW system started in 1991, following a proof-of-principle phase which demonstrated the basic concept of a millimeter-wave radar target acquisition and fire-control radar operating on a helicopter platform in conjunction with a fire-and-forget radar-guided missile. The intent of the weapons system is to destroy ground and low-level airborne targets under conditions of low visibility, at night, in poor weather (rain, fog, and snow), and in battlefield obscurants and smoke. The combination of a small, light weight, high resolution sensor for both the

Fig. 7. AN/MPS-29 MMW short-range surveillance and target acquisition system.

Fig. 8. LONGBOW Ka-band radar mounted on an AH-64 Apache helicopter.

radar and the missile seeker and the requirement for operation in adverse weather resulted in the selection of millimeter wavelengths for the frequency band of operation.

Especially important to the development of the LONGBOW radar/missile seeker is the gallium arsenide monolithic millimeter integrated circuits (MMIC) required in the missile/radar transceiver. LONGBOW Ka-band Power Amplifier Modules (PAM) are now in production at TRW's Space and Electronics Group in Redondo Beach, California. "A total of 16 MMIC chips, fabricated from TRW's 0.15 Micron Power High Electron Mobility Transistor (Power HEMT) technology, are used in the fabrication of a PAM."[20] Each of the power chips shown in Fig. 9 operates in the Ka-band and delivers a nominal output of one watt.

Full-scale LONGBOW production is expected to begin in 1996 with the first units employed in the AH-64 Apache helicopter. Depending on Army procurement and funding decisions, the RAH-66 Commander Helicopter may also receive LONGBOW weapon systems.

4.1.2. Battlefield Combat Identification System (BCIS)

In every major hostile action, the U.S. Army has experienced significant fratricide losses; i.e., casualties from friendly fire. This was especially true in Operation Desert Storm. Unfortunately, this problem is expected to worsen as "the gap continues to widen between the range at which weapons can engage targets and the range at which targets can be positively identified as friend or enemy,[39] and as new, more advanced autonomous weapons that can acquire and engage targets beyond visual range continue to be developed and become operational.

Fig. 9. Ka-band Power Amplitude Modules used in the LONGBOW radar.

In an effort to identify technologies to assist in reducing fratricide, the Army tested seven different "near-term" combat identification systems at Fort Bliss, Texas during April, May and June 1992. According to the report describing the results of the tests and the evaluation of the competing technologies, "the millimeter question and answer technology works best in all environments; meets or exceeds user requirements for range, identification accuracy, and time to identify; has a low vulnerability to enemy countermeasures and exploitation; offers the greatest potential growth; and (is) easy to integrate as a 'bold-on' to tank turrets."[40]

As a result of these tests and other supporting studies and assessments of the operational requirements and available technology, the Army initiated a program in late summer of 1993 to develop a Battlefield Combat Identification System or BCIS. As indicated above, this system is based on a millimeter-wave question and answer approach, and while not a radar system in the classic sense, (it is more appropriately termed a "secondary radar"), this Army program and the LONGBOW radar development constitute two of the Army's major, current R&D programs involving MMW technology. Both systems represent major investments of R&D funds and future commitments on the part of the Army to MMW-based solutions to major Army operational problems.

Although much of the specifics of the BSIC is classified, the basic operational concept is shown in Fig. 10, and a possible "bolt-on" installation on an Abrams MIAI Tank is shown in Fig. 11. The link between the shooter and the potential target is a MMW transmission involving an interrogation (question) by the shooter and a coded response (answer) from the potential target. Performance in the various degraded and contaminated environmental conditions outlined in Fig. 10 were the key to the selection of MMW over the other competing technologies. Another key was the availability of much of the required technology "off-the-shelf," or in the form of MMIC's as illustrated in the block diagram of the BCIS RF Transceiver shown in Fig. 12.

4.1.3 Brilliant Anti-Tank Submunition (BAT)

A unique new weapons system currently under development is the Army's Brilliant Anti-Tank (BAT), self-guided submunition that is intended to autonomously locate, attack, and destroy moving tanks and other armored vehicles. The current version of BAT (shown in Fig. 13) uses both an acoustic sensor and a two-color infrared seeker.[41] To overcome the limitation of the current BAT sensors to detection of only moving targets, the Army is investigating replacement of one of the two sensors with millimeter-wave and/or imaging infrared technology in various seeker concepts.[42]

4.2 Phenomenology/Instrumentation Systems

The requirement for basic propagation, reflectivity, and phenomenological data parallels and supports the development of new, advanced MMW sensors. In fact, the technology employed in the instrumentation systems used to gather such data must, in many ways, be more advanced than that of the actual MMW sensors since the basic data must cover a range of operational parameters that exceed not only the current system performance parameters but the measured data must establish the performance limitations of future systems. Accordingly, radar systems used to gather basic phenomenological data,

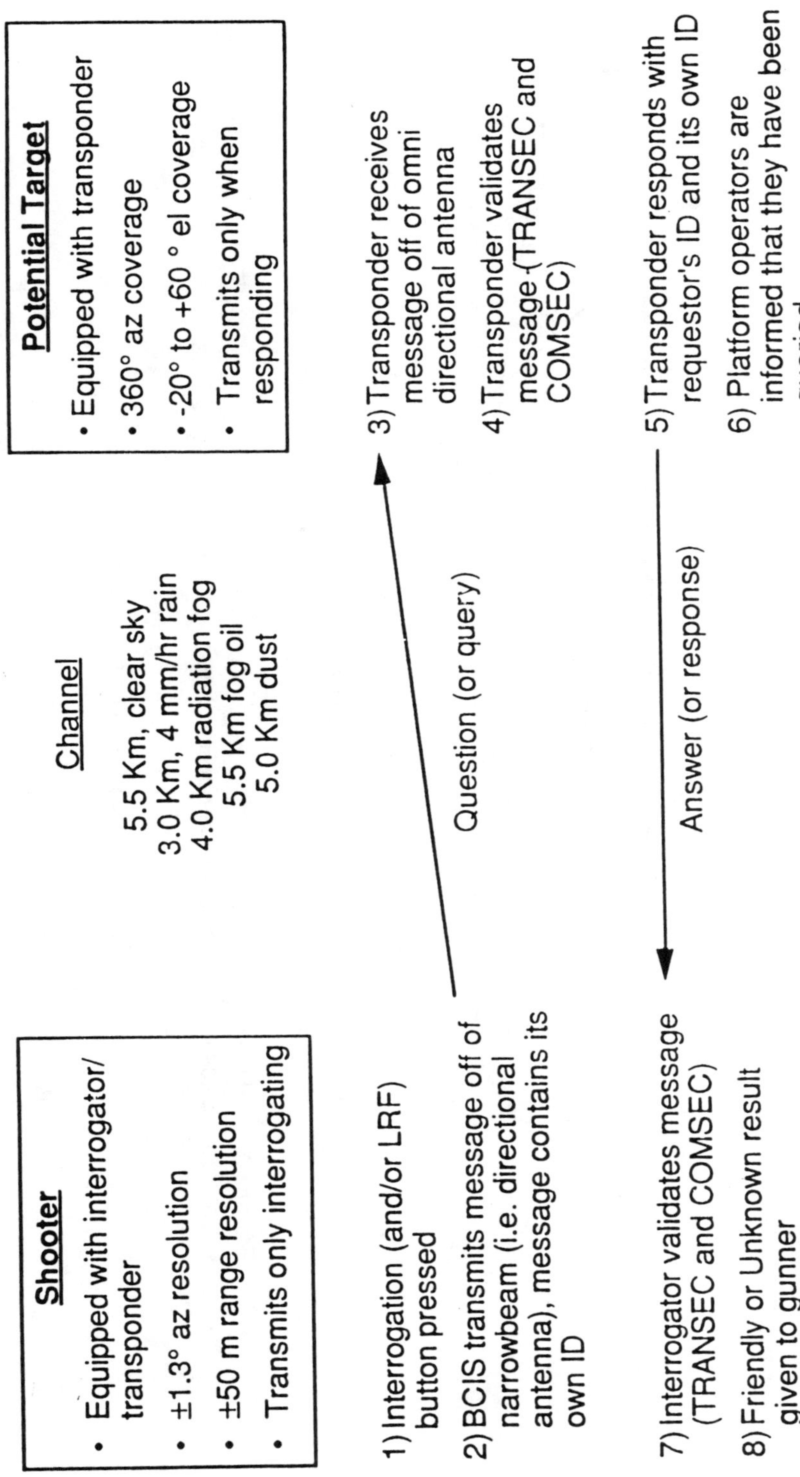

Fig. 10. BCIS Basic Operational Concept

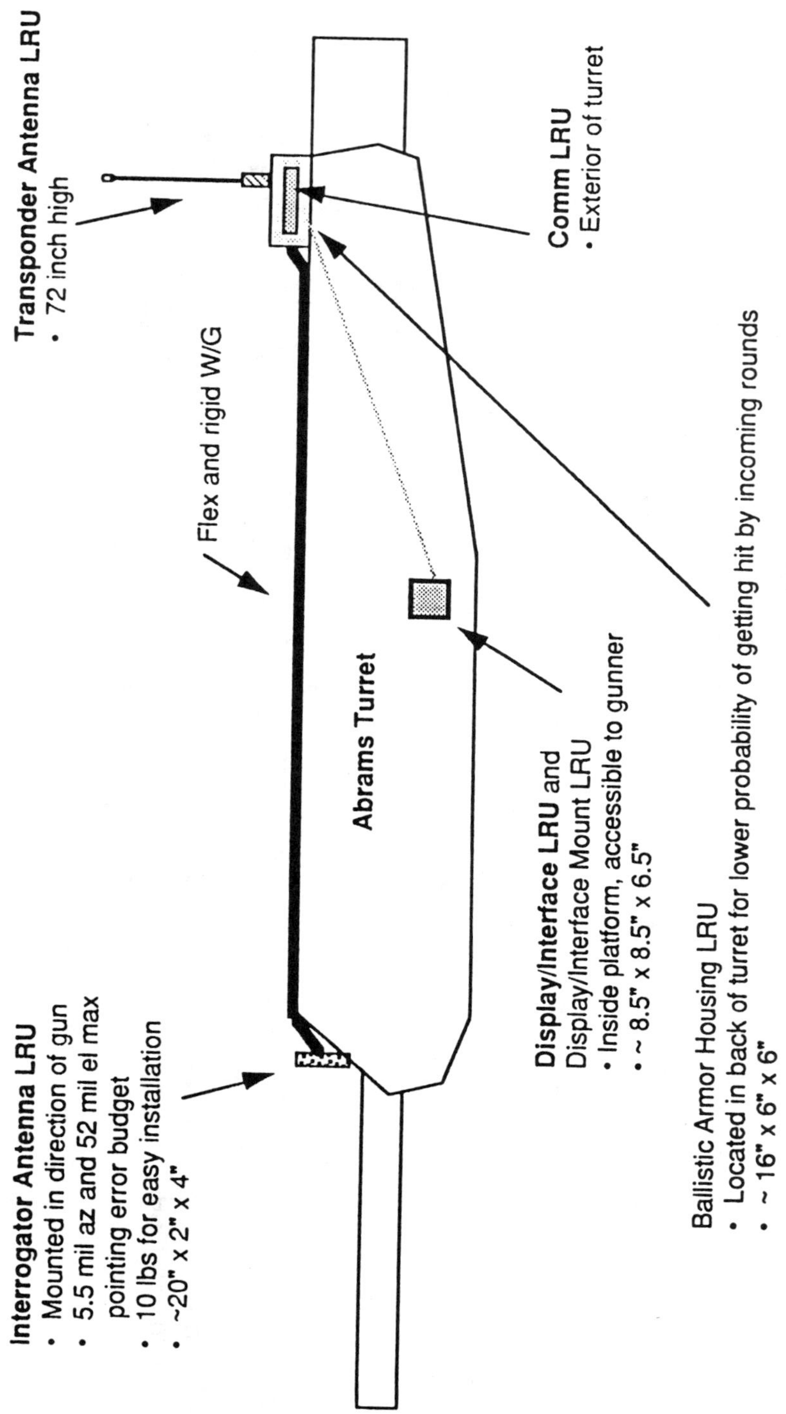

Fig. 11. BCIS Installation on Abrams Tank

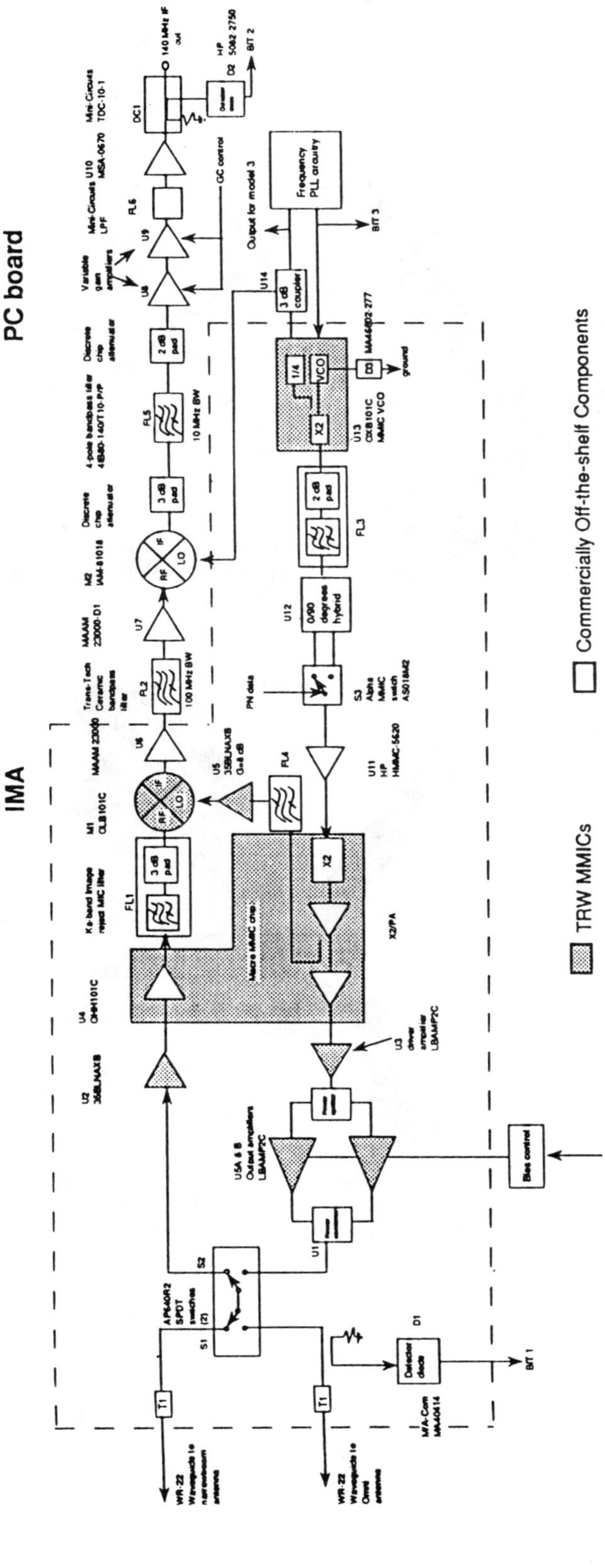

Fig. 12. BCIS RF Transceiver Block Diagram

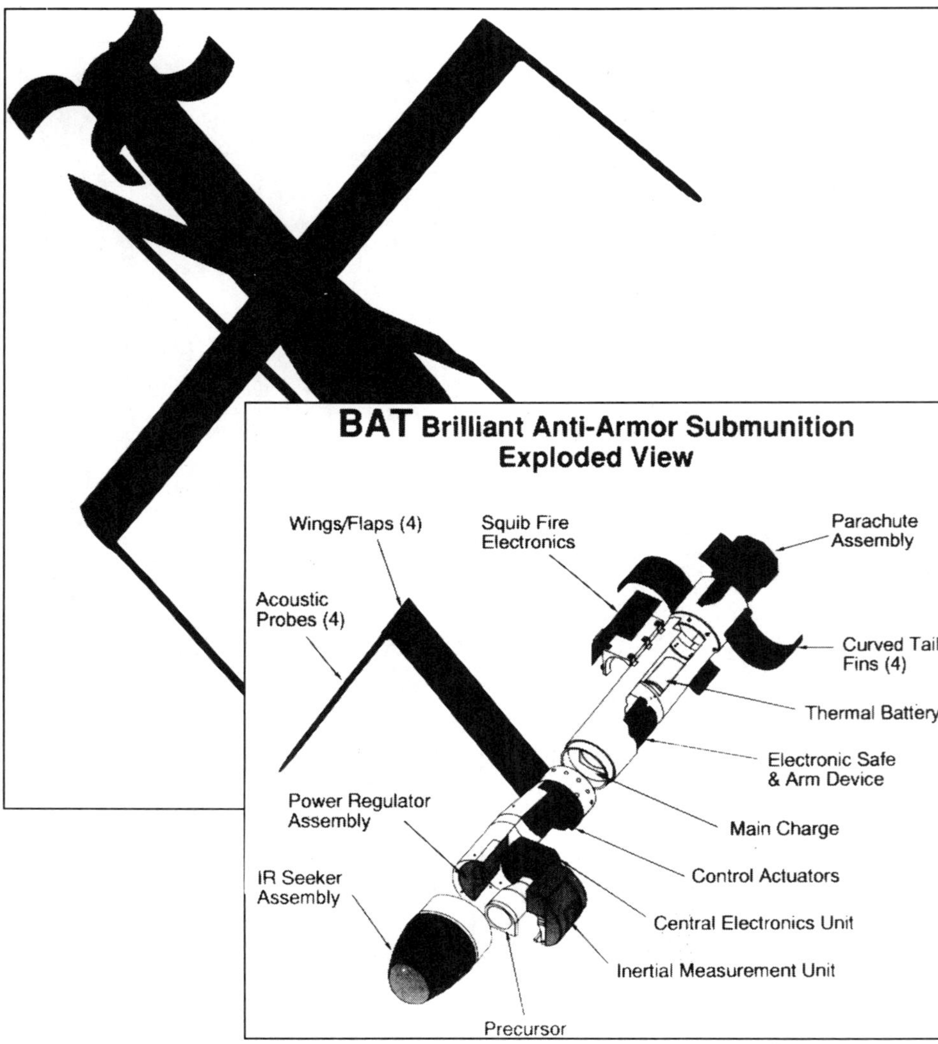

Fig. 13. Current Configuration of the IR/Acoustic Brilliant Anti-armor (BAT) Submunition

especially at the millimeter wavelengths, may actually include more advanced technology and have performance parameters which exceed those found in developmental systems in the same frequency range.

Three such instrumentation radars are briefly discussed in this section.

4.2.1 HIPCOR-95

One of the most advanced, and flexible, MMW instrumentation radars was developed during the late 1980's for the U.S.

Army Missile Command by the Georgia Tech Research Institute. Called HIPCOR-95 for high-power, coherent radar operating at 95 GHz, this radar incorporated the heretofore difficult to achieve characteristics of combined high power with coherent operation in an integrated, MMW instrumentation radar operating at 95 GHz. Furthermore, the radar is unique in that it combines both an Extended Interaction Klystron Amplifier (EIKA) and a TWT amplifier in one master oscillator, power amplifier coherent chain. The radar is also dual polarized, with two operational modes: high power (1-2 kW peak) and moderate power (~ 80 W peak). Coherent frequency agility over 2 GHz bandwidth is available along with two pulsewidths of 50 and 10 ns allowing range resolution of approximately 4-6 inches with off-line processing.

The salient features of the HIPCOR-95 instrumentation radar are given in Table 3 and a basic block diagram is shown in Fig. 14. The radar has recently been modified to include both monopulse operation and an MTI processor. It is currently being used for low-angle propagation and multipath studies by U.S. Army MICOM personnel.[36]

Table 3 HIPCOR-95 Salient Features

Coherent	I and Q Outputs
Polarization Agile (Transmit)	H/V, RHC/LHC
Dual Polarized (Receive)	Co-Polarization Cross-Polarization
Operational Modes:	
High Power (Moderate Bandwidth)	2 kW 1 W (Average)
Moderate Power (Wide Bandwidth)	80 W 0.1 W (Average)
Coherently Frequency Agile	2 GHz BW at Moderate Power Level
Pulse Width	50, 10 ns
High Range Resolution	0.5 ft (Off-Line Processing) at Moderate Power Level
Multiple Range Samples	1 per Channel (6 Total)
Monopulse	Single Polarization
MTI Processor	
Track Processor	Range - only

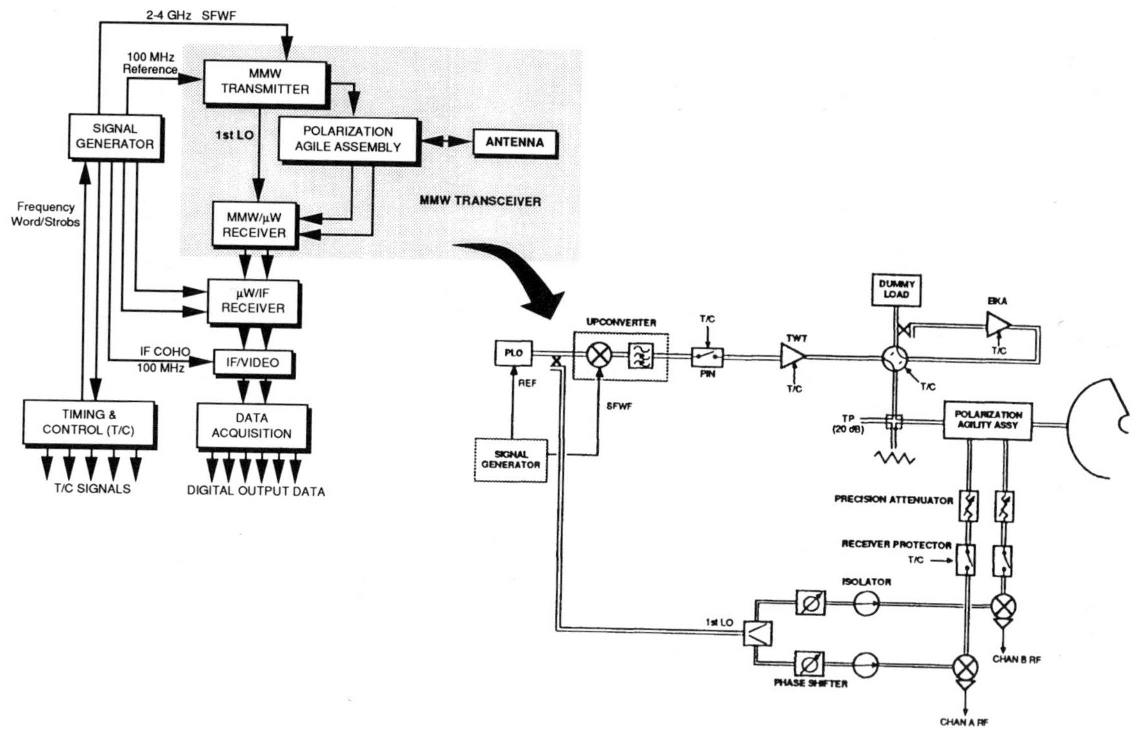

Fig. 14. HIPCOR-95 MMW transceiver block diagram.

4.2.2 Advanced Detection Technology Sensor (ADTS)

One of the most advanced MMW instrumentation radar systems currently in operation is the Advanced Detection Technology Sensor developed by MIT Lincoln Laboratory for the Advanced Research Projects Agency (ARPA).[43, 44] This radar incorporates some of the latest MMW technology into an instrumentation-quality airborne synthetic aperture imaging radar and associated ground processing equipment which is capable of producing calibrated, fully polarimetric ground reflectivity data of targets and clutter. The radar can achieve range and cross range resolution to 1 foot by 1 foot at a slant range of 7 kM.[44]

The ADTS radar is mounted in a specially designed radome pod beneath the fuselage of a Gulfstream G1 aircraft, as shown in Fig. 15 a and b. Table 4 gives the major system parameters and some performance characteristics for the ADTS radar.

(a)

(b)

Fig. 15 a and b. ADTS radar configuration on Gulfstream aircraft.

Table 4 ADTS Major System Parameters

Frequency	33.56 GHz
RF Bandwidth	600 MHz
Resolution:	
Range	1 ft
Azimuth (SAR)	1 ft
Beamwidth	2 degrees
Polarization Isolation	30 dB
Sensitivity	
(SAR Mode)	S/N = 10 dB for -30 dBsm RCS @ 7 km range
Transmitter Power	84 Watts Peak
Receiver Noise Figure	6.8 dB
Antenna Peak Sidelobe	-27 dB
Waveform	Linear FM (deramp on Receive)

This radar, coupled with the ground processor and optimal processing algorithms developed by the MIT/LL staff, can produce truly outstanding imagery. An example of that imagery is shown in Fig. 16. Here a 1 meter by 1 meter SAR image of a rural farm area is compared with the actual photograph of that same area.[44] Notice the detail and highlighting in the SAR image. Information and data derived from such images are used to develop algorithms for separating, more effectively, ground clutter from true targets.

Fig. 16 a. Photograph Image

Fig. 16 (b). SAR Image

Fig. 16 a and b. Comparison of Photographic and ADTS SAR Images.

4.2.3 95-ARMS

In contrast to the airborne SAR imaging system described above, the Georgia Tech Research Institute, in conjunction with the U.S. Air Force at Eglin AFB, Florida, has recently developed an airborne, real aperture MMW instrumentation radar called 95-ARMS for 95 GHz airborne radar measurement systems. Instead of a SAR processor, this radar incorporates a real aperture rapid scanning pencil beam antenna pattern to collect coherent clutter and target-in-clutter data from an airborne C-130 platform. Major performance parameters of the 95-ARMS are given in Table 5.

Table 5 95 GHz ARMS Sensor Parameters

Transmitter	Frequency	W Band (Fixed or Agile) 92 GHz
	RF Bandwidth	1,000 MHz (Coherent)
	Frequency Step Size	8 MHz (Variable)
	Peak Power	15 Watts
	Pulse Width	70 ns
	PRF	110 kHz
Antenna	Type	8 Inch Rotating Mirror/Dielectric Lens
	Beam Width	0.8 Degrees Pencil Beam (2 Way, 3 dB)
	Gain	43 dB
	Sidelobes	-20 dB
	Scan Angle	± 20 Degrees
	Scan Rate	400 Degrees Per Second (Variable)
	Acceleration	2000 Degrees Per Second Squared
Polarization	Transmit	Circular: RHC, LHC; Linear: H, V
	Receive	Simultaneous Dual Polarized: Parallel and Cross I, Q, & Log

4.3 Commercial or Non-Military

At a time when it appears that the military market and funding to support MMW radar development is declining, commercial or non-military applications for MMW radar seem to be increasing. Evidence of this is represented in various conferences where sessions on automotive electronics, biomedical applications, and plasma diagnostics and fusion applications are included, with the automotive electronics session normally the most directly related to MMW radar. Rather than repeating here a survey of current civilian or non-military millimeter wave radar applications which has been very well covered recently in several articles by Johnston,[45, 46] this paper will simply identify the following general categories of applications which represent the areas of current significant activity. Additional discussion of commercial markets for MMW radar is given in a paper by Schneiderman.[47]

- <u>Automobile Applications</u>
  Currently the largest area of MMW radar commercial activity with most of the emphasis on anti-collision radars. MMW radars have recently been installed on a fleet of commercial buses.

- <u>Meteorological Applications</u>
  Weather monitoring, severe weather warning, and wind-shear and microburst detection represent the areas of current activity. Also, MMW radar has been suggested for a satellite-based precipitation and cloud-top monitoring system.

- <u>Aircraft Collision Avoidance and Wire Detection</u>
  Millimeter wave length radar has been evaluated for the detection and avoidance of high-power transmission lines and wires, especially by helicopters.

- <u>Automonous Landing Systems</u>
  MMW radar is one of the sensors under evaluation in the Synthetic Vision program to allow aircraft to land under difficult low-visibility conditions.

- <u>Industrial Control/Instrumentation</u>
  Remote sensing of industrial processes under difficult environments by MMW radar is also receiving attention currently.

5. MMW RADAR FUTURE DIRECTIONS

This paper has presented a current survey and assessment of MMW technology, especially as this technology base relates to radar applications. Recent MMW radar research efforts have also been summarized along with examples of millimeter wavelength radar hardware development programs. This previous discussion provides the necessary background from which possible future directions for the application and development of millimeter wavelength radar and radar hardware technology can be predicted with some measure of confidence. The projection of future directions for MMW radar can be subdivided into two categories — new directions based on 1) technology push and 2) applications pull.

5.1 Technology Push

The following technology developments and advances in current technology are expected to drive or "push" MMW radar development s over the next several years.

- Fully Coherent, High Power Transmitters
- Packaging Improvements and Component Development (Smaller, Lighter)
  - Integrated Devices (MIMIC, HEMP/PHEMP)
  - Hybrid Components
  - Quasi-Optical Technology
- Electronic Scanning
- Low Noise/Low Loss Subsystems
- Solid-State Phased Array
- Power Combining
- Cost Reduction

5.2 Applications Pull

Certain radar applications where MMW systems offer significant and sometimes unique advantages will "pull" MMW radar developments forward also. Some of these potential MMW applications are listed below.

- Space Sensors
- Point Surveillance and Fire Control
- IFF, Target Recognition, and Classification
- Electronic Warfare
- Dual Mode
  - MMW/μW
  - MMW/EO-IR
- Civil Transportation (Cars, Ships, Aircraft)
- Industrial Instrumentation
- Dual Use
  - Military/Commercial
  - Commercial/Scientific
- Lower Cost

Note that lower cost and cost reduction are so important that they are included in both the technology driven and applications driven lists even though cost is not explicitly a member of either category.

It is the intersection of the technology push with the applications pull where significant future development of MMW radar will occur. For example the combination of fully coherent, high power transmitters and electronic scanning, with low noise/low loss subsystems will drive the development of MMW radar to point surveillance and fire control applications, while packaging improvements and component development coupled with cost reduction will support more extensive applications in the civil transportation and industrial areas.

6. REFERENCES

1. E.K. Reedy and J.C. Wiltse, "Fundamentals of Millimeter-Wave Radar Systems," Chapter 7, Aspects of Modern Radar, Artech House, 1988.

2. J. Pressner, "The Influence of the Atmosphere on Passive Radiometric Measurements," AGARD Conference Reprint No. 245, Millimeter and Sub-Millimeter Wave Propagation and Circuits, 1978.

3. P. Bhartia and I. Bahl, Millimeter Wave Engineering and Application, John Wiley and Sons, New York, 1984.

4. J.M. Osterwalder and T.W. Nyland, "Nonintrusive Cryogenic Propellant Sensing with Millimeter Wave/EM Beams," SPIE Infrared and Millimeter Wave Engineering, Vol. 1874, 1993.

5. D. Yi-Yuan, L. Xiao-Bo, and H. Shung-Ji, "Millimeterwave Real Time Three Dimensional Imaging, Its Clutter Rejection Properties and Application to the Metallurgical Industry," 19th European Microwave Conference, pp. 1081-1085, 1989.

6. W.H. Lockyear, "Stock Components Support MM-Wave Power Standard," Microwaves & RF, pp. 61-65
July 1991.

7. L.B. Spence and R.J. Temkin, "A Very High Power, Wide Bandwidth Imaging Radar Design," 1993 IEEE National Radar Conference, pp. 182-185.

8. V.L. Granatstein and I. Alexeff, "High-Power Microwave Sources," Artech House, Norwood, MA, 1987.

9. J. Benford and J. Swegle, "High-Power Microwaves," Artech House, Norwood, MA, 1992.

10. A.W. Fliflet, R.B. McCowan, C.A. Sullivan, D.A. Kirkpatrick, S.H. Gold, and W.M. Manheimer, "Development of High Power CARM Oscillators," Nuclear Instruments and Methods in Physics Research A285, pp. 233-238, 1989.

11. A.C. DiRienzo, G. Bekefi, C. Chen and J.S. Wurtele, "Experimental and Theoretical Studies of a 35 GHz Cyclotron Autoresonant Maser Amplifier," Physics of Fluids, pp. 1755-1765, July 1991.

12. E.V. George and D. Prosnitz, "Radar FEL (RAFEL)," briefing at Lawrence Livermore National Laboratory, October 1991.

13. H.C. Limburg, Dr. O. Sauseng, R.R. Hull, Dr. J.A. Davis, C. Goodman, and G. Lee, "Development of a 10 Watt 95 GHZ Helix TWT," Tri-Service Millimeter-Wave Workshop, p. 24, October 27-29, 1993.

14. P.L. Walchi, "High Power Pulsed Millimeter-Wave Helix TWT," Tri-Service Millimeter-Wave Workshop, p. 26, October 27-29, 1993.

15. C.R. Seashore, "Integrated Circuit Impact on Millimeter -Wave Sensors," 1991 National Telesystems Conference-NTC 1991, pp. 363-366.

16. J.W. Rawles, "MIMIC Program Shifts to High Gear," Defense Electronics, pp. 49-56, May 1991.

17. D.J. Kozakoff and J.C. Wiltse, "Overview of System Engineering Methods and Objectives," SPIE Vol. 1874 Infrared and Millimeter-Wave Engineering, pp. 2-10, 1993.

18. R. Carandang, J. Yonaki, W.L. Jones, R.E. Kasody, W. Lam, and L.C.T. Liu, "A K-Band HEMT Low Noise Receive MMIC for Phased Array Applications," 1991 IEEE MTT-S Digest, pp. 521-524.

19. J. Berenz, M. LaCon, and M. Luong, "Single Chip Ka-Band Transceiver," 1991 IEEE MMT-S Digest, pp. 517-520.

20. "MMIC-Based Power Amplifier Modules for LONGBOW Transceiver," Defense Electronics, September 1993.

21. P. Bhartia, K.V.S. Rao, and R.S. Tomar, "Millimeter-Wave Microstrip and Printed Circuit Antennas," Artech House, Norwood, MA, 1991.

22. J.E. Garrett and J.C. Wiltse, "Fresnel Zone Plate Antennas," International Journal of Infrared and Millimeter Waves, Plenum Press, Vol. 12, pp. 195-220, March 1991.

23. G.V. Eleftheriades, W.Y. Ali-Ahmad, and G.M. Rebeiz, "A 20-dB Quasi-Integrated Horn Antenna," IEEE Microwave and Guided Wave Letters, Vol. 2, No. 2, pp. 73-75, February 1992.

24. S.C. Shrivastava and A.K. Tiwari, "Dielectric Corrugated for Millimeter-Wave Applications," 4th IEEE Region 10 International Conference-TENCON, pp. 218-220, 1989.

25. G.N. Hulderman, "Five-Port Hydbrid-Integrated MMW Monopulse Tracking Sensor," SPIE Vol. 1874 Infrared and Millimeter-Wave Engineering, pp. 155-159, 1993.

26. J. C. G. Lesurf, "Millimeter-Wave Optics, Devices & Systems," Adam Hilger, Bristol, England, 1990.

27. P. F. Goldsmith, "Quasi-Optics in Radar Systems," Microwave Journal, pp. 79-98, January 1991.

28. W.D. Fitzgerald, "A 35-GH Beam Waveguide System for the Millimeter-Wave Radar," The Lincoln Laboratory Journal, Vol. 5, pp. 245-271, Number 2, 1992.

29. U. Aulenbacher and R. Barg, "Target and Clutter Modeling for Millimeter-Wave," SPIE Infrared and Millimeter Wave Engineering, Vol. 1874, 1993.

30. L.M. Novak, S.R. Hesse, "Optimal Polarizations For Radar Detection and Recognition of Targets in Clutter," 1993 IEEE National Radar Conference, pp. 76-78, April 1993.

31. W. Parnell and T.L. Lane, "Development of a Multispectral Airborne Instrumentation System," 16th International Conference on Infrared and Millimeter Waves, pp. 1514-1522, August 1991.

32. N.C. Currie, R.D. Hayes, and R.N. Trebits, "Millimeter-Wave Radar Clutter," Artech House, Norwood, MA, 1992.

33. F.T. Ulaby and M. Craig Dobson, "Handbook of Radar Scattering Statistics for Terrain", Artech House, Norwood, MA, 1989.

34. W.L. Cassaday, L.F. Moore and M.T. Tuley, "Full Polarization Scattering Matrix Ka-Band Measurements of Tactical Ballistic Missiles," 37th Tri-Service Radar Symposium, pp. 101-108, June 1992.

35. J.A. Bruder and J.A. Saffold, "Multipath Effects on Low-Angle Tracking at Millimeter-Wave Frequencies", IEEE Proceedings-F, Vol. 138, No. 2, pp. 172-184, April 1991.

36. M.R. Christian and J.B. Winter, "High Range Resolution Low Angle Track Measurements," Tri-Service Millimeter Wave Workshop, p. 51, October 27-29, 1993.

37. D.F. Bond, "Army Evaluating LONGBOW, Weights Broader Program," Aviation Week and Space Technology, May 11, 1992.

38. "LONGBOW Transceivers Pass Key Test," Defense Electronics, February 1993.

39. Charles L. Malone, "Army Science Board 1992 Summer Study," Land Warfare Combat Identification; Office of the Assistant Secretary of the Army, Washington, D.C., December 1992.

40. "Army Study Says Millimeter-Wave Design Best for Combat ID Requirements," Inside the Army, March 1, 1993.

41. K. Atwal, "Piercing the Enemy Heart," Defense, May 1993.

42. "Improved BAT Submunition Planned for New AT ACMS," Aviation Week and Space Technology, December 13/20, 1993.

43. J.C. Henry, "The Lincoln Laboratory 35 GHz Airborne SAR Imaging Radar System," Proceedings of the 1991 National Telesystems Conference, Atlanta, Georgia, March 26, 27, 1991.

44. L.M. Novak, et al, "Optimal Processing of Polarimetric Synthetic - Aperture Radar Imaging," Lincoln Laboratory, Journal Vol. 3, Number 2, 1990.

45. S.L. Johnston, "Contemporary Millimeter Wave Radar," Tri-Service Millimeter Wave Workshop, Atlanta, Georgia, October 27-29, 1993.

46. S.L. Johnston, "Modern Ka- Band/Millimeter Wave Radar," International Conference on MMW and Far-Infrared Technology, Beijing, June 1990.

47. R. Schneiderman, "Millimeter-Waves Find Commercial Markets," Microwaves and RF, July 1991.

Millimeter-wave Fresnel zone plate antennas

James C. Wiltse

Georgia Tech Research Institute
Atlanta, Georgia 30332-0801

ABSTRACT

Fresnel zone plate antennas have seen extensive research in recent years. Both transmission and reflection types have been analyzed and tested, and found to give excellent performance. Parameters such as far-field patterns, gain, efficiency, frequency dependence, aberrations, scan capability, focal region properties, and off-axis performance have been examined. While most zone plates are planar, curved (spherical or paraboloidal) versions have been found to offer certain advantages, such as sharpness of focus or larger field of view. In general, millimeter-wave results are emphasized.

1. INTRODUCTION

The transmission Fresnel zone plate is a device that produces lens-like focusing and imaging of electromagnetic waves. These effects are produced by diffraction, rather than refraction. The zone plate transforms a normally incident plane wave into a converging wave, concentrating the radiation field to a small focal region. Conceived a century ago, it has been used often in optical systems and sparingly in millimeter-wave applications for the past three decades. Only recently, however, have its performance characteristics been investigated in detail.

Fresnel zone plates provide performance comparable to, or in some cases, better than lenses or parabolic reflector antennas and have been used in transmission or in reflector configurations [1-3]. In general, Fresnel zone plates may be lower in cost, simpler, lighter in weight, have lower loss and wider fields of view, while still providing the same beamwidth and comparable sidelobe performance to that obtainable from a lens or paraboloid reflector of equal aperture. There are two basic types, the partially opaque (in which alternate zones are blocked) and the phase-correcting (in which all zones are utilized). Figure 1 shows some examples. The partially opaque type only uses about half of the available aperture and is very inefficient. Nonetheless, it is sometimes employed at millimeter wavelengths because it is so easy to build. At millimeter wavelengths the focal lengths and diameters are often comparable in size, whereas in the optical range very long focal lengths are often used (e.g., 600 times the plate diameter). Also, optical zone plates may have large numbers of annular zones (perhaps 200 to 300), while at millimeter wavelengths there may be only a few zones (e.g., 5 to 30). This is illustrated in Figures 2-4. The number of zones increases with diameter (D) for a fixed focal length (F) and with smaller F/D for a given diameter. [3].

2. BACKGROUND AND HISTORY

Much of the early history is summarized in References 1 and 2, published in 1991. They also give the design equations for determining the radii of the annular zones of a circular aperture, as well as the depth of cut for a dielectric plate. These equations may be used for half-wave, quarter-wave, or smaller phase-correcting steps, and the dependence of efficiency on the degree of phase correction is also given. Figure 5 illustrates the typical depths of cut in three different plastic materials for the half-wave (phase correcting) zone plate. The first usage of zone plates at millimeter wavelengths was reported in 1961 [4]. Since that time scores of general papers have been published, but many of these deal with partially-opaque zone plates or with optical applications, and therefore may not be of interest to the millimeter-wave user. The literature contains references to phase-correcting zone plates that have been designed and tested at frequencies from K_a-band to 230 GHz (see Table I). Some of these have employed better than half-wave phase correction, such as quarter-wave correction. An extensive bibliography on zone plates was published in 1994 [5]. In these investigations a number of technical requirements and needs have been addressed. They include far-field patterns (beamwidth, peak sidelobe level, average sidelobe level), gain, bandwidth and frequency dependence, aberrations, knowledge of the focal region fields, off-axis performance, field of view, and efficiency or effectiveness when compared with a lens. In spite of a significant amount of this prior work, there is no unified collection of results, but examples will be discussed below.

The first significant description of microwave zone plates appears to have occurred in the NRL report authored by Maddaus [7]. Although described as an interim report, it was a relatively detailed treatment containing theoretical analyses, as well as measured results on antenna radiation patterns at 23.6 GHz. The report seems to have received very little circulation and few references to it have been found.

In 1961, Van Buskirk and Hendrix reported the results of analysis and measurements on a zone plate made for X-band [8]. They demonstrated that a partially opaque zone plate has

Table I. Examples of Zone Plates at Numerous Millimeter-Wave Frequencies

FREQUENCY & DATE	INVESTIGATOR(s)	ZONE PLATE TYPE (PHASE)
34.6 GHz 1982	Lazarus, Pantoja, Novak & Semekh	Transmission, Partially Opaque
35 1993	Garrett & Wiltse	Transmission; Reflecting Off-Axis ($\lambda/4$)
35 1985	Wiltse	Planar-Dielectric ($\lambda/2$)
36-38 1983	Thornton & Strozyk	Transmission ($\lambda/8$)
54-58 1983	Thornton & Strozyk	Transmission ($\lambda/8$)
60 1994	Yu Ji & Masaharu Fujita	Reflecting ($\lambda/2$)
90.5 1967	Weibel & Dressel	Transmission ($\lambda/2$)
94 1988	Huder & Menzel	Reflecting ($\lambda/2$)
140, 210, 280 1961	Sobel, Wentworth, Wiltse	Transmission ($\lambda/2$)
230 1992	Gouker & Smith	Reflecting ($\lambda/2$)
235 1962	Cohn, Wentworth, Wiltse	Transmission ($\lambda/2$)

a focus for reflection as well as transmission. If the opaque rings or zones are made from conducting materials, they can be used as sources of reflection. When a ground plane is used one quarter wavelength behind the plate, both foci are brought into coincidence. Thus, the equivalent half-wave phase-correcting zone plate is produced in a reflecting configuration. Since then numerous millimeter-wave investigations have been conducted, and Table I lists the majority of them.

3. TECHNICAL CHARACTERISTICS

Far-field patterns have been measured and/or calculated by several groups [1,4, 6-12], and in general the patterns are similar to those from a lens having a comparable aperture distribution. That is, beamwidths are approximately the same, as are principal sidelobe levels, Secondary sidelobes are typically 20 to 30 dB below the main beam. Attempts have been made to find a simple, closed-form expression for the far-field pattern, but normally one must perform a computer-numerical solution for the Kirchoff integral [9]. Gain is somewhat less than for a lens, but this depends upon the number of phase steps utilized, and that has been analyzed in Reference 11, which also shows that far-out sidelobes are 35 to 45 dB below the main beam (see also references 1 and 12).

The beamwidth and first sidelobes are determined primarily by the amplitude taper across the aperture, and so are similar to those obtained for a lens with the same taper. But because of the phase steps in the aperture (deviations from uniform phase), the far-out sidelobes are higher than for the lens, and this of course is why the gain of the zone plate is lower. The analysis by Baggen and Herben [11] is particularly informative because they plot the far-field patterns for half-opaque and for phase correcting (half-wave and quarter-wave) plates having F/D values of 1.9 or 0.52. Figure 6, taken from Reference 11, shows their results for the smaller F/D, which has the better sidelobe performance. This is in agreement with the results of Guo and Barton [12], who also found that a smaller F/D gave lower sidelobes. Figure 6 also shows that the on-axis gain of the half-wave correcting plate is 3 dB lower than that of the quarter-wave plate, while the gain of the half-opaque plate is 9 dB down from the quarter-wave design.

In 1988 Huder and Menzel [10] reported on a reflecting zone plate operating at 94 GHz. Their measured radiation pattern had a clean mainlobe (1.6-degree half-power beamwidth) and a sidelobe level of -19 dB. This probably would be lower except for feed blockage, since the feed was on-axis. To improve the blockage problem, an off-axis feed arrangement may be utilized. Previous experience has shown that an off-axis angle up to about 15 to 20 degrees may be used without serious degradation of the pattern [3,9], although the correct design for an off-axis feed would require the annular zones to be non-confocal ellipses, rather than concentric circles [1,3,5,13].

To examine this off-axis behavior we have conducted tests at 35 GHz on a transmission quarter-wave corrected zone plate and a reflecting quarter-wave plate using a feed at 17 degrees off axis. Both employ circular zones and were fabricated from Rexolite. Figure 7 shows the cross-section of the transmission version and Figure 8 is a plot of its measured H-plane radiation pattern. Figure 9 illustrates the geometry for the off-axis case, while Figure 10 gives the plate design and Figure 11 shows the measured far-field pattern (H-plane). The feed is at -34 degrees. It may be seen that in the latter case the main beam is not as clean (and is broadened below the -10 dB level), while sidelobe levels are higher. Off-axis fed reflecting zone plates employing non-confocal elliptical zones provide much better patterns [1,13].

The bandwidth and frequency dependence are related to the focal length over diameter (F/D) and the number of zones [1,9,12], but percentage bandwidths of 12% to 20% are typical. The main effect is the change in the axial location of the focal point. A large number of investigations have dealt with the knowledge of the fields in the vicinity of the focal region [14-19]. Obviously, the zone plate does not focus energy to a "point", but produces a diffraction pattern (similar to that of a lens), with a prominent main beam surrounded by sidelobes and second-order foci. It is important to know the structure of these fields in a detailed quantitative manner. Understanding this structure gives information about off-axis performance, aberrations, and scan capability.

A very comprehensive investigation in this area was conducted by Sanyal and Singh [16]. They measured the fields in the axial and transverse directions around the focal area and did it for a range of frequencies (8.2 to 10.4 GHz) for both a semi-opaque simple zone plate and a phase-correcting example. However, it was discovered by the present author that their results are flawed because the design of the phase-correcting plate was in error. The more recent references [17-19] have provided further information about the focal region fields, but it is still an area that needs further quantitative theoretical and measured results.

The Sanyal and Singh case [16] warrants further comment, because it appeared to be such a thorough, detailed study. First, they developed a theoretical explanation of the fields in the vicinity of the focal region, and then made extensive measurements of the axial and transverse field variations near the focus, for a wide range of frequencies. They reported no improvement in going from a transmission half-opaque zone plate to a "phase correcting" plate. However, the depth of cut for their X-band example was 1.0 cm, whereas it should have been 2.67 cm. The groove depth corresponded to a phase retardation of 67.5 degrees rather than the expected 180 degrees needed for half-wave correction. Therefore the resultant patterns should be those of two superimposed opaque-transparent plates out of phase by 112.5 degrees. Adding two vectors of equal magnitude with relative phase given above should yield a resultant vector of a magnitude of approximately 1.11 or only 0.9 dB larger than each vector, compared to the 6 dB expected. The phase reversing zone plate can be thought of as two partially opaque zone plates in parallel. The first zone plate has alternating transmissive zones with no phase correction. The remaining zones are blocked (i.e., opaque). The second zone plate has alternating transmissive zones that introduce a $\lambda/2$ phase correction with opaque zones in between. Note that the second zone plate has transmissive zones where the first zone plate has opaque zones and vice versa.

The resulting vector diagram at the focal point due to the first zone plate spans between 0 degrees and 180 degrees (choosing an arbitrary phase reference). Therefore, the total radiation can be summed into one vector at 90 degrees representing the first zone plate. The second zone plate would have a vector sum at 270 degrees if there were no phase correction; however, with a $\lambda/2$ phase correction, the second zone plate has a vector at 90 degrees. The incorrect phase correction will cause it to point at some other angle and the two vectors will not add maximally resulting in a lower intensity focal point.

Sanyal and Singh also proposed a simple numerical method for estimating the axial variation of the Fresnel zone plate focal region. In many calculations it is reasonable to assume that the field contributions from each zone are approximately equal at the focal point. This may be assumed since the slight effect of the obliquity factor is countered by the ever so slight increase of zone plate area as one moves to the outer zones on the plate. Thus, for an incident plane wave, the outer zones which are a somewhat longer distance from the designed focal point are subtending a correspondingly greater amount of incident energy. Sanyal and Singh extended this assumption as one moves away from the designed focal point along the zone plate axis. They effectively ignored the obliquity factor at any point along the axis. The authors even went beyond the assumption of equal zonal contributions along the axis. For their axial pattern calculations, they assumed that the magnitudes of the zonal field contributions do not change as one moves away from the focus while remaining on the plate's axis.

This calculation only involves the addition of vectors of equal magnitudes interacting via their relative phases which vary along the axis. At the designed focus, all of the zonal contributions are assumed to have a phase of zero. In other words the zone field vectors add up in a completely constructive fashion at the focal point. For points out along the axis, the phases of the contributions from the zones change in correspondence with the change in path-length difference with respect to the focus. These numerical calculations yield curves which make some sense intuitively. However, when the calculations are tried, for example, for zone plates as presented by such authors as Guo, Baybulatov [14], and others, the results differ significantly from the more accurate integrations presented by the above-mentioned authors for calculation of the axial field. The ignoring of the obliquity factor likely is not the culprit as long as one is making calculations close to the focal point. The source for the discrepancies is likely due to the lack of recalculation (integration) of each zonal contribution along the axis.

An analysis of the zone plate focal region for an arbitrary phase difference between adjacent zones has been presented by Jones [17] for the phase-correcting case. Such analysis could be applied to Sanyal and Singh's case, in which their dielectric zone plate had an incorrect groove depth. Jones's analysis could be applied to their arbitrary phase difference in order to verify the experimental results obtained by Sanyal and Singh.

Guo, Barton and Wright [18] calculate the lateral and axial field intensity patterns for the zone plate focal region. They confirm the result of Sussman [15] with regard to the positioning of the multiple axial peaks, or foci. Where F is the focal distance from the zone plate, such secondary foci are shown to be positioned at F/3, F/5, and F/7 from the zone plate. It is confirmed that the principal focal peak width decreases with a decreasing F/D ratio. This is to be expected since a relatively greater diameter zone plate aperture should yield more effective focusing. In the area of scan performance of zone plates, Baggen, Jeronimus, and Herben have carried out a theoretical analysis to compare the zone plate with the parabola [20]. Unfortunately, only the half-opaque zone plate is considered.

Most zone plates are planar, and, in fact, this is often touted as an advantage (i.e., simple construction, less weight and volume, lower cost, and usefulness in a microelectronics planar environment). Nonetheless, there have been several investigations of spherical and paraboloidal plates. An extensive series of papers was published by authors in India [21-25], but because they were published in Indian journals, most researchers have been unaware of them and they have seldom been referenced by American or European authors. (The Indian authors similarly did not reference the relevant Western publications). These and other references [26,27] have given quantitative information about the characteristics of curved zone plates, which offer advantages such as a better axial focusing property, wider field of view, and aerodynamic shapes. The several papers by Day and Khastgir are excellent because they compare the spherical and paraboloidal cases with the planar version. This has provided a basis for the hardware application of paraboloidal zone plates for signal reception from multiple satellites (Delmas, et al [27]).

4. SUMMARY

Although a large amount of research has been conducted on Fresnel zone plate antennas, there is no unified summary of the theory or experimental results. Nonetheless, it is well-established that these antennas provide excellent performance. It is hoped that the extensive list of references will be useful to workers in the field.

5. ACKNOWLEDGEMENT

The author would like to acknowledge the assistance of Mr. James E. Garrett. This work was supported by the U.S. Army Research Office. Opinion, interpretations, conclusions and recommendations are those of the author and not necessarily endorsed by the U.S. Army.

6. REFERENCES

1. J.C. Wiltse and J.E. Garrett, "The Fresnel Zone Plate Antenna", Microwave Journal, Vol. 34, Jan. 1991, pp. 101-114.

2. J.E. Garrett and J.C. Wiltse, "Fresnel Zone Plate Antennas at Millimeter Wavelengths", International Journal of Infrared and Millimeter Waves, Vol. 12, March, 1991, pp. 195-220.

3. C.A. Barrett and J.C. Wiltse, "Design Parameters for Zone Plate Antennas", IEEE Antennas and Propagation Society International Symposium, July 18-25, 1992, pp. 608-611.

4. F. Sobel, F.L. Wentworth, J.C. Wiltse, "Quasi-Optical Surface Waveguide and Other Components for the 100 to 300 GHz Region", I.R.E. Transactions on Microwave Theory and Techniques, Vol. MTT-9, November, 1961, pp. 512-518.

5. J.C. Wiltse, "Characteristics of Phase-Correcting Fresnel Zone Plates and Elliptical Waveguides", Final Report on U.S. Army Research Office Grant No. DAAL 03-91-G-0114, February, 1994.

6. A. Boivin, A. Dion, and H.P. Koenig, "Etude Experimental de la Diffraction Des Micro-Ondes Par Des Ouvertures a Symetrie de Revolution", Canadian Journal of Physics, Vol. 34, pp. 166-178, 1956.

7. I. Maddaus, "Fresnel Zone Plate Antenna", Naval Research Lab., Washington D.C., Report R-3293, June, 1948.

8. L.F. Van Buskirk and C.E. Hendrix, "The Zone Plate as a Radio Frequency Focusing Element", IRE Transactions on Antennas and Propagation, Vol. AP-9, pp. 319-320, May, 1991.

9. D.N. Black and J.C. Wiltse, "Millimeter-Wave Characteristics of Phase-Correcting Fresnel Zone Plates", IEEE Transactions on Microwave Theory and Techniques, Vol. MTT-35, no. 12, pp. 1122-1128, December, 1987.

10. B. Huder and W. Menzel, "Flat Printed Reflector

Antennas for MM-Wave Applications", Electronics Letters, Vol. 25, p. 318, March, 1988.

11. L.C.J. Baggen, and M.H.A.J. Herben, "Design Procedure for a Fresnel-Zone Plate Antenna", International Journal of Infrared and Millimeter Waves, Vol. 14, no. 6, pp. 1341-1352, June, 1993.

12. Y.J. Guo and S.K. Barton, "On the Sidelobe Performance of Fresnel Zone Plate Antennas", 1992 IEEE Antennas and Propagation Symposium, Chicago, IL, pp. 2175-2178, June 1992.

13. Y.J. Guo, I.H. Sassi, and S.K. Barton, "Multilayer Offset Fresnel Zone Plate Reflector", IEEE Microwave and Guided Wave Letters, Vol. 4, pp. 196-198, June 1994.

14. F.Kh. Baybulatov, I.V. Minin, and O.V. Minin, "Investigation of the Focusing Properties of a Fresnel-Zone Plate", Soviet Journal of Communications Technology and Electronics,no. 1, pp. 1-7, 1986. (Trans. from Radioteknika i Elektronika, No. 9, 1985, pp. 1681-1688)

15. M. Sussman, "Elementary Diffraction Theory of Zone Plates", American Journal of Physics, Vol. 28, pp. 394-398, 1960.

16. G.S. Sanyal and M. Singh, "Fresnel Zone Plate Antenna", Journal of the Institution of Telecommunication Engineers India), Vol. 14, pp. 265-281, 1968.

17. A.R. Jones, "The Focal Properties of Phase Zone Plates", British Journal of Applied Physics (Journal of Physics D.), Vol. 2, pp. 1789-1791, September 16, 1969.

18. Y.J. Guo, S.K. Barton, and T.M.B. Wright, "Focal Field Distribution of Fresnel Zone Plate Antennas", Proc. of the 7th International Conf. on Antennas and Propagation (ICAP), York, UK, April 1991, pp. 6-8.

19. K.K. Dey and P. Khastgir, "A Study of the Characteristics of a Microwave Spherical Antenna", International Journal of Electonics, Vol. 35, no. 2, 1973, pp. 97-103.

20. L.C.J. Baggen, C.J.J. Jeronimus, and M.H.A.J. Herben,

"The Scan Performance of the Fresnel Zone Plate Antenna: A Comparison with the Parabolic Reflector Antenna", Microwave and Optical Technology Letters, Vol. 6, p. 769, October, 1993.

21. K.K. Dey and P. Khastgir, " Comparative Focusing Properties of Spherical and Plane Microwave Zone Antennas", International Journal of Electronics, Vol. 35, No. 4, 1973, pp. 497-506.

22. K.K. Dey and P. Khastgir, "A Theoretical Study of the Axial Field Amplitude of Microwave Paraboloidal, Spherical and Plane Zone Plate Antennas", Journal of the Institution Electronics and Telecommunication Engineers, Vol. 19, no., 12, pp. 697-700, 1973.

23. K.K. Dey and P. Khastgir, "A Generalized Method of Zone Construction for Plane Zone Plate Antennas", International Journal of Electronics, Vol 36, no. 1, 1974, pp. 91-96.

24. S.C. Chakravarty, P. Khastgir, and K.K. Dey, "Microwave Fresnel Zone Plates with Non-Uniform Aperture Excitation", Journal of the Institution of Electronics and Telecommunication Engineers, Vol. 23, no. 3, 1977, pp. 145-147.

25. P. Khastgir, K.M. Bhowick, and K.K. Dey, "Analysis of the Off-axis Defocus of Microwave Zone Plates", Indian Journal of Pure and Applied Physics, Vol. 16, February, 1978, pp. 96-101.

26. W.A. Kleinhans, "Aberrations of Curved Zone Plates and Fresnel Lenses", Applied Optics, Vol. 16, no. 6, pp. 1701-1704, June 1977.

27. J-J. Delmas, S. Toutain, G. Landrac, and P. Cousin, "TDF Antenna for Multisatellite Reception Using 3D Fresnel Principle and Multilayer Structure", 1993 AP-S Antennas and Propagation Symposium Digest, Ann Arbor, MI, June 28-July 2, 1993.

Figure 1. RELATION BETWEEN LENSES AND ZONE PLATE

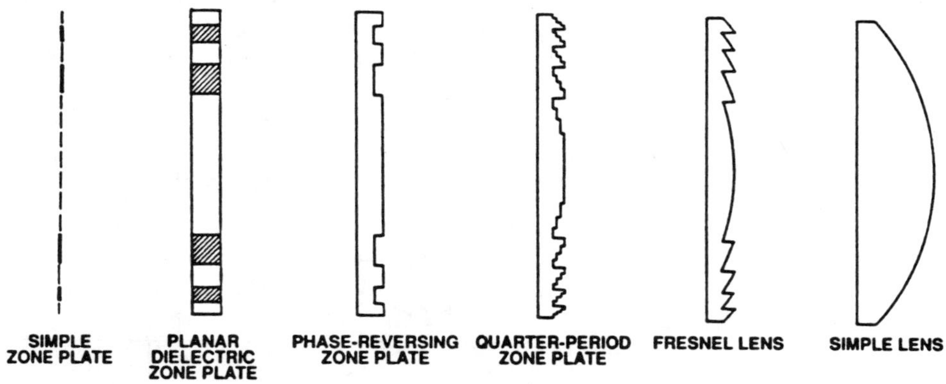

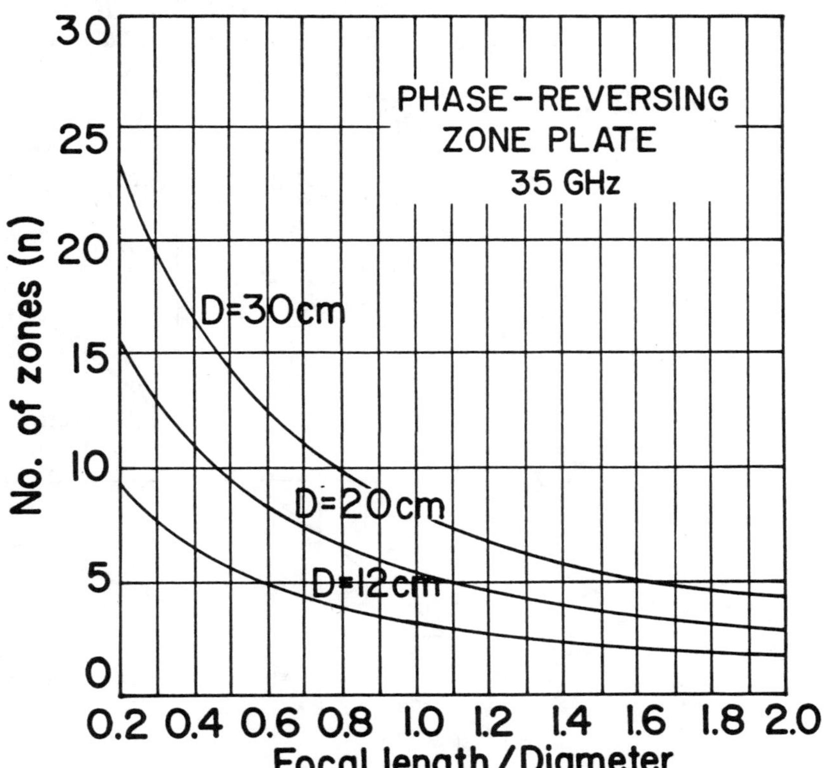

Figure 2. Number of zones as a function of focal length over diameter.

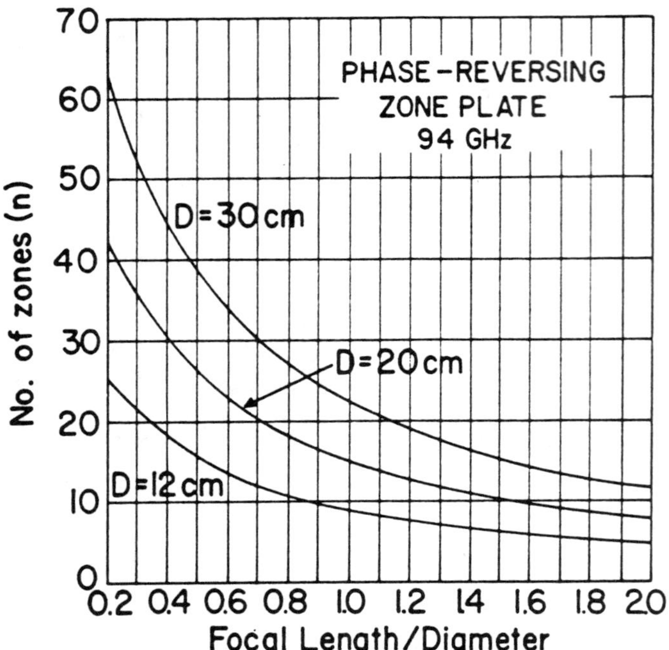

Figure 3. Number of zones as a function of focal length over diameter at 94 GHz.

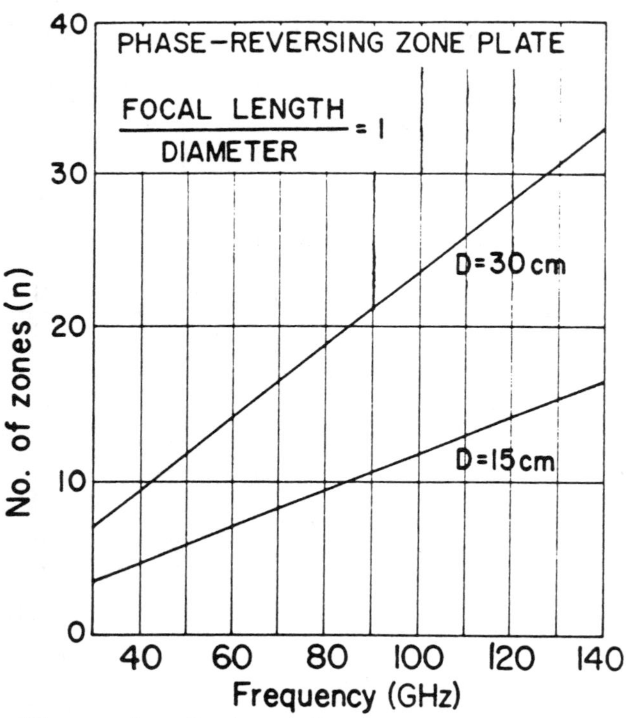

Figure 4. Number of zones as a function of frequency for two diameters.

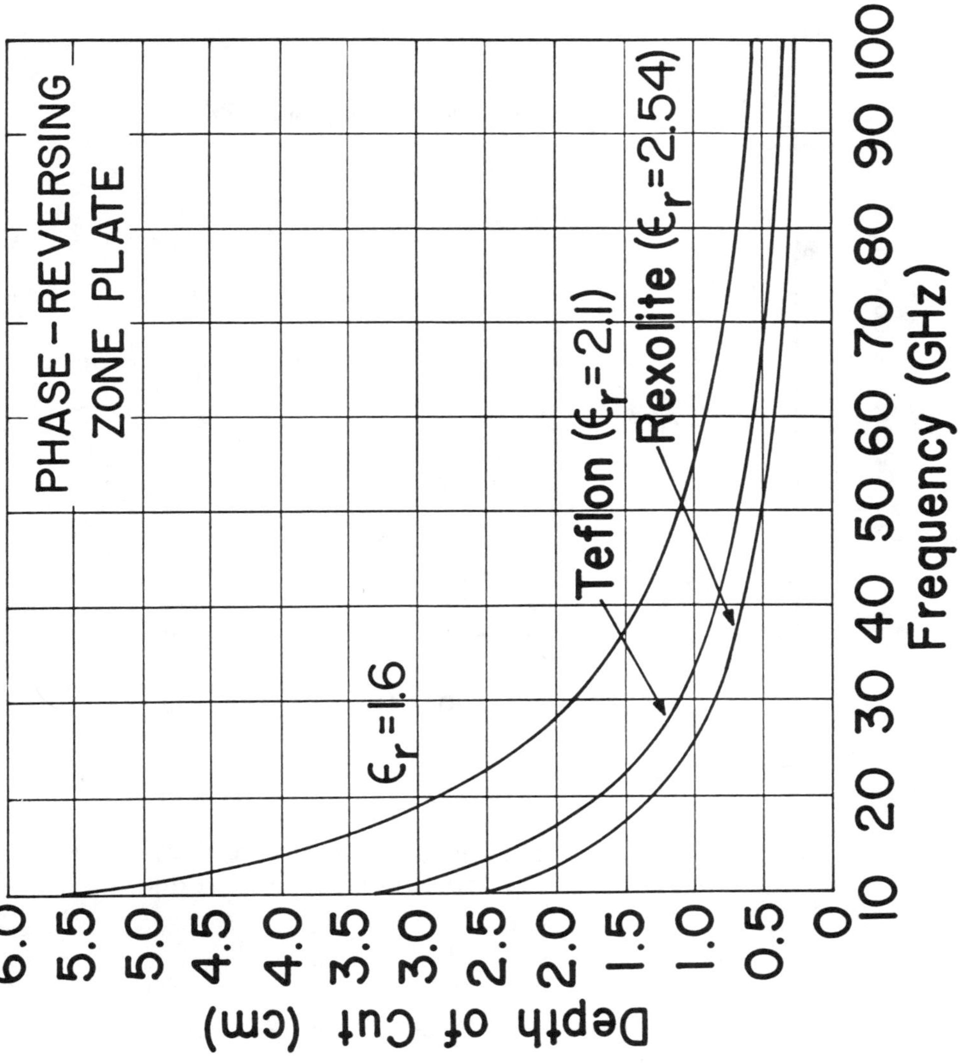

Figure 5. Depth of Cut vs. Frequency

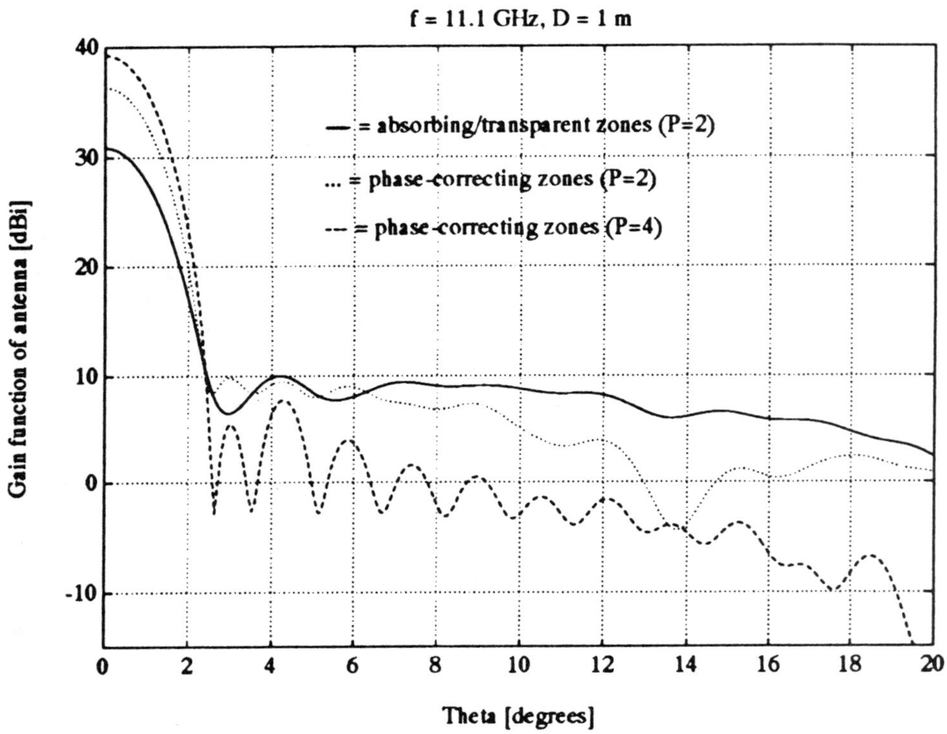

Figure 6. Radiation patterns of three gain-optimized zone plates (F/D=0.52), from Reference 11.

r_1 = 1.265 inches	r_6 = 3.134 inches	d = 0.1406 inches
r_2 = 1.793 inches	r_7 = 3.392 inches	3.57 mm
r_3 = 2.202 inches	r_8 =3.634 inches	
r_4 = 2.548 inches	r_9 = 3.863 inches	r ~ $\pm$ 0.05 inches
r_5 = 2.854 inches		$\pm$ 0.127 cm
		could be tolerated as an error

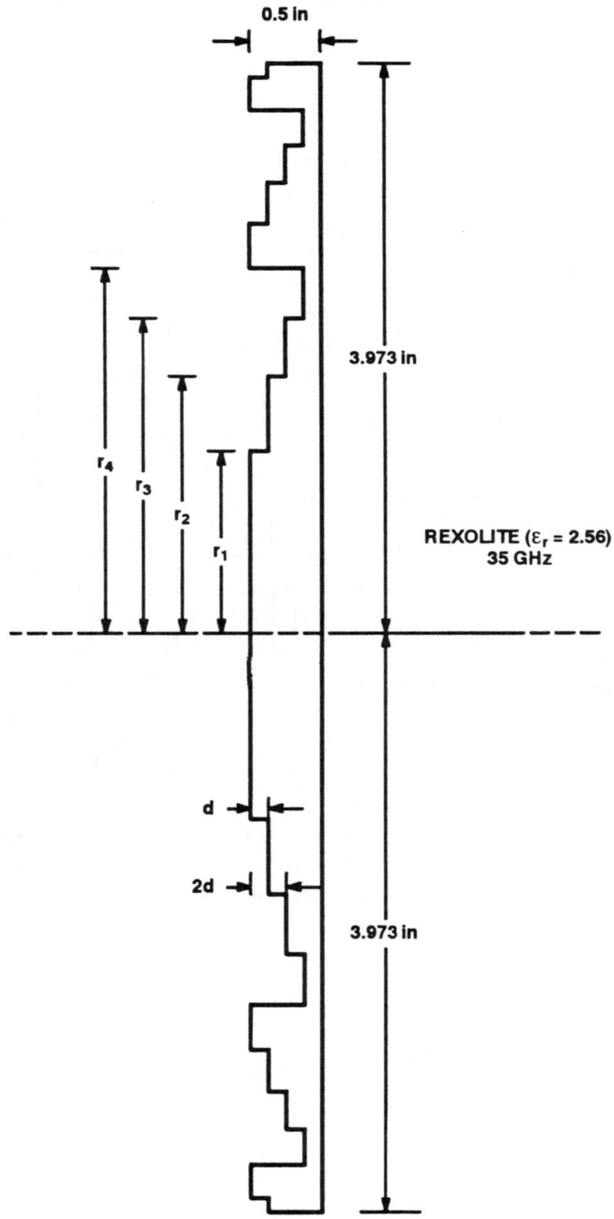

Transmitting Zone Plate

Figure 7. Cross-section of a 35GHz quarter-wave zone plate.

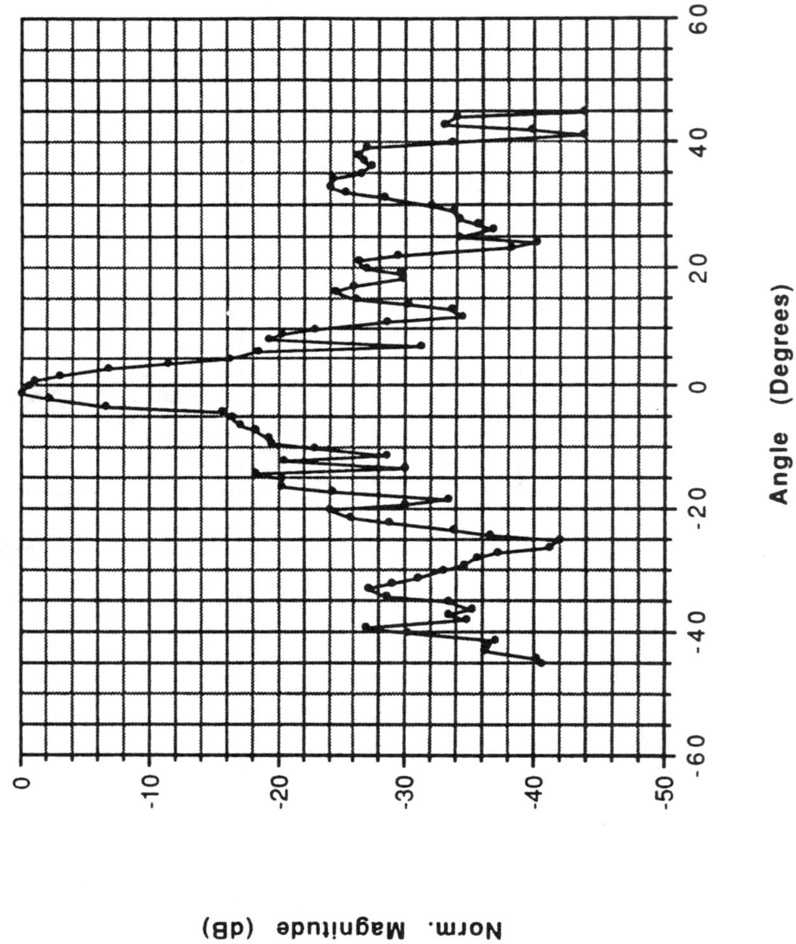

Figure 8. Pattern of 8in Diameter Transmission Zone Plate (35 GHz)

OFF-AXIS FEED CONFIGURATION

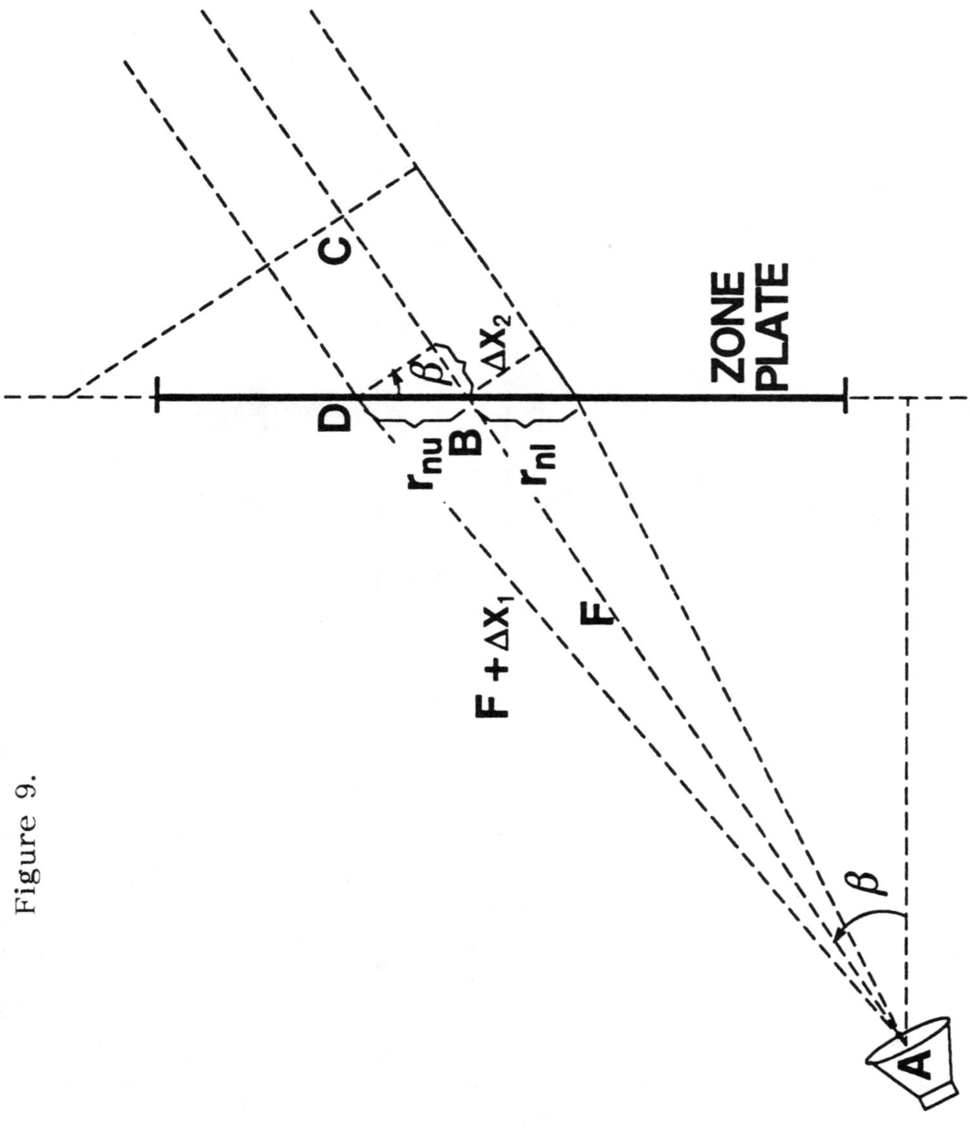

Figure 9.

r_1 = 1.265 inches	r_6 = 3.134 inches	d = 0.0703 inches
r_2 = 1.793 inches	r_7 = 3.392 inches	1.79 mm
r_3 = 2.202 inches	r_8 =3.634 inches	
r_4 = 2.548 inches	r_9 = 3.863 inches	
r_5 = 2.854 inches		

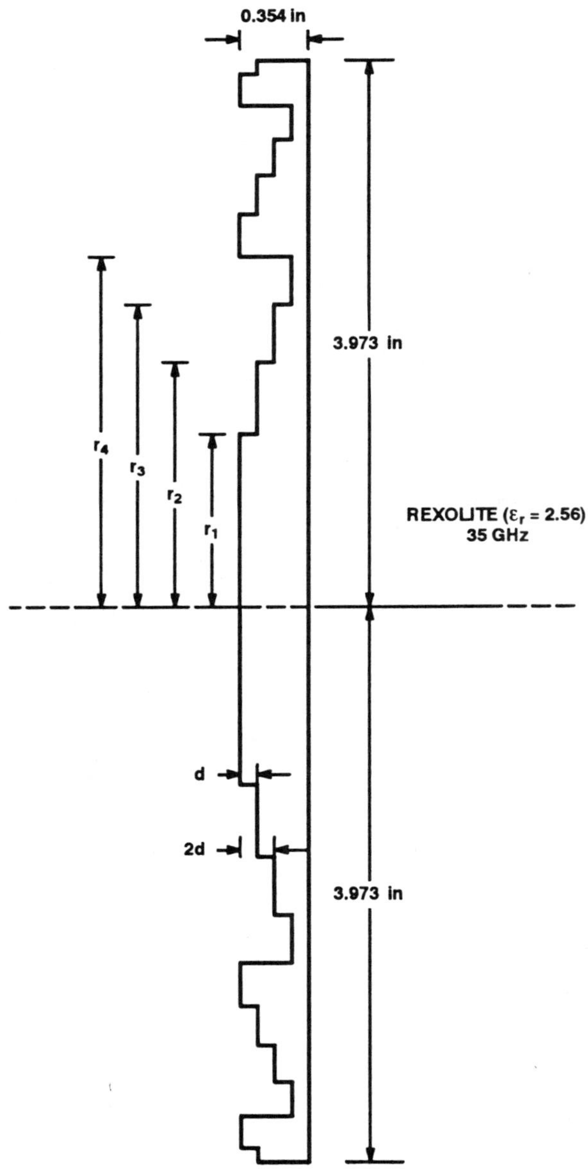

Reflecting Zone Plate

Figure 10. Reflector zone plate for 35GHz (quarter - wave correction).

Figure 11. **Pattern of 8in Diameter Reflecting Off-Axis Zone Plate (35 GHz)**

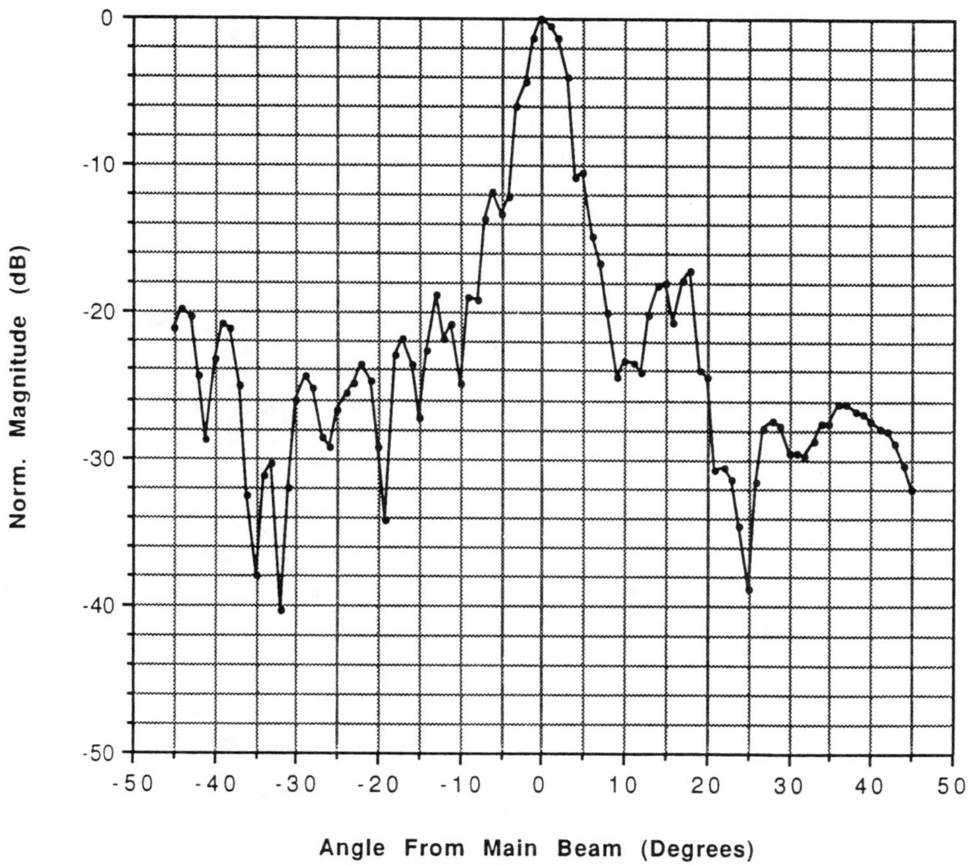